Lecture Notes in Mathematics   2090

Editors:
J.-M. Morel, Cachan
B. Teissier, Paris

For further volumes:
http://www.springer.com/series/304

Fondazione C.I.M.E., Firenze

C.I.M.E. stands for *Centro Internazionale Matematico Estivo*, that is, International Mathematical Summer Centre. Conceived in the early fifties, it was born in 1954 in Florence, Italy, and welcomed by the world mathematical community: it continues successfully, year for year, to this day.

Many mathematicians from all over the world have been involved in a way or another in C.I.M.E.'s activities over the years. The main purpose and mode of functioning of the Centre may be summarised as follows: every year, during the summer, sessions on different themes from pure and applied mathematics are offered by application to mathematicians from all countries. A Session is generally based on three or four main courses given by specialists of international renown, plus a certain number of seminars, and is held in an attractive rural location in Italy.

The aim of a C.I.M.E. session is to bring to the attention of younger researchers the origins, development, and perspectives of some very active branch of mathematical research. The topics of the courses are generally of international resonance. The full immersion atmosphere of the courses and the daily exchange among participants are thus an initiation to international collaboration in mathematical research.

C.I.M.E. Director	C.I.M.E. Secretary
Pietro ZECCA	Elvira MASCOLO
Dipartimento di Energetica "S. Stecco"	Dipartimento di Matematica "U. Dini"
Università di Firenze	Università di Firenze
Via S. Marta, 3	viale G.B. Morgagni 67/A
50139 Florence	50134 Florence
Italy	Italy
e-mail: zecca@unifi.it	e-mail: mascolo@math.unifi.it

For more information see CIME's homepage: http://www.cime.unifi.it

CIME activity is carried out with the collaboration and financial support of:

- INdAM (Istituto Nazionale di Alta Matematica)

- MIUR (Ministero dell'Università e della Ricerca)

Martin Burger • Andrea C.G. Mennucci
Stanley Osher • Martin Rumpf

Level Set and PDE Based Reconstruction Methods in Imaging

Cetraro, Italy 2008

Editors: Martin Burger
Stanley Osher

Martin Burger
Institute for Computational
 and Applied Mathematics
University of Münster
Münster, Germany

Andrea C.G. Mennucci
Scuola Normale Superiore
Pisa, Italy

Stanley Osher
Department of Mathematics
University of California
Los Angeles, CA, USA

Martin Rumpf
Institute for Numerical Simulation
University of Bonn
Bonn, Germany

ISBN 978-3-319-01711-2   ISBN 978-3-319-01712-9 (eBook)
DOI 10.1007/978-3-319-01712-9
Springer Cham Heidelberg New York Dordrecht London

Lecture Notes in Mathematics ISSN print edition: 0075-8434
                             ISSN electronic edition: 1617-9692

Library of Congress Control Number: 2013950139

Mathematics Subject Classification (2010): 94A08, 35Q94, 68U10, 65K10, 35J20, 45Q05, 53C44

© Springer International Publishing Switzerland 2013

This work is subject to copyright. All rights are reserved by the Publisher, whether the whole or part of the material is concerned, specifically the rights of translation, reprinting, reuse of illustrations, recitation, broadcasting, reproduction on microfilms or in any other physical way, and transmission or information storage and retrieval, electronic adaptation, computer software, or by similar or dissimilar methodology now known or hereafter developed. Exempted from this legal reservation are brief excerpts in connection with reviews or scholarly analysis or material supplied specifically for the purpose of being entered and executed on a computer system, for exclusive use by the purchaser of the work. Duplication of this publication or parts thereof is permitted only under the provisions of the Copyright Law of the Publisher's location, in its current version, and permission for use must always be obtained from Springer. Permissions for use may be obtained through RightsLink at the Copyright Clearance Center. Violations are liable to prosecution under the respective Copyright Law.

The use of general descriptive names, registered names, trademarks, service marks, etc. in this publication does not imply, even in the absence of a specific statement, that such names are exempt from the relevant protective laws and regulations and therefore free for general use.

While the advice and information in this book are believed to be true and accurate at the date of publication, neither the authors nor the editors nor the publisher can accept any legal responsibility for any errors or omissions that may be made. The publisher makes no warranty, express or implied, with respect to the material contained herein.

Printed on acid-free paper

Springer is part of Springer Science+Business Media (www.springer.com)

Preface

The present volume collects the lecture notes of courses given at the CIME Summer School in Cetraro in September 2008. The school appeared to be a highly successful event with around 50 participants from all over the world, enjoying both the scientific content of the courses and the atmosphere and opportunities for discussions among participants and with the speakers.

Mathematical imaging and inverse problems are two of the fastest growing disciplines in applied and interdisciplinary mathematics, with a strong overlap since many image analysis and reconstruction problems are effectively formulated as inverse problems. A key issue in such problems is to preserve and analyze edges or shapes in the image. Those are taken care by the methods covered in the summer school, which had the particular aim of providing a unified picture to methods of image reconstruction and analysis, on the one hand, and techniques for shape reconstruction and analysis, on the other hand. The key steps connecting those are methods based on level set representations and geometric partial differential equations, which effectively operate on the level sets (respectively, discontinuity set) of an image as well as on a single shape. The subject of lectures were chosen to cover these aspects and provide an outlook to other relevant topics:

- *Computational methods for nonlinear PDE and L1 techniques*: Stanley Osher (UCLA).
- *Total variation and related methods*: Martin Burger (Münster).
- *The use of level set methods in inverse problems*: Oliver Dorn (Manchester).
- *Variational methods in image matching and motion extraction: Medical and biological applications*: Martin Rumpf (Bonn).
- *Metrics of curves in shape optimization and analysis*: Andrea C.G. Mennucci (SNS Pisa).

In addition, few young participants already involved in research related to the course topics were given an opportunity to present their work in shorter talks.

The lecture series by Stanley Osher gave a general overview to modern computational methods for image reconstruction and analysis, including Bregman iterations, sparsity-based methods with ℓ^1-type functionals, and local and nonlocal total variation methods for image denoising.

Martin Burger further expanded on total variation techniques and discussed the analysis of total variation methods, further details on Bregman iterations and inverse scale space methods, and also applications in biomedical imaging.

The lectures by Oliver Dorn gave an overview of level set methods for the reconstruction of shapes in inverse problems. He discussed basics about level set representation of shapes and the construction of computational methods to solve appropriate variational formulations, including the necessary shape calculus. A major part of the lecture series was devoted to several applications modeled by inverse problems for partial differential equations.

Martin Rumpf moved from single images to pairs of images. The major tasks then consist of finding deformations of the image domains that make the images look as similar as possible. The corresponding objectives are either matching if interested in resulting images, and motion estimation if interested in the deformation. Extensions of the framework to shapes and their matching respectively averaging were presented as well.

Andrea C.G. Mennucci provided an overview of metrics on curves, linking this topic originating in differential geometry to very applied problems in image processing and shape optimization. He showed that by choosing different metrics methods for tasks such as object segmentation in images can be improved.

In the lecture notes most topics of the first two lectures were unified in a chapter providing a comprehensive introduction to the zoo of total variation and related methods. More specialized topics related to photon count data frequently arising in imaging were unified with the contents of participant talks by Christoph Brune and Alex Sawatzky in the second chapter. The third chapter covers the contents of Martin Rumpf's lecture series, the fourth chapter the ones of Andrea C.G. Mennucci's. We hope that readers find the material as interesting as we do and enjoy the breadth of covered topics as well as the depth of detail in the single chapters.

Münster, Germany   Martin Burger
Los Angeles, CA   Stanley Osher

Contents

A Guide to the TV Zoo 1
Martin Burger and Stanley Osher

EM-TV Methods for Inverse Problems with Poisson Noise 71
Alex Sawatzky, Christoph Brune, Thomas Kösters,
Frank Wübbeling, and Martin Burger

Variational Methods in Image Matching and Motion Extraction 143
Martin Rumpf

Metrics of Curves in Shape Optimization and Analysis 205
Andrea C.G. Mennucci

A Guide to the TV Zoo

Martin Burger and Stanley Osher

Abstract Total variation methods and similar approaches based on regularizations with ℓ^1-type norms (and seminorms) have become a very popular tool in image processing and inverse problems due to peculiar features that cannot be realized with smooth regularizations. In particular total variation techniques had particular success due to their ability to realize cartoon-type reconstructions with sharp edges. Due to an explosion of new developments in this field within the last decade it is a difficult task to keep an overview of the major results in analysis, the computational schemes, and the application fields. With these lectures we attempt to provide such an overview, of course biased by our major lines of research.

We are focusing on the basic analysis of total variation methods and the extension of the original ROF-denoising model due various application fields. Furthermore we provide a brief discussion of state-of-the art computational methods and give an outlook to applications in different disciplines.

1 Introduction

Reconstructing and processing images with appropriate edges is of central importance in modern imaging. The development of mathematical techniques that preserve or even favour sharp edges has become a necessity and created various interesting approaches. The two most succesful frameworks are two variational approaches: total variation models on the one hand and models with explicit edges

---

M. Burger (✉)
Institute for Computational and Applied Mathematics, University of Münster, Münster, Germany
e-mail: martin.burger@www.de

S. Osher
Department of Mathematics, UCLA, Los Angeles, USA

in the framework of Mumford and Shah [140] on the other. The latter usually lead to various difficulties in the analysis (cf. [138]) and numerical realization due to the explicit treatment of edges and arising nonconvexity, consequently they have found limited impact in practical applications beyond image segmentation. In these lectures notes we will discuss various developments for total variation methods, which can be formulated as convex variational methods. A whole zoo of approaches to the modelling, analysis, numerical solution, and applications has developed in the last two decades, through which we shall try to develop a guide.

The starting point of total variation (TV) methods has been the introduction of a variational denoising model by Rudin et al. [163], consisting in minimizing total variation among all functions within a variance bound

$$TV(u) \to \min_u \quad \text{subject to} \quad \int_\Omega (u-f)^2 dx \le \sigma^2. \tag{1}$$

Introducing a Lagrange multiplier λ it can be shown that this approach is equivalent to the unconstrained problem of minimizing

$$E_{ROF}(u) := \frac{\lambda}{2} \int_\Omega (u-f)^2 dx + TV(u), \tag{2}$$

in the following often referred to as the ROF model. In subsequent years this model was generalized for many imaging tasks and inverse problems and found applications in different areas.

We will provide a more detailed motivation for total variation regularization in a rather general setup in Sect. 2. In Sects. 3–5 we provide an overview of various aspects in the analysis of total variation regularization. In Sect. 6 we discuss the concepts of Bregman iterations and inverse scale space methods, which allow to compensate systematic errors of variational methods, e.g. contrast loss in the case of TV regularization, and gave another boost to research in this area in recent years. In Sects. 7 and 8 we discuss some variants of the models, with changes concerning the fidelity term in Sect. 7 and the regularization term in Sect. 8. In Sect. 9 we discuss some approaches for the numerical solution of the variational problems. Section 10 is devoted to geometric aspects of total variation minimization and the relaxation of segmentation problems into convex models for functions of bounded variation. We then proceed to applications, which we mainly incorporate as further links to literature in Sect. 11. Finally we present some open questions in the modelling and analysis of TV methods in Sect. 12.

2 The Motivation for TV and Related Methods

In the following we provide basic motivations for the general setup of TV methods with respect to forward operators, data fidelity terms, and regularization.

2.1 MAP and Penalized ML Estimation

In order to obtain a general variational model including total variation and similar penalties we resort to the Bayesian approach for computing solutions to an operator equation

$$Au = fv \tag{3}$$

in the presence of stochastic effects such as noise (cf. e.g. [110, 114, 129]). In a classical log-likelihood estimation technique one computes a solution by minimizing the negative log-likelihood of observing f under u, i.e.,

$$u_{LL} = \arg\min_u (-\log p(f|u)), \tag{4}$$

where $p(f|u)$ denotes an appropriate probability density for observing f given u. This can usually be identified with the probability density of the noise, e.g. in the frequently investigated case of additive noise

$$f = Au + \eta, \tag{5}$$

where η is a stochastic perturbation, i.e. noise, we find

$$p(f|u) = p_\eta(f - Au).$$

The computationally efficient part of the Bayesian approach is to compute the MAP (maximum a-posteriori probability) estimate, by

$$u_{MAP} = \arg\min_u (-\log p(u|f)). \tag{6}$$

The posterior probability density is obtained from Bayes formula

$$p(u|f) = \frac{p(f|u) p_0(u)}{\tilde{p}_0(f)}, \tag{7}$$

where p_0 denotes the prior probability for u and $\tilde{p}_0$ is the prior probability for f. Since the latter will only contribute a constant term in the minimization of the negative logarithm of the prior probability, $\tilde{p}_0(f)$ is not important for the MAP estimate. We can rewrite the MAP estimation in a similar form to log-likelihood estimation as

$$u_{MAP} = \arg\min_u (-\log p(f|u) - \log p_0(u)). \tag{8}$$

The negative logarithm of the prior probability density hence acts as a penalty or regularization functional, which creates a detailed link to total variation methods. Since the second term penalizes nonsmooth functions in addition to the logarithmic likelihood, this approach is also called penalized maximum likelihood (ML) method.

Since we assume that images of low total variation have a higher prior probability than those with high total variation, it seems obvious that p_0 should be a monotonously decreasing function of $TV(u)$. For related methods we of course need to replace TV by the appropriate prior, we consequently shall write the general functional J instead of TV. The natural choice in probability is a Gibbs form (cf. [94])

$$p_0(u) \sim e^{-\beta J(u)} \tag{9}$$

with some constant $\beta > 0$. We can then rewrite the MAP estimation as

$$u_{MAP} = \arg\min_u (-\log p(f|u) + \beta J(u)). \tag{10}$$

Rescaling to

$$\lambda H(u, f) := -\frac{1}{\beta} p(f|u) \tag{11}$$

where λ is a parameter depending on β and possibly the noise (see below), we see that MAP estimation yields a minimization problem of the form

$$\lambda H(u, f) + J(u) \to \min_u. \tag{12}$$

For specific imaging tasks and specific noise models the form of $H(u, f)$ can be written down, usually by straight-forward calculations. We shall now and in the following consider a linear image formation model $u \mapsto Au$ and a Gaussian additive noise η. We assume that A is a bounded linear operator to some Banach space. This means that for the "exact" image $\hat{u}$ the data are generated from

$$f = A\hat{u} + \eta, \tag{13}$$

where η follows a normal distribution with variance σ^2 and zero mean. This means that (formally)

$$p(f|u) = p_{Gauss}(f - Au) \sim \exp\left(-\frac{\|Au - f\|^2}{2\sigma^2}\right).$$

Hence we can define $\lambda := \frac{1}{\beta\sigma^2}$ and obtain the MAP estimation from the variational problem

$$\frac{\lambda}{2}\|Au - f\|^2 + J(u) \to \min_u. \tag{14}$$

We shall investigate non-Gaussian noise models leading to nonquadratic fidelity terms in the variational problem in Sect. 7.

2.2 Imaging Tasks

Above we have simply modelled image formation by a linear operator A mapping between some Banach (or ideally Hilbert) spaces. In our case the preimage space will consist of functions of bounded variation on a domain $\Omega \subset \mathbb{R}^d$ (usually $d = 2, 3$). In the following we give an overview how various imaging tasks can be formulated as such (cf. e.g. [65]):

- *Denoising:* In the case of denoising (with Gaussian noise) the given (noisy) image is measured as an element of the Hilbert space $L^2(\Omega)$. The clean image is usually modelled as an element of a smaller function space X, e.g. a Sobolev space $W^{1,p}(\Omega)$, a Besov space, or in our context usually $BV(\Omega) \cap L^2(\Omega)$. The operator A is an embedding operator into $L^2(\Omega)$ and hence in particular bounded.
- *Deblurring:* In the case of deblurring, A is modelled as an integral operator of the form

$$(Au)(x) = \int_\Omega k(x, y) u(y) \, dy \qquad (15)$$

in most cases with a convolution kernel $k(x, y) = \tilde{k}(x - y)$. For typical kernels A is a bounded operator from $L^1(\Omega)$ or $L^2(\Omega)$ to $L^2(\Omega)$.
- *Decompositon:* In an image decomposition model one aims to separate certain components of the image, e.g. smooth ones from texture. A typical model for decomposition into two parts is an operator

$$A : X_1 \times X_2 \to Y, \quad A(u_1, u_2) = u_1 + u_2. \qquad (16)$$

A possible choice of the output space $Y \supset X_i$ is $L^2(\Omega)$, but if one seeks decompositions also into oscillatory components such as texture a larger space such as $H^{-1}(\Omega)$ is needed.
- *Inpainting:* The aim of inpainting is to restore respectively extend an image given in $\Omega \setminus D$ into the inpainting domain D. Hence, the operator A is an embedding operator from a function space on $\Omega \setminus D$ into a function space on Ω, which is again linear and bounded.
- *Medical image reconstruction:* In medical imaging, the image formation model depends on the specific modality (cf. e.g. [141]). Examples are the Radon/X-ray transform (in computerized tomography and PET) or samples of the Fourier transform (in MRI), in both cases linear bounded operators. Certain modalities (e.g. electron and optical tomography) need to be modelled by nonlinear operators, which are still compact however.
- *Image segmentation:* In edge-based image segmentation one seeks a decomposition of the image into the smooth parts of the image in some function space and in particular an edge set $\Gamma \subset \Omega$ of finite $(n-1)$-dimensional Hausdorff-measure. If the edge set is the boundary of some subset of Ω such as in object-based segmentation, it is often replaced by a function of bounded variation taking only the values zero (outside this subset) and one (inside).

2.3 Regularization Functionals

In the following we consider the variational model

$$E(u) = \frac{\lambda}{2} \|Au - f\|^2 + J(u) \tag{17}$$

and discuss some implications of different choices of J.

A simple choice of a regularization functional is a quadratic energy of the form

$$J(u) = \frac{1}{2} \int_\Omega |Du|^2 \, dx, \tag{18}$$

where $D : X \to L^2(\Omega)^m$ is a linear operator, e.g. a gradient in $X = H^1(\Omega)$, $m = d$. We assume that A is a bounded linear operator on X. Then the optimality condition for (17) with regularization (18) becomes

$$0 = E'(u) = \lambda A^*(Au - f) + D^* Du. \tag{19}$$

Hence we see that for D^*D being invertible

$$u = \lambda (D^*D)^{-1} A^*(f - Au).$$

Consequently $u \in \mathcal{R}((D^*D)^{-1}A^*)$, the range of the smoothing operator $(D^*D)^{-1}A^*$, and hence the reconstructed image will be blurred. Consider e.g. the denoising case where $D = \nabla$ and $A : H^1(\Omega) \to L^2(\Omega)$ is the embedding operator, then

$$-\Delta u = \lambda(f - u) \in L^2(\Omega).$$

By elliptic regularity we thus expect $u \in H^2(\Omega)$, i.e. the image is oversmoothed and there are no edges in particular. On the other hand consider $X = L^2(\Omega)$, D being the identity, and A is an integral operator as in deblurring or image reconstruction. Then u is in the range of A^*, which is again an integral operator (with kernel $k(y, x)$). This again implies smoothness of u and the nonexistence of edges.

An obvious next step would be to replace the square in the regularization (18) by a smooth strictly convex function R, i.e. to choose

$$J(u) = \frac{1}{2} \int_\Omega R(Du) \, dx, \tag{20}$$

in (17). The corresponding optimality condition becomes

$$0 = E'(u) = \lambda A^*(Au - f) + D^*(R'(Du)). \tag{21}$$

The linearization of this equation is given by

$$\lambda A^* A v + D^*(R''(Du)Dv) = g. \tag{22}$$

If R is smooth and strictly convex, then the "diffusion" coefficient $R''(Du)$ is strictly positive and the linearized operator has analogous smoothing properties as (19). Consequently we need to expect a similar smoothing behaviour as in the quadratic case.

As a consequence one might argue that only a regularization functional based on R being not strictly convex can indeed prevent oversmoothing and should be able to maintain edges. As we shall see this is indeed the case for total variation. In order to gain a formal understanding it is instructive to consider $R(u) = \frac{1}{p}|u|^p$ for p approaching 1. The corresponding optimality condition becomes

$$0 = E'(u) = \lambda A^*(Au - f) + D^*(|Du|^{p-2}Du). \tag{23}$$

For $p > 1$ a large "gradient" Du will have an impact on the differential operator, since the Euclidean norm of $|Du|^{p-2}Du$ is $|Du|^{p-1}$. In the limit $p \to 1$ this behaviour changes and the involved vector field $|Du|^{-1}Du$ has Euclidean norm one no matter how large Du becomes. Hence, large (infinite) "gradients" Du will not have particular impact on $D^*(|Du|^{p-2}Du)$.

Visually the improvement when choosing $p = 1$ compared to larger values (e.g. $p = 2$) can be observed in Figs. 1 and 2. The noisy versions of the images are used as input f for two variational denoising methods, a linear scheme (on the lower left)

$$\frac{\lambda}{2}\int_\Omega (u-f)^2 \, dx + \int_\Omega |\nabla u|^2 \, dx \to \min_u \tag{24}$$

equivalent to

$$\lambda(u - f) - \Delta u = 0, \tag{25}$$

and the ROF model (on the lower right)

$$\frac{\lambda}{2}\int_\Omega (u-f)^2 \, dx + \int_\Omega |\nabla u| \, dx \to \min_u. \tag{26}$$

In both cases, a visually optimal value of λ is chosen. The linear scheme leads to significant blurring and non-sharp images, while the ROF model succeeds in computing reconstructions with sharp edges, which seem much more appealing to the human eye (cf. [100] for a further interpretation of this effect). Comparing the ROF reconstructions in Figs. 1 and 2 one observes that the quality of the ROF solution is better in the case of the blocky image, which ideally corresponds to the properties of total variation as we shall also see below. In the case of a natural image the denoising is of lower visual quality (still clearly outperforming the

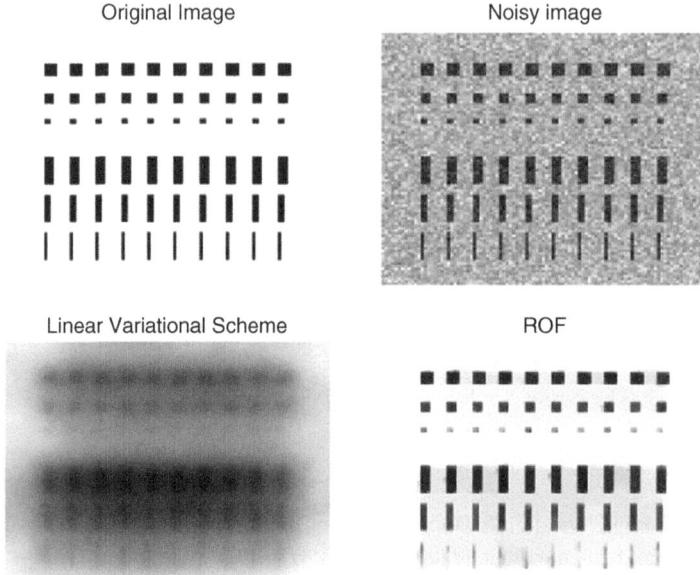

Fig. 1 A blocky image and its noisy version (*top*), reconstruction with linear scheme (quadratic regularization) and with the ROF model

linear scheme) due to various oscillatory patterns (texture) appearing in the image (compare e.g. the flower bush in the background). In such cases the ROF model still manages to reconstruct a cartoon of the image, i.e. its major structure without small-scale textures. Further techniques can subsequently be applied to find textures in the residual.

A similar reasoning holds for schemes enforcing sparsity. Suppose $(\psi_j)_{j \in \mathbb{N}}$ is an orthonormal basis of a separable Hilbert space X and

$$B : \ell^2(\mathbb{N}) \to X, \ \mathbf{c} = (c_j)_{j \in \mathbb{N}} \mapsto \sum c_j \psi_j \qquad (27)$$

is the map from coefficients to elements in X. Then one can rephrase linear methods via the variational problem (Fig. 3)

$$\frac{\lambda}{2} \|AB\mathbf{c} - f\|^2 + \frac{1}{2} \sum_{j=1}^{\infty} \omega_j |c_j|^2, \qquad (28)$$

with (ω_j) being a suitable sequence of weights (e.g. $\omega_j = 1$ corresponds to the original squared norm in X). The optimality condition for (28) becomes

$$\omega_j c_j = -(B^* A^* (AB\mathbf{c} - f)) \cdot e_j, \qquad (29)$$

Fig. 2 A natural image (lecturers of the summer school) and its noisy version (*top*), reconstruction with linear scheme (quadratic regularization) and with the ROF model

where e_j is the j-th unit vector. Hence the decay of the coefficients c_j is mainly determined by the properties of the adjoint operator B^*A^* and the weights ω_j. There is no particular reason for achieving a sparse (in particular finite) weight sequence, one will rather encounter a slowly decaying one. If one generalizes the regularization to

$$\frac{\lambda}{2}\|AB\mathbf{c} - f\|^2 + \frac{1}{p}\sum_{j=1}^{\infty}\omega_j|c_j|^p, \qquad (30)$$

the optimality condition becomes

$$\omega_j c_j |c_j|^{p-2} = -(B^*A^*(AB\mathbf{c} - f)) \cdot e_j. \qquad (31)$$

For $p > 1$ there still exists a unique simple monotone relation between c_j and the right-hand side, so the decay behaviour is still dominated by ω_j and B^*A^*. Again the limit $p \to 1$ gives an interesting change, since for $c_j \to 0$ the limit of the left-hand side can be quite arbitrary. Thus the case $p = 1$ (or $p < 1$) will be particularly interesting also in such situations.

Fig. 3 From *left* to *right*: clean image, denoising with the isotropic ROF model with three different scale parameters λ (decreasing from *left* to *right*)

2.4 Total Variation and Variants

The total variation of a function is formally defined as

$$TV(u) = \int_\Omega |\nabla u|\, dx, \tag{32}$$

a definition which makes sense if $u \in W^{1,1}(\Omega)$. Since neither the structure of $W^{1,1}(\Omega)$ is very convenient (it is not the dual of a Banach space) nor it contains piecewise constant functions, it is reasonable to consider a slightly larger space which has the desired properties, namely the space of functions of bounded variation. An exact definition of the total variation is (cf. [95])

$$TV(u) := \sup_{\substack{g \in C_0^\infty(\Omega;\mathbb{R}^d) \\ \|g\|_\infty \leq 1}} \int_\Omega u \nabla \cdot g\, dx. \tag{33}$$

Here we use the vectorial norm

$$\|g\|_\infty := \operatorname*{ess\,sup}_{x \in \Omega} \sqrt{g_1(x)^2 + \ldots + g_d(x)^2}, \tag{34}$$

variants in the choice of the vector norm will be discussed below. The space of functions of bounded variation is then defined as

$$BV(\Omega) = \{\, u \in L^1(\Omega) \mid TV(u) < \infty \,\}. \tag{35}$$

We shall later verify that BV has indeed the desired properties, see also [6, 86, 95] for detailed discussions of functions of bounded variation.

The total variation as defined so far was isotropic, in the sense that it is invariant with respect to rotations in the argument x. This is often an advantage, since the result becomes independent of the rotation of the input image. Variants of total variation are basically obtained by changing the norm of the vector g in (33) from the standard Euclidean norm to different vector norms

A Guide to the TV Zoo

Fig. 4 From *left* to *right*: noisy image, denoising with the isotropic ROF model (Euclidean vector norm), denoising with the anisotropic ROF Model (supremum vector norm), and a ROF model with adapted anisotropy. From [30]

$$TV(u) := \sup_{\substack{g \in C_0^\infty(\Omega;\mathbb{R}^d) \\ \gamma(g) \leq 1 \text{ a.e.}}} \int_\Omega u \, \nabla \cdot g \, dx, \qquad (36)$$

where $\gamma : \mathbb{R}^d \to \mathbb{R}$ is a nonnegative one-homogeneous functional satisfying a triangle-inequality. Of course the arising total variation norms are equivalent and give the same space $BV(\Omega)$ in the above definition. However, in the reconstruction the anisotropy of vector norms different from the Euclidean one can have a significant impact (cf. [30, 83]). While the isotropic total variation (based on the Euclidean norm) favours rounded edges in reconstructions, anisotropic definitions can favour different structures, e.g. corners or rectangular structures. This behaviour is illustrated in Fig. 4.

3 Existence, Uniqueness, and Optimality

In the following we investigate the well-posedness of the general model

$$\frac{\lambda}{2} \|Au - f\|^2 + TV(u) \to \min_{u \in BV(\Omega)}. \qquad (37)$$

This problem has been analyzed first by Acar and Vogel [2], using strong L^p topologies that are suitable due to compact embedding results for $BV(\Omega)$. We shall here present a different analysis based on the weak* topology of $BV(\Omega)$. In order to avoid technicalities in the analysis we first eliminate the mean value of u, which later allows to define an equivalent norm on those functions with mean zero. We introduce the corresponding subspace

$$BV_0(\Omega) = \{ u \in BV(\Omega) \mid \int_\Omega u \, dx = 0 \}. \qquad (38)$$

In the remainder of the paper we shall assume that $A\mathbf{1} \neq 0$, where $\mathbf{1}$ is the constant function taking the value 1 everywhere. We further assume

$$\langle Av, A\mathbf{1}\rangle = 0 \qquad \forall\, v \in BV_0, \tag{39}$$

which is true for most of the operators we consider (e.g. denoising, inpainting, deblurring, decomposition). The main results still hold true if (39) is not satisfied but with additional technical effort related to the mean value of u.

Lemma 3.1. *Let $\lambda > 0$. Then the minimizer of (37) is of the form*

$$u = v + \frac{\langle f, A\mathbf{1}\rangle}{\|A\mathbf{1}\|^2}\mathbf{1} \tag{40}$$

with $v \in BV_0(\Omega)$.

Proof. Each $u \in BV(\Omega)$ can be written as $u = v + c\mathbf{1}$ with mean value $c \in \mathbb{R}$ and $v \in BV_0(\Omega)$. Since $TV(u) = TV(v)$, we can rewrite the functional to be minimized using (39) as

$$\frac{\lambda}{2}\|Au - f\|^2 + TV(u) = \frac{\lambda}{2}\|cA\mathbf{1} - f\|^2 + \frac{\lambda}{2}\|Av\|^2 - \lambda\langle Av, f\rangle + TV(v).$$

Hence in order to compute the optimal c we can simply minimize $\|cA\mathbf{1} - f\|^2$ with respect to $c \in \mathbb{R}$, which yields the desired result. □

As a result of Lemma 3.1, we can now reduce the minimization over $BV_0(\Omega)$, shifting f to $f - \frac{\langle f, A\mathbf{1}\rangle}{\|A\mathbf{1}\|^2}A\mathbf{1}$. With abuse of notation we also call the shifted data f and consider the minimization

$$E(u) = \frac{\lambda}{2}\|Au - f\|^2 + TV(u) \to \min_{u \in BV_0(\Omega)}. \tag{41}$$

3.1 Coercivity

At the first glance we observe that the total variation is bounded if the functional in (41) is bounded. Thus we seek coercivity in the norm of $BV(\Omega)$. A first useful property of BV-spaces is embedding into Lebesgue spaces (cf. [86, 95]):

Lemma 3.2. *Let $\frac{q}{q-1} \geq d$, then*

$$BV_0(\Omega) \hookrightarrow L^q(\Omega).$$

The embedding is compact if $\frac{q}{q-1} > d$.

From Lemma 3.2 we see that in any dimension there exists $q > 1$ with $BV_0(\Omega) \hookrightarrow L^q(\Omega)$, even with compact embedding. If the operator A is defined on such an L^q-space, the analysis could be carried out in the strong topology of $L^q(\Omega)$, an approach used e.g. in [2]. We take a different approach using a different topology in $BV_0(\Omega)$ directly. The main result we shall use to verify coercivity of E is the following classical theorem (cf. e.g. [122]):

Theorem 3.3 (Banach-Alaoglu). *Let X be the dual of some Banach space Z. Then each bounded set in X is precompact in the weak-$*$ topology.*

Hence we need to define a weak-$*$ topology on $BV_0(\Omega)$, respectively find a space whose dual $BV_0(\Omega)$ is. For this sake we define a normed space

$$Z_0 = \{ \nabla \cdot g \mid g \in C_0^\infty(\Omega; \mathbb{R}^d) \}, \qquad (42)$$

with norm (the norm properties can be checked in a straight-forward way)

$$\|p\|_Z = \inf_{g \in C_0^\infty(\Omega;\mathbb{R}^d), \nabla \cdot g = p} \|g\|_{L^\infty}. \qquad (43)$$

Its completion in this norm is denoted by

$$Z := \overline{Z_0}. \qquad (44)$$

We now find $BV_0(\Omega)$ as the dual of Z:

Proposition 3.4. *$BV_0(\Omega)$ can be identified with the dual space of Z defined by* (44)

Proof. First of all we observe that for each $u \in BV_0(\Omega)$ we can construct a linear functional in Z_0 given by

$$\ell_u : p \mapsto \int_\Omega u p \, dx.$$

For $p = \nabla \cdot g_p$ we have

$$|\ell_u(p)| = |\int_\Omega u \nabla \cdot g_p \, dx| \leq \|g_p\|_{L^\infty} \sup_{g \in C_0^\infty(\Omega; \mathbb{R}^d), \|g\|_{L^\infty} \leq 1} \int_\Omega u \nabla \cdot g \, dx = \|g_p\|_{L^\infty} TV(u).$$

Taking the infimum over all such g_p we conclude

$$|\ell_u(p)| \leq \|p\|_Z \|u\|_{BV}.$$

Thus, ℓ_u is bounded on Z_0 and can therefore be extended in a unique way to a bounded linear functional on Z. Moreover, for $u_1 \neq u_2$ it is easy to see that $\ell_{u_1} \neq \ell_{u_2}$, otherwise

$$TV(u_1 - u_2) = \sup_{g \in C_0^\infty(\Omega;\mathbb{R}^d), \|g\|_{L^\infty} \leq 1} (\ell_{u_1}(g) - \ell_{u_2}(g)) = 0.$$

This implies that we can identify $BV_0(\Omega)$ with a subspace of Z^*.

On the other hand we see that for q_* sufficiently large we have $L_0^{q_*}(\Omega) \hookrightarrow Z$, where

$$L_0^{q_*}(\Omega) = \{\, v \in L^{q_*}(\Omega) \mid \int_\Omega v\, dx = 0 \,\}.$$

This can be shown e.g. by regularity results for the homogeneous Neumann problem for the Poisson equation (note that functions in $L_0^{q_*}(\Omega)$ satisfy the solvability criterion). Thus we obtain the opposite inclusion for the dual spaces, $Z^* \subset L_0^q(\Omega)$ with $q = \frac{q_*}{q_*-1}$. Hence, each functional $\ell \in Z^*$ can be identified with a functional of the form

$$\ell(p) = \int_\Omega v\, p\, dx$$

for some $v \in L_0^q(\Omega)$. The boundedness of the linear functional ℓ further implies $TV(v) < \infty$, hence $v \in BV_0(\Omega)$. Thus, we can also identify Z^* with a subspace of $BV_0(\Omega)$. □

In order to complete our analysis of coercivity, we need to replace the BV-norm by the seminorm, i.e., the total variation, since only the latter is bounded on sub-level sets of E. Hence we prove a Poincare-Wirtinger inequality on $BV_0(\Omega)$:

Lemma 3.5. *The total variation is an equivalent norm on $BV_0(\Omega)$.*

Proof. It is obvious that

$$\|u\|_{BV} = \|u\|_{L^1} + TV(u) \geq TV(u)$$

and it remains to show that there exists $c > 0$ with $\|u\|_{BV} \leq cTV(u)$ for all $u \in BV_0(\Omega)$. We prove this assertion by contradiction and hence assume that for each $n \in \mathbb{N}$ there exists $u_n \in BV_0(\Omega)$ with $\|u_n\|_{BV} > n\,TV(u_n)$. Now let $v_n := \frac{u_n}{\|u_n\|_{BV}}$, then $TV(v_n) < \frac{1}{n}$, i.e. $TV(v_n) \to 0$. Since $\|v_n\|_{BV} = 1$, we conclude $\|v_n\|_{L^1} \to 1$. Since v_n is uniformly bounded in $BV_0(\Omega)$, there exists a weak-* convergent subsequence v_{n_k} with some limit v. Due to the compact embedding of $BV_0(\Omega)$ into $L^1(\Omega)$, this sequence converges strongly in $L^1(\Omega)$ and thus

$$\|v\|_{L^1} = \lim_k \|v_{n_k}\|_{L^1} = 1.$$

By the lower semicontinuity of the total variation in the weak-* topology (see below) we further conclude

$$TV(v) \leq \liminf_k TV(v_n) = 0,$$

A Guide to the TV Zoo

hence $TV(v) = 0$, which implies that v is constant and since $v \in BV_0(\Omega)$ even that $v \equiv 0$. The latter contradicts $\|v\|_{L^1} = 1$. □

We are finally able to state the main coercivity result as a direct consequence of the properties shown above:

Lemma 3.6. *The sub-level set*

$$\mathcal{M}_C = \{\, u \in BV_0(\Omega) \mid E(u) \leq C \,\} \tag{45}$$

is precompact in the weak-* topology of $BV_0(\Omega)$.

We mention that the weak-* topology in BV always implies strong convergence in $L^1(\Omega)$, hence we can work with sequences in all arguments in the following.

3.2 Lower Semicontinuity

Besides coercivity, a major ingredient is the lower semicontinuity of the functional E in (41). Since a sum of two functionals is lower semicontinuous if both of them are, we verify this property separately, starting with the total variation:

Proposition 3.7. *The total variation is weak-* lower semicontinuous on $BV_0(\Omega)$.*

Proof. Let $u_n \rightharpoonup^* u$ and let $\psi_k \in C_0^\infty(\Omega; \mathbb{R}^d)$ with $\|\psi_k\|_{L^\infty} \leq 1$ such that

$$\int_\Omega u \nabla \cdot \psi_k \, dx \to TV(u).$$

Then we have due to the weak-* convergence

$$\int_\Omega u \nabla \cdot \psi_k \, dx = \lim_n \int_\Omega u_n \nabla \cdot \psi_k \, dx$$

$$= \liminf_n \int_\Omega u_n \nabla \cdot \psi_k \, dx$$

$$\leq \liminf_n TV(u_n).$$

Taking now the limit $\psi_k \to \infty$ we find

$$TV(u) = \lim_k \int_\Omega u \nabla \cdot \psi_k \, dx \leq \liminf_n TV(u_n),$$

hence TV is weak-* lower semicontinuous. □

For the fitting functional we need to make a basic assumption on the operator A in order to obtain weak-* lower semicontinuity, namely a range property of the adjoint operator

$$\mathcal{R}(A^*) \subset Z. \tag{46}$$

Note that for A being a bounded linear operator on $BV_0(\Omega)$, $\mathcal{R}(A^*)$ is always a subset of the dual space (cf. [122]) of $BV_0(\Omega)$. The additional assumption that $\mathcal{R}(A^*)$ lies in the smaller predual space is a regularity assumption on the operator A.

Lemma 3.8. *For $\lambda > 0$ the functional $u \mapsto \lambda \|Au - f\|^2$ is weak-* lower semicontinuous on $BV_0(\Omega)$.*

Proof. Since a positive multiple and a square preserve lower semicontinuity of a positive functional it suffices to show that $u \mapsto \|Au - f\|$ is weak-* lower semicontinuous. By the dual characterization of a Hilbert space norm (cf. [122]) we have

$$\|Au - f\| = \sup_{\varphi, \|\varphi\|=1} \langle Au - f, \varphi \rangle.$$

Now we can again choose a sequence φ_k with norm equal one such that $\langle Au - f, \varphi_k \rangle \to \|Au - f\|$. Since $A^*\varphi_k \in Z$ we can employ weak-* convergence of u_n to see

$$\begin{aligned}
\langle Au - f, \varphi_k \rangle &= \langle u, A^*\varphi_k \rangle - \langle f, \varphi_k \rangle \\
&= \lim_n \langle u_n, A^*\varphi_k \rangle - \langle f, \varphi_k \rangle \\
&= \liminf_n \langle u_n, A^*\varphi_k \rangle - \langle f, \varphi_k \rangle \\
&= \liminf_n \langle Au_n - f, \varphi_k \rangle \\
&\leq \liminf_n \|Au_n - f\|.
\end{aligned}$$

Again with the limit $k \to \infty$ we obtain weak-* lower semicontinuity. □

3.3 Existence

With the above prerequisites we can now show the existence of a minimizer of E by a standard variational technique:

Theorem 3.9. *There exists a minimizer of E in $BV_0(\Omega)$, i.e. a solution of (41).*

Proof. The proof technique is a standard combination of coercivity and lower semicontinuity, which we give here for completeness. Since

$$+\infty > \frac{\lambda}{2}\|f\|^2 = E(0) \geq \inf_u E(u) \geq 0 > -\infty,$$

the infimum of E is finite and hence there exists a minimizing sequence u_k with $E(u_k) \to \inf_u E(u)$. For k sufficiently large, this minimizing sequence lies in the sub-level set $E(u) \leq C$ with $C = E(0) + 1$, which is precompact in the weak-* topology. Hence u_k has a weak-* convergent subsequence, again denoted by u_k without loss of generality. If $\hat{u}$ is the weak-* limit of u_k, then the lower semicontinuity implies

$$E(\hat{u}) \leq \liminf_k E(u_k) = \inf_u E(u),$$

thus $\hat{u}$ is a solution of (41). □

3.4 Uniqueness

The uniqueness of (41) is related to the convexity of the functional. Since both the quadratic fitting term and the total variation are convex, (41) is a convex variational problem, hence the following general result applies (cf. [81]).

Lemma 3.10. *The set of minimizers of* (41) *is convex.*

As a consequence we see that we can at least uniquely pick a certain solution by a different criterion, e.g. by minimizing some squared norm of the set of minimizers. An obvious candidate is the solution of minimal L^2-norm among all minimizers of (41). Since in this case a strictly convex functional is minimized over a convex set, the solution is unique.

A general uniqueness result can only be formulated if A is injective:

Theorem 3.11. *Let A have trivial nullspace and $\lambda > 0$. Then* (41) *has a unique minimizer.*

Proof. If A has trivial nullspace, then the second variation of $F(u) = \frac{\lambda}{2}\|Au - f\|^2$ is given by

$$F''(u)(v, v) = \lambda \|Av\|^2,$$

which is positive for $v \neq 0$. Hence the functional is strictly convex, which implies that the minimizer is unique. □

3.5 Optimality Conditions

In order to obtain optimality conditions for the variational problem (41) we need a general notion of derivative of the functional E since the total variation is not differentiable in a classical sense. However, since the total variation is convex, we can adopt the multivalued notion of a *subdifferential* (cf. [81]). The subdifferential of a convex functional $J : X \to \mathbb{R} \cup \{+\infty\}$ at $u \in X$ (here with $X = BV_0(\Omega)$ is given by

$$\partial J(u) = \{\, p \in X^* \mid \langle p, v - u \rangle \leq J(v) - J(u), \forall\, v \in X \,\}. \tag{47}$$

A subgradient $p \in \partial J(u)$ can be identified with the slope of a plane (of codimension one) in $X \times \mathbb{R}$ through $(u, J(u))$ that lies under the graph of J.

From the definition of the subdifferential it is straight-forward to see that $\hat{u}$ is a minimizer of J if and only if $0 \in \partial J(\hat{u})$. Due to convexity the first-order optimality condition is not only necessary, but also sufficient. It thus remains to compute the subdifferential of E to characterize minimizers of (41). Since the quadratic fitting term is Frechet-differentiable, we can decompose the subdifferential of E into this derivative and the subdifferential of the total variation (cf. [81]). Thus, the optimality condition for a minimizer u of E becomes

$$\lambda A^*(Au - f) + p = 0, \quad p \in \partial TV(u). \tag{48}$$

It thus remains to characterize the subdifferential of the total variation, which is a rather difficult task. We start with a general property of one-homogeneous functional (i.e. $J(tx) = |t| J(x)$ for all $t \in \mathbb{R}$)

Lemma 3.12. *Let $J : X \to \mathbb{R} \cup \{+\infty\}$ be a convex one-homogeneous functional. Then*

$$\partial J(u) = \{\, p \in X^* \mid \langle p, u \rangle = J(u), \langle p, v \rangle \leq J(v), \forall\, v \in X \,\}.$$

Proof. Using $v = 0$ and $v = 2u$ in the definition of a subgradient we find

$$-\langle p, u \rangle \leq -J(u)$$

and

$$\langle p, u \rangle \leq J(2u) - J(u) = 2J(u) - J(u) = J(u).$$

Thus, $\langle p, u \rangle = J(u)$ and the assertion follows. □

In the case of the total variation we see that for each subgradient the dual norm is bounded by

A Guide to the TV Zoo

$$\|p\| = \sup_{v \in BV_0(\Omega), TV(v)=1} \langle p, v \rangle \leq \sup_{v \in BV_0(\Omega), TV(v)=1} TV(v) = 1.$$

Hence, we see that

$$\partial TV(u) = \{\, p \in BV_0(\Omega)^* \mid \|p\| \leq 1, \langle p, u \rangle = TV(u) \,\}. \tag{49}$$

From the structure of the dual space of $BV_0(\Omega)$ we see that for each $p \in BV_0(\Omega)^*$ with $\|p\| \leq 1$ there exists $g \in L^\infty(\Omega; \mathbb{R}^d)$ with $p = \nabla \cdot g$ (the opposite is not true in general). Hence, the optimality condition can be stated equivalently as

$$\lambda A^*(Au - f) + \nabla \cdot g = 0, \quad \|g\|_{L^\infty} \leq 1, \quad \langle \nabla \cdot g, u \rangle = TV(u). \tag{50}$$

For detailed investigations of subgradients of the total variation and properties of the dual space and its norm (often called G-norm) we refer to [15, 134, 143, 146, 179].

4 Examples

In the following we present some examples that can be computed exactly in order to provide further insight into the behaviour of total variation regularization in general and the ROF-functional (2). In this way one obtains further insight into structural properties of solutions, but also into remaining deficiencies of total variation regularization such as staircasing and loss of contrast. Since solutions cannot be found in closed form in general, the only way to obtain exact solutions for some functions f is to guess the form of the solution u and verify that $\lambda(f - u)$ is a subgradient of the total variation at u. Of course some general structural results, in particular for piecewise constant data (cf. [58, 162]) and numerical experiments can be a good guideline to find such solutions. The solutions that can be found in the easiest way are related to eigenvalues, more precisely to the nonlinear eigenvalue inclusion problem $\mu u \in \partial TV(u)$, which has been discussed in [26, 27] and effectively used in [134] for a celebrated example of an exact solution of the ROF functional, namely for Ω being the whole $\mathbb{R}^d$ and

$$f = \chi_{B_R(0)} = \begin{cases} 1 & \text{if } |x| < R \\ 0 & \text{else.} \end{cases} \tag{51}$$

The solution of this problem is given by

$$u = \left(1 - \frac{2}{\lambda R}\right)^+ \chi_{B_R(0)}, \tag{52}$$

i.e. u is proportional to f and thus also the subgradient p, hence the solution is based on an eigenfunction of the total variation. For further exact solutions we refer to [27, 178].

Note that for each function $f \in BV(\Omega)$ satisfying $\mu f \in \partial TV(f)$ for some $\mu \geq 0$ we can explicitly construct the unique minimizer of the ROF functional, which is of the form $u = cf$ for some $c \geq 0$. With this form the optimality condition gives for $\mu < \lambda$

$$cf - f + \frac{\mu}{\lambda} f = 0,$$

hence $c = 1 - \frac{\mu}{\lambda}$ yields a solution. If $\mu \geq \lambda$ then we can choose $c = 0$ to obtain the solution, which follows from a general results shown below: For λ sufficiently small (with explicit bound depending on f) $u \equiv 0$ is always the unique minimizer.

We shall in the following work on a bounded domain, for simplicity considering $d = 1$ and $\Omega = [0, 1]$, which allows some elementary computations, since the dual norm of p is simply $\|P\|_\infty$ with P being the primitive of p with $P(0) = 0$. We do not only provide the explicit examples but also check the computations verifying them, since this might be useful for research on other ℓ^1-type regularizations in the future. Multidimensional examples can be constructed in a rotationally symmetric setup by completely analogous methods.

4.1 Indicator Functions

We start with an example related to indicator functions of subintervals $[a, b]$ related to Meyer's example (cf. [134]) with input data from (51). Since we are considering a bounded image domain and not the whole space we need to subtract a normalizing constant to achieve $\int_\Omega f \, dx = 0$, i.e.

$$f = \alpha \chi_{[a,b]} - \alpha(b - a). \tag{53}$$

We are looking for a subgradient $p = \mu f$, which is characterized by $\|P\|_\infty \leq 1$ and

$$TV(f) = 2\alpha(1 + b - a) = \langle p, f \rangle = \mu \int_0^1 f(x)^2 \, dx = \mu \alpha^2 (1 - b + a)(b - a)(1 + b - a), \tag{54}$$

which can be solved for μ to obtain

$$\mu = \frac{2}{\alpha(b - a)(1 - b + a)}. \tag{55}$$

A Guide to the TV Zoo 21

The primitive is a piecewise linear function given by

$$P(x) = \begin{cases} -\mu\alpha(b-a)x & \text{if } x \leq a, \\ -\mu\alpha(b-a)x + \mu\alpha(x-a) & \text{if } a < x < b, \\ -\mu\alpha(b-a)x + \mu\alpha(b-a) & \text{if } x \geq b. \end{cases}$$

The absolute value of P will attain its maximum at $x = a$ or $x = b$, hence $\|P\|_\infty \leq 1$ is equivalent to the two conditions

$$\frac{2a}{1-b+a} \leq 1 \quad \text{and} \quad \frac{2-2b}{1-b+a} \leq 1, \tag{56}$$

which is obviously satisfied for $a = 1 - b$, i.e. the interval $[a, b]$ is placed symmetrically. Hence, any function of this type is a solution of the nonlinear eigenvalue problem. We also see the scale aspect in the eigenvalue μ, the small eigenvalues (large scales) are related to a high product $\alpha(b-a)$, which somehow characterizes the size of the step in f (the additional factor $1 - b + a$ can be interpreted as the effect of the finite domain).

4.2 Staircasing

It has been observed frequently in numerical experiments that total variation methods inherit staircasing phenomena, i.e. they often tend to produce stair-like step functions instead of smoothly increasing ones for 1D signals, respectively blocky structures for 2D images. In order to obtain a better understanding why staircasing is favoured, we consider two explicit examples in the following. Unfortunately it is not possible to construct an analytical example of noise that exhibits staircasing due to Lipschitz regularity results for minimizers of the ROF functional (cf. [59]), but at least we can give some hints in the right direction. Note also that is was believed for a long while that real staircasing in the sense of piecewise constant regions separated by discontinuities is a typical structure in ROF-denoising, however the recent results in [59] rather indicate that the piecewise constant regions are often rather separated by steep continuous parts instead.

We start with a stair-like structure

$$f = \alpha \sum_{k=1}^{N-1} \chi_{(k/N,1]} - \alpha \frac{N-1}{2}. \tag{57}$$

for N being an even number. The total variation of f equals $\alpha(N-1)$ and thus the a subgradient $p = \mu f$ needs to satisfy

$$\alpha(N-1) = \mu \int f^2\, dx = \frac{2\mu}{N} \sum_{k=0}^{N/2-1} \left(\frac{\alpha(N-1)}{2} - k\alpha\right)^2 = \frac{\mu\alpha^2}{12}(2N^2 - 2N - 1),$$

which yields

$$\mu = \frac{12(N-1)}{\alpha(2N^2 - 2N - 1)}. \tag{58}$$

The primitive is given by

$$P(x) = \mu\alpha \sum_{k=1}^{N-1} \left(x - \frac{k}{N}\right)^+ - \mu\alpha\frac{N-1}{2} x,$$

and $|P|$ attains its maximum at $x = \frac{1}{2}$ as

$$|P(\tfrac{1}{2})| = \mu\alpha \sum_{k=1}^{N/2} \left(\frac{1}{2} - \frac{k}{N}\right) - \mu\alpha\frac{N-1}{4} = \mu\alpha\frac{N}{8} = \frac{3N(N-1)}{2(2N^2 - 2N - 1)}.$$

Thus $\|P\|_\infty \leq 1$ and hence μf is a subgradient, i.e. a solution of the eigenvalue problem.

The second example is a linearly growing function

$$f(x) = \alpha(x - \frac{1}{2}), \tag{59}$$

which due to staircasing we do not expect to solve the eigenvalue problem. The candidate for a minimizer of the ROF functional is of the form

$$u(x) = \max\{-\beta, \min\{\beta, f(x)\}\}. \tag{60}$$

We thus have to verify that $p = \lambda(f - u)$ is a subgradient in $\partial TV(u)$. First of all we compute β from the relation

$$2\beta = TV(u) = \int_0^1 pf\, dx = \lambda\alpha\beta \left(\frac{1}{2} - \frac{\beta}{\alpha}\right)^2,$$

which yields

$$\beta = \alpha \left(\frac{1}{2} - \sqrt{\frac{2}{\lambda\alpha}}\right)^+.$$

The primitive $|P|$ attains its maximum at $x = \frac{1}{2} - \frac{\beta}{\alpha}$ with $|P(x)| = 1$. Thus $p = \lambda(f - u) \in \partial TV(u)$ and hence u is a minimizer of the ROF model, and hence f is partly replaced by piecewise constant structures.

5 Asymptotics and Stability Estimates

In the following we discuss some asymptotic properties in terms of the Lagrange parameter λ as well as stability estimates with respect to λ and the data f. Such asymptotics have first been investigated by Acar and Vogel [2], error estimates are due to [47, 72].

5.1 Asymptotics in λ

We shall now discuss the asymptotic behaviour of minimizers of (41) with respect to λ. For simplicity we assume that A has trivial nullspace, hence the minimizer is unique, and we shall denote the functional E and the minimizer for specific λ as E^λ and u^λ, respectively. The simplest asymptotic for the variational problem (41) is the one for $\lambda \to 0$, for which of course we expect the minimization of the total variation as the asymptotic problem:

Theorem 5.1. *For $\lambda \downarrow 0$ the minimizers of (41) satisfy $u^\lambda \to 0$ in the strong topology of $BV_0(\Omega)$.*

Proof. By the minimization property of u^λ we have

$$\frac{\lambda}{2}\|Au^\lambda - f\|^2 + TV(u^\lambda) = E^\lambda(u^\lambda) \leq E^\lambda(0) = \frac{\lambda}{2}\|f\|^2.$$

Hence $TV(u^\lambda)$ converges to zero (even with order λ). Since TV is an equivalent norm on $BV_0(\Omega)$ the convergence is also in the norm topology. □

The above convergence result and its proof can be applied for various regularization functionals, not only for total variation. A specific feature of total variation (and other ℓ^1-type functionals) is that the convergence to zero arises already for finite λ. The key is to investigate the norm in the dual space of BV, which can be characterized as

$$\|p\|_* = \inf_{g \in L^\infty(\Omega;\mathbb{R}^d), \nabla \cdot g = p} \|g\|_\infty. \tag{61}$$

Meyer [134] showed that

Theorem 5.2. *Let $\lambda \|A^* f\|_* \leq 1$. Then $u \equiv 0$ is a minimizer of (41).*

Proof. Since

$$\partial J(0) = \{ p \in BV^* \mid \|p\|_* \leq 1 \}$$

we obtain

$$p = 0 - \lambda A^* f \in \partial J(0),$$

hence 0 satisfies the optimality condition and is a minimizer. □

More involved and also more interesting is the asymptotic for $\lambda \to \infty$. Since the fitting term dominates in this asymptotic one expects convergence towards a solution of $Au = f$, which can however only be true if $f \in \mathcal{R}(A)$. In this case convergence can be verified in the weak-* topology as we shall see below. Before we provide a general monotonicity property:

Proposition 5.3. *The map $\lambda \mapsto \|Au^\lambda - f\|$ is nonincreasing and the map $\lambda \mapsto TV(u^\lambda)$ is nondecreasing. If $A^* f \neq 0$, then the decrease respectively increase are strict for λ sufficiently large.*

Proof. Let $\mu > \lambda$, then from the definition of u^λ and u^μ as minimizers of the regularized functional we find

$$\frac{\lambda}{2}\|Au^\lambda - f\|^2 + TV(u^\lambda) \leq \frac{\lambda}{2}\|Au^\mu - f\|^2 + TV(u^\mu)$$

and

$$\frac{\mu}{2}\|Au^\mu - f\|^2 + TV(u^\mu) \leq \frac{\mu}{2}\|Au^\lambda - f\|^2 + TV(u^\lambda).$$

By adding these inequalities and simple rearrangement we find $\|Au^\mu - f\| \leq \|Au^\lambda - f\|$. Dividing the first inequality by λ and the second by μ and subsequent analogous comparison yields

$$TV(u^\lambda) \leq TV(u^\mu).$$

We finally show the strict decrease (increase of TV) by contradiction. Assuming $\|Au^\mu - f\| = \|Au^\lambda - f\|$, we conclude from the first two inequalities $TV(u^\lambda) \leq TV(u^\mu)$. Hence u^λ is also a minimizer for the functional with parameter μ and from the optimality conditions we see that there exist $p, q \in \partial TV(u^\lambda)$ such that

$$0 = \lambda A^*(Au^\lambda - f) + p = \mu A^*(Au^\lambda - f) + q,$$

which implies

$$p = \frac{\lambda}{\mu} q.$$

Since both p and q are subgradients of the total variation, we have

$$TV(u^\lambda) = \langle q, u^\lambda \rangle = \frac{\lambda}{\mu} \langle q, u^\lambda \rangle,$$

which can only hold for $u^\lambda = 0$ since $\frac{\lambda}{\mu} < 1$. $u^\lambda = 0$ can only be a solution if

$$A^* f = \frac{1}{\lambda} p.$$

Since $\|p\| \leq 1$ for a subgradient, this can only be true if $\|A^* f\| \leq \frac{1}{\lambda}$, which contradicts $A^* f \neq 0$ if λ is sufficiently large. □

For exact data, i.e. f in the range of A, we can show the following weak convergence result:

Theorem 5.4. *Let $f \in \mathcal{R}(A)$, then every subsequence of (u^λ) has a weak-* convergent subsequence. Every weak-* accumulation point is a solution of $Au = f$ with minimal total variation.*

Proof. Let $\hat{u} \in BV_0(\Omega)$ be an arbitrary solution of $Au = f$. Then by the definition of u^λ we have

$$\frac{\lambda}{2} \|Au^\lambda - f\|^2 + TV(u^\lambda) \leq TV(\hat{u}).$$

Hence $\|Au^\lambda - f\|$ is of order $\lambda^{-1/2}$ and converges to zero, while $TV(u^\lambda) \leq TV(\hat{u})$ is uniformly bounded. From the latter we infer the existence of a weak-* convergent subsequence. For such a weak-* convergent sequence, which we denote by $u_k := u^{\lambda_k}$ with limit $\bar{u}$ we find from the lower semicontinuity of the fitting term

$$\|A\bar{u} - f\| \leq \liminf_k \|Au_k - f\| \leq \liminf_k \sqrt{\frac{2 TV(\hat{u})}{\lambda}} = 0$$

and from the lower semicontinuity of the total variation

$$TV(\bar{u}) \leq \liminf_k TV(u_k) \leq TV(\hat{u}).$$

Thus, $\bar{u}$ is a solution of $Au = f$ with minimal total variation. □

In the more relevant case of noisy data, i.e. f being a perturbation of $g \in \mathcal{R}(A)$, the convergence has to be interpreted relative to the noise, more precisely the estimated variance of the noise,

$$\sigma^2 := \|f - g\|^2. \tag{62}$$

As usual for ill-posed problems (cf. [82]) convergence is achieved only if λ does not converge to infinity too fast relative to the decrease of the noise level.

Theorem 5.5. *Let $g \in \mathcal{R}(A)$ and f^σ be noisy data with variance σ^2, i.e. satisfying (62). Moreover, let δ_σ be an estimate of the noise variance such that there exists $c > 0$ with*

$$\delta_\sigma \geq c\sigma \qquad \forall\, \sigma > 0. \tag{63}$$

Let the Lagrange parameter be chosen such that $\lambda = \lambda(\delta_\sigma) \to \infty$ as $\delta_\sigma \to 0$ such that $\lambda(\delta_\sigma)\delta_\sigma^2 \to 0$. Then every subsequence of (u^λ) has a weak-$*$ convergent subsequence and every weak-$*$ accumulation point is a solution of $Au = g$ with minimal total variation.

Proof. In the following we always write λ meaning $\lambda(\delta_\sigma)$ for brevity. Let $\hat{u}$ be any solution of $Au = g$. From the definition of the minimizer we have

$$\frac{\lambda}{2}\|Au^\lambda - f^\sigma\|^2 + TV(u^\lambda) \leq \frac{\lambda}{2}\|f^\sigma - g\|^2 + TV(\hat{u}).$$

Using the noise variance and the properties of the estimator we find

$$\frac{\lambda}{2}\|Au^\lambda - f^\sigma\|^2 + TV(u^\lambda) \leq \frac{\lambda}{2}\sigma^2 + TV(\hat{u}) \leq \frac{\lambda}{2c}\delta_\sigma^2 + TV(\hat{u}).$$

Since the right-hand side converges to $TV(\hat{u})$, the total variation of u^λ is uniformly bounded and $\|Au^\lambda - f^\sigma\|$ converges to zero, from which the convergence properties can be infered as in the proof of Theorem 5.4. □

5.2 Stability Estimates

Error estimation for variational models like the ROF-model was an open problem in imaging and inverse problems for a rather long period, which is due to the difficulty to find an appropriate error measure. Intuitively it becomes obvious quite soon that the total variation norm (or seminorm) cannot be an appropriate measure, since it penalizes small visual differences too strongly.

Moreover, the above asymptotic analysis gives only weak-$*$ convergence, but no strong convergence in $BV_0(\Omega)$, hence also error estimation seems out of reach. In [47] a generalized Bregman distance was introduced as an error measure, which allowed to derive suitable estimates consistent with the norm estimates in the case of penalization with a squared norm. The estimation in Bregman distances has been widely accepted since then and extended to various other situations (cf. [27, 28, 88, 98, 99, 107, 127, 160, 161]). We shall here mainly recall the results obtained in [47], and start by defining Bregman distances.

Definition 5.6. Let $J : X \to \mathbb{R} \cup \{+\infty\}$ be a convex functional. For each $p \in \partial J(u)$,

$$D_J^p(v, u) = J(v) - J(u) - \langle p, v - u \rangle \tag{64}$$

is a generalized Bregman distance between u and v. Moreover, for $q \in \partial J(v)$

$$D_J^{q,p}(v, u) = \langle p - q, u - v \rangle \tag{65}$$

is called a symmetric Bregman distance between u and v.

The Bregman distance is not a distance in the classical sense, since it is not symmetric in its original form, and it is not strict, i.e. $D_J^p(v, u) = 0$ is possible if $v \neq u$. For the total variation in particular contrast changes are not detectable in the Bregman distance, i.e. for a smooth monotone map F the functions u and $F(u)$ have zero distance (for any choice of subgradients). On the other hand differences in edges can be measured well in Bregman distances for the total variation (cf. [47, 50]), which is of particular interest for total variation techniques.

The stability of the variational problem in terms of the data f is given by the following result:

Theorem 5.7. *Let u_i, $i = 1, 2$ be the minimizers of* (41) *with data f_i. Then there exists $p_i \in \partial TV(u_i)$ such that the estimate*

$$\lambda \|Au_1 - Au_2\|^2 + 2D_{TV}^{p_1,p_2}(u_1, u_2) \leq \lambda \|f_1 - f_2\|^2 \tag{66}$$

holds.

Proof. Since the u_i are minimizers, they satisfy the optimality condition

$$\lambda A^*(Au_i - f_i) + p_i = 0, \qquad p_i \in \partial TV(u_i).$$

Subtracting them and taking the duality product with $u_1 - u_2$ we find

$$\lambda \langle A^*(Au_1 - Au_2), u_1 - u_2 \rangle + \langle p_1 - p_2, u_1 - u_2 \rangle = \lambda \langle A^*(f_1 - f_2), u_1 - u_2 \rangle.$$

Using the definition of the adjoint we can rewrite the first term on the left-hand side as $\lambda \|Au_1 - Au_2\|^2$ and we estimate the right-hand side as

$$\langle A^*(f_1 - f_2), u_1 - u_2 \rangle = \langle f_1 - f_2, A(u_1 - u_2) \rangle \leq \|f_1 - f_2\| \|Au_1 - Au_2\|$$

$$\leq \frac{1}{2} \|f_1 - f_2\|^2 + \frac{1}{2} \|Au_1 - Au_2\|^2.$$

Inserting these relations and multiplying by two we obtain (66). □

Besides stability at fixed λ, quantitative estimates concerning the asymptotics in λ are of interest. Due to the ill-posedness of the operator equation with A additional conditions on the solution are needed. In particular the existence of a Lagrange parameter for the constrained problem

$$TV(u) \to \min_{u \in BV_0(\Omega), Au=g}, \qquad (67)$$

characterizing the solution of minimal total variation, is a standard regularity assumption. This can be formulated as the source condition (cf. [47, 72])

$$A\hat{u} = g, \qquad \hat{p} = A^*w \in \partial TV(\hat{u}). \qquad (68)$$

Theorem 5.8. *Let u^λ be the minimizer of E and let $\hat{u}$, $\hat{p}$ satisfy (68). Then there exists $p \in \partial TV(u^\lambda)$ such that the estimate*

$$\frac{\lambda}{2} \|Au^\lambda - A\hat{u}\|^2 + D_{TV}^{p,\hat{p}}(u^\lambda, \hat{u}) \le \frac{1}{\lambda}\|w\|^2 + \lambda\|g - f\|^2. \qquad (69)$$

holds.

Proof. We start from the optimality condition for u^λ and subtract $(\hat{p} + \lambda A^*g)$ on both sides to have

$$\lambda A^*(Au^\lambda - A\hat{u}) + p - \hat{p} = -\hat{p} + \lambda A^*(f - g).$$

The duality product with $u^\lambda - \hat{u}$ yields

$$\lambda \|Au^\lambda - A\hat{u}\|^2 + D_{TV}^{p,\hat{p}}(u^\lambda, u) = -\langle \hat{p}, u^\lambda - \hat{u}\rangle + \lambda \langle f - g, Au^\lambda - A\hat{u}\rangle.$$

For the first term on the right-hand side we insert (68) and apply Young's inequality to obtain

$$-\langle \hat{p}, u^\lambda - \hat{u}\rangle = -\langle w, Au^\lambda - A\hat{u}\rangle \le \frac{\lambda}{4}\|Au^\lambda - A\hat{u}\|^2 + \frac{1}{\lambda}\|w\|^2,$$

and for the second term we directly apply Young's inequality, which finally yields the desired estimate. □

The error estimate (69) consists of two parts, a decreasing term of order $\frac{1}{\lambda}$ with a constant depending on the source condition (thus on the smoothness of $\hat{u}$) and a second term simply depending on the noise variance. Balancing the terms gives an indication how to choose the optimal λ at given noise, which is of course not directly possible in practice since $\|w\|$ is not known.

6 Bregman Iterations and Scale Spaces

Although the ROF-model (2) had great success in applied imaging, there is still one deficiency remaining, which is loss of contrast in the reconstruction compared to the original image. This aspect was highlighted by Meyer [134], who showed that an application of the ROF-model (2) to the characteristic function of a ball (f from (51)) results in a shrunk version as the minimizer, with the shrinkage proportional to $\frac{1}{\lambda}$ (cf. Sect. 4). A fundamental reason for the loss of contrast is a systematic error of variational regularization methods, since in the case of exact data $A\hat{u} = f$ we obviously have

$$J(u) \leq J(\hat{u}) - \frac{\lambda}{2}\|Au - f\|^2 < J(\hat{u})$$

for (17), since typically $Au \neq f$. For J being the total variation, this means that the total variation of the reconstruction is smaller than the total variation of the exact solution, which results in the above mentioned contrast loss.

An approach to overcome this issue has been presented in [148], we here start with a derivation in the case of the ROF-model (2). Assume that u_1 is the minimizer of (2) for fixed λ chosen rather to oversmooth (hence there should be no significant noise part in u_1). Then the main deficiency will be that the residual

$$v_1 = f - u_1 \tag{70}$$

still contains too much signal—actually enough to restore quite detailed information about the original image as demonstrated in [117]. Since this part is obviously shrunk too strongly by the ROF-model one might try and add the "noise" back to the image, so that in another run of the minimization this compensates the loss of contrast. This means a new image u_2 is computed from minimizing

$$J_2(u) := \frac{\lambda}{2}\int_\Omega (u - f - v_1)^2 dx + TV(u). \tag{71}$$

This minimization can be interpreted in an alternative way if we take a look at the optimality condition for (2), which together with the definition of v_1 implies

$$p_1 = \lambda v_1 = \lambda(f - u_1) \in \partial TV(u_1). \tag{72}$$

Hence, we can rewrite J_2 as

$$J_2(u) = \frac{\lambda}{2}\int_\Omega (u - f)^2 dx + TV(u) - \langle p_1, u\rangle + \langle p_1, f + v_1\rangle. \tag{73}$$

Ignoring the constant term $\langle p_1, f + v_1 \rangle$ we observe that minimizing J_2 is equivalent to minimize the fitting functional penalized by the Bregman distance to the last iterate u_1 (with subgradient p_1).

As a consequence of the above reasoning we can generalize the one-step contrast correction introduced via (73) to an iterative scheme. We start with u_0 and $p_0 \in \partial TV(u_0)$ and then subsequently compute a sequence u_k via

$$u_{k+1} = \arg\min_{u \in BV(\Omega)} \left[\frac{\lambda}{2} \int_\Omega (u-f)^2 dx + D_{TV}^{p_k}(u, u_k) \right], \quad p_k \in \partial TV(u_k). \tag{74}$$

This iterative refinement scheme introduced and analyzed in [148] turns out to be equivalent to the so-called *Bregman iteration* introduced in [37] for quite general, but continuously differentiable functionals. The more general form of the Bregman iteration for computing a minimizer of

$$J(u) \to \min_u \quad \text{subject to } u \in \arg\min_v H(v, f) \tag{75}$$

is given by

$$u_{k+1} = \arg\min_{u \in BV(\Omega)} \left[H(u, f) + D_J^{p_k}(u, u_k) \right], \quad p_k \in \partial J(u_k). \tag{76}$$

The problem with convex fitting term $H(., f)$ can be generalized to convex and possibly nondifferentiable J, but in this case some regularity of H is still needed to guarantee well-definedness of the iteration and suitable properties of subgradients.

6.1 Interpretations of the Bregman Iteration

In the following we further discuss different interpretations of the Bregman iteration in the case of a quadratic fitting functional (Gaussian noise) and general convex regularization, i.e., we consider

$$J(u) \to \min_u \quad \text{subject to } u \in \arg\min_v \frac{\lambda}{2} \|Av - f\|^2, \tag{77}$$

where A is a linear operator. The corresponding Bregman iteration is defined by

$$u_{k+1} = \arg\min_{u \in BV(\Omega)} \left[\frac{\lambda}{2} \|Au - f\|^2 + D_J^{p_k}(u, u_k) \right], \quad p_k \in \partial J(u_k). \tag{78}$$

We observe that the optimality condition for the minimizer in (78) yields

$$\lambda A^*(Au_{k+1} - f) + p_{k+1} = p_k, \tag{79}$$

which can also be seen as an update rule for the dual variable. Such a dual interpretation of the Bregman iteration will be derived in the following.

Primal-Dual Iteration

If the data f lie in the domain of A, then we can write (77) equivalently as

$$J(u) \to \min_{u} \quad \text{subject to } Au = f. \tag{80}$$

The associated Lagrangian for this problem is

$$L_0(u, w) = J(u) + \langle w, Au - f \rangle \tag{81}$$

which can be augmented without changing minimizers to

$$L(u, w) = J(u) + \frac{\lambda}{2} \|Au - f\|^2 + \langle w, Au - f \rangle \tag{82}$$

and the optimality conditions for a saddle point (if it exists, which is a regularity property not automatic for A being compact) are given by

$$0 \in \partial J(u) + \lambda A^*(Au - f) + A^* w = \partial_u L(u, w),$$
$$0 = Au - f = \partial_w L(u, w).$$

A primal dual scheme iterates on both optimality conditions, e.g. by using a quadratic penalty on w. This yields

$$0 \in \partial J(u_{k+1}) + \lambda A^*(Au_{k+1} - f) + A^* w_k,$$
$$0 = Au_{k+1} - f = \tau(w_{k+1} - w_k).$$

From the first relation we see that w_{k+1} is related to the dual variable in the Bregman iteration by $p_{k+1} = -A^* w_{k+1}$. With $\lambda = \frac{1}{\tau}$ and applying A^* to the second relation, we see that this primal-dual scheme is exactly equivalent to the Bregman iteration, cf. (79). The iteration in the variable w_k can be interpreted as a generalization of the original motivation (70), it describes the update of the residuals. It also delivers an alternative way to realize the Bregman iteration as in (71), the minimizer can be computed from minimizing

$$u_{k+1} = \arg\min_{u \in BV(\Omega)} \left[\frac{\lambda}{2} \|Au - f + \frac{1}{\lambda} w_k\|^2 + J(u) \right], \tag{83}$$

such that again the problem can be realized as the original variational model just changing the input data.

The primal-dual interpration also induces an alternative to the Bregman iteration, by considering an Uzawa-type iteration with an explicit treatment of the constraint

$$0 \in \partial J(u_{k+1}) + \lambda A^*(Au_{k+1} - f) + A^* w_k,$$
$$0 = Au_k - f = \tau(w_{k+1} - w_k).$$

This scheme is equivalent to the iteration

$$u_{k+1} = \arg\min_{u \in BV(\Omega)} \left[\lambda \langle Au, Au_k - f \rangle + D_J^{p_k}(u, u_k) \right], \tag{84}$$

which linearizes the fitting term and was hence called *linearized Bregman iteration* in [55–57] in applications to compressed sensing (J being an ℓ^1-norm). The scheme was also called *Landweber-type method* in an application to total variation methods (even allowing A nonlinear) in [20, 21]—due to analogies with classical iteration schemes for inverse problems.

Updated Bayesian Prior

In the Bayesian setting we computed the MAP estimate by maximizing the posterior probability density

$$p(u|f) \sim p(f|u) p_0(u). \tag{85}$$

The prior density is usually centered at zero, i.e., $u \equiv 0$ maximizes p_0. Given the knowledge from the solution of (17) one could however update the prior probability distribution. Instead of having a probability centered at zero, it makes more sense to use a probability at least locally centered at u_1, where u_1 is the solution of (17). A simple way to do so is to shift the log likelihood by an affinely linear term, namely the first-order Taylor approximation around u_1, in order to obtain a new prior log likelihood and probability density, respectively. The shifted prior probability

$$p_1(u) \sim \exp\left(\log p_0(u) - \log p_0(u_1) + \frac{p_0'(u_1)(u-u_1)}{p_0(u_1)} \right) \tag{86}$$

now has a maximum at u_1 instead of zero. This process of shifting the prior probability is repeated during the Bregman iteration.

In the case of total variation we even find that the prior probability p_1 has a maximum at each multiple of u_1, since $\partial TV(u_1) = \partial TV(cu_1)$ for any $c \in \mathbb{R}$. This means that the contrast is left competely open by the new prior probability, a reason why the Bregman iteration can improve upon contrast losses.

Dual Interpretation

If A^*A is invertible, the Bregman iteration can be interpreted as a dual ascent scheme, preconditioned with $(A^*A)^{-1}$. Defining the standard convex conjugate (dual functional)

$$J^*(p) = \sup_u [\langle p, u \rangle - J(u)], \tag{87}$$

we find $u_{k+1} \in \partial J^*(p_{k+1})$ under appropriate conditions, since $p_{k+1} \in \partial J(u_{k+1})$. Hence, (79) can be rewritten as

$$(A^*A)^{-1}(p_{k+1} - p_k) \in \lambda A^* f - \lambda \partial J^*(p_{k+1}) \tag{88}$$

Hence, the Bregman iteration in the dual variable computes

$$p_{k+1} = \arg\max_p \left(-\frac{1}{2\lambda} \|(A^*A)^{-1/2}(p_{k+1} - p_k)\| + \langle p, A^* f \rangle - J^*(p) \right). \tag{89}$$

This is just a proximal point algorithm (with special norm) for the dual problem

$$\langle p, A^* f \rangle - J^*(p) \to \max_p. \tag{90}$$

A similar relation holds for the linearized Bregman iteration, in this case we find

$$(A^*A)^{-1}(p_{k+1} - p_k) \in \lambda A^* f - \partial \lambda J^*(p_k), \tag{91}$$

i.e., we recover a dual (sub)gradient ascent scheme.

Geometric Interpretation for Total Variation

In the special case of J denoting the total variation we can add a further interpretation based on a geometric viewpoint. The formal version of the Bregman distance is

$$D_J^p(v, u) = \int_\Omega (|\nabla u| - \frac{\nabla v}{|\nabla v|} \nabla u) \, dx = \int_\Omega (\frac{\nabla u}{|\nabla u|} - \frac{\nabla v}{|\nabla v|}) \cdot \nabla u \, dx. \tag{92}$$

The difference $\frac{\nabla v}{|\nabla v|} - \frac{\nabla u}{|\nabla u|}$ can be interpreted as the difference in normals to level sets or edge sets. Hence the Bregman distance measures the alignment of level and edge sets, and in the iteration (76) there is no further penalty for those edge sets aligned with the ones of the previous iterates. In particular there is no penalty if the edge sets are aligned, but the height of the jump is different. This allows the Bregman iteration to correct contrast losses inherent in (2). The idea of matching normal fields has also been used previously, e.g. in pan-sharpening of color and spectral images (cf. [22]), the connection to the Bregman distance has been established recently (cf. [136]).

6.2 Convergence Analysis

In the following we provide a convergence analysis of the Bregman iteration (78). For simplicity we assume $u_0 = 0$ and $p_0 = 0 \in \partial J(u_0)$, an extension to arbitrary initial values satisfying the consistency condition $p_0 \in \partial J(u_0)$ is quite obvious. Our basic assumption is again the existence of an exact solution $\hat{u}$ satisfying $A\hat{u} = g$ and f is a noisy version of the exact data. We do not further discuss the well-definedness of the iterates u_k, since due to (83) this issue is equivalent to existence and uniqueness of the original variational problem for arbitrary data, which has been discussed above. Besides monotone decrease of the residual, the fundamental property of the Bregman iteration is that the Bregman distance to the exact solution is decreasing up to some point, with a dissipation governed by the least-squares functional.

Lemma 6.1. *Let u_k be a sequence generated by the Bregman iteration (78). Then the inequalities*

$$\|Au^{k+1} - f\| \leq \|Au^k - f\| \tag{93}$$

and

$$D_J^{p_{k+1}}(\hat{u}, u_{k+1}) + \frac{\lambda}{2}\|Au_{k+1} - f\|^2 + D_J^{p_k}(u_{k+1}, u_k) \leq D_J^{p_k}(\hat{u}, u_k) + \frac{\lambda}{2}\|g - f\|^2 \tag{94}$$

hold for all $k \geq 0$.

Proof. Due to the positivity of the Bregman distance and since $D_J^{p_k}(u_k, u_k) = 0$, we directly obtain (93) from comparing the functional values in (78) for the minimizer u_{k+1} and for u_k. We further have

$$\begin{aligned}
D_J^{p_{k+1}}(\hat{u}, u_{k+1}) &- D_J^{p_k}(\hat{u}, u_k) + D_J^{p_k}(u_{k+1}, u_k) \\
&= \langle p_{k+1} - p_k, u_{k+1} - \hat{u} \rangle \\
&= -\lambda \langle Au_{k+1} - f, Au_{k+1} - g \rangle \\
&= -\lambda \|Au_{k+1} - f\|^2 + \lambda \langle Au_{k+1} - f, g - f \rangle \\
&\leq -\frac{\lambda}{2}\|Au_{k+1} - f\|^2 + \frac{\lambda}{2}\|g - f\|^2,
\end{aligned}$$

where we have inserted the optimality condition (12) for $p_{k+1} - p_k$. □

Of particular importance for the convergence of the scheme is the boundedness of the total variation, which will again yield weak-* convergence of subsequences. In the case of noisy data the uniform bound can only be guaranteed for a finite number of iterations, which can be used determine the stopping index:

Lemma 6.2. *Let u_k with subgradients $p_k \in \partial J(u_k)$ be a sequence generated by (78), then*

$$J(u_m) \leq 5J(\hat{u}) + 2\lambda m \|f - g\|^2 \qquad (95)$$

for all $m \in \mathbb{N}$.

Proof. Summing (94) and (79) we obtain

$$D_J^{p_m}(\hat{u}, u_m) + \frac{\lambda}{2} \sum_{k=1}^{m} \|Au_k - f\|^2 \leq D_J^{p_0}(\hat{u}, u_0) + \frac{\lambda m}{2} \|g - f\|^2$$

respectively

$$p_m = -\lambda \sum_{k=1}^{m} A^*(Au_k - f).$$

From the first inequality we obtain using nonnegativity of $D_J^{p_m}(\hat{u}, u_m)$ that

$$\|Au_m - f\|^2 \leq \frac{1}{m} \sum_{k=1}^{m} \|Au_k - f\|^2 \leq \frac{2}{\lambda m} J(\hat{u}) + \|g - f\|^2$$

and from the identity for p_m we estimate

$$J(u_m) - J(\hat{u}) \leq \langle p_m, u_m - \hat{u} \rangle = -\lambda \sum_{k=1}^{m} \langle Au_k - f, Au_m - g \rangle$$

$$\leq \lambda \sum_{k=1}^{m} \|Au_k - f\| (\|Au_m - f\| + \|f - g\|)$$

$$\leq \lambda \sum_{k=1}^{m} \|Au_k - f\|^2 + \frac{\lambda m}{2} \left(\|Au_m - f\|^2 + \|f - g\|^2 \right)$$

$$\leq 4J(\hat{u}) + 2\lambda m \|f - g\|^2,$$

which yields the assertion. □

Hence in the case of exact data ($f = g$) we have $J(u_m) \leq 5J(\hat{u})$ and thus uniform boundedness of the iterates. Thus we have the following result by standard reasoning as above:

Theorem 6.3 (Convergence for Exact Data). *Let u_k with subgradients $p_k \in \partial J(u_k)$ be a sequence generated by (78) and let $f = g$. Then there exists a weak-* convergent subsequence and every weak-* accumulation point solves $Au = g$. If A has trivial nullspace, then the sequence u_k converges to the unique solution $\hat{u}$ of $A\hat{u} = g$ in the weak-* topology.*

In the case of noisy data, boundedness is obtained if $m\|f - g\|^2 \leq C$ for some constant C. This provides a criterion for the choice of a stopping index k_* in dependence of the noise level $\|f - g\|$:

Theorem 6.4 (Semi-Convergence for Noisy Data). *Let u_k^f with subgradients $p_k^f \in \partial J(u_k^f)$ be a sequence generated by (78) for specific noisy data f and let $k_* = k_*(f)$ be a stopping index satisfying*

$$k_*(f)\|f - g\|^2 \leq C \tag{96}$$

for some constant $C > 0$. Then, if f_n is a sequence with $\|f_n - g\| \to 0$, there exists a weak-* convergent subsequence of $\{u_{k_*(f_n)}^{f_n}\}$ and every weak-* accumulation point solves $Au = g$. If A has trivial nullspace, then the sequence $u_{k_*(f_n)}^{f_n}$ converges to the unique solution $\hat{u}$ of $A\hat{u} = g$ in the weak-* topology.

The convergence analysis can be refined further: In [148] an a-posterior stopping via the discrepancy principle, i.e.,

$$k_*(f) = \min\{ k \mid \|Au_k^f - f\| \leq \tau\|g - f\| \} \tag{97}$$

for $\tau > 1$ fixed, has been shown to yield convergence. In [50] error estimates for the Bregman iteration have been derived under the source conditions discussed above.

6.3 Inverse Scale Space Methods

In general it seems favourable to choose a rather small λ in the Bregman iteration, i.e. each iteration step is actually oversmoothing. In this way one slowly iterates to a point where a stopping criterion is satisfied and should obtain a reasonable reconstruction at this point of stopping. If λ is too large one might end up with a final reconstruction of bad quality just since the last iteration step went too far. The most extreme case for large λ could be that the Bregman iteration stops after only one step, i.e. at the solution of (17) for large λ—hence it cannot lead to an improvement. This reasoning indicates to study the limit $\lambda \to 0$ with a simultaneous increase of iteration steps, i.e. $t = k\lambda$ being constant. We can observe the result of this limit by rewriting (79) as

$$\frac{p(t+\lambda) - p(t)}{\lambda} = A^*(f - Au(t+\lambda)), \tag{98}$$

where we define $u(k\lambda) := u_k$ and $p(k\lambda) := p_k$. We see that the Bregman iteration is just the implicit Euler discretization of a flow, the so-called *inverse scale space flow*

$$\partial_t p = A^*(f - Au), \quad p \in \partial J(u). \tag{99}$$

This flow has been derived in [49] and called *inverse scale space flow*, since it coincides in the case of quadratic regularization with the inverse scale space methods in [169]. The wording inverse is due to the opposite behaviour to classical scale space methods or diffusion filters (cf. [194]). While scale space methods start with the noisy image and smooth increasingly with proceeding time (coarsening of scales), inverse scale space methods start with the coarsest scale (e.g. the mean value of the image intensity) and roughen the image with increasing time (refinement of scales). Inverse scale space methods are computationally more involved than classical diffusion filtering techniques, but there are several advantages: First of all, inverse scale space methods can be applied to very general imaging tasks such as deblurring or other inverse problems, while scale space techniques are restricted mainly to denoising-type problems. Secondly, appropriate stopping rules can be derived for inverse scale spaces such as the discrepancy principle

$$t_*(f) = \inf\{\, t \mid \|Au(t) - f\| \leq \tau \|g - f\| \,\}. \tag{100}$$

Finally, the quality of reconstructions obtained with inverse scale space methods seems to be better compared to scale space methods in particular in the case of total variation and ℓ^1-regularization, both visually and with respect to some error measures (cf. [49, 53]).

Here we shall neither provide details on the analysis of the well-posedness of the flow (cf. [51] for the challenging case of total variation) nor on the convergence analysis (cf. [49]) or error estimates (cf. [50]), which are analogous to the case of Bregman iterations.

It is interesting to consider a relaxation of the inverse scale space method introduced in [49], with an auxiliary variable w:

$$\epsilon \partial_t u = \gamma A^*(f - Au) + w - p \tag{101}$$

$$\partial_t w = A^*(f - Au) \tag{102}$$

$$p \in \partial J(u) \tag{103}$$

with small parameters ϵ and γ. In the limit $\epsilon \to 0$ and $\gamma \to 0$ one expects that $w - p$ tends to zero, and hence one recovers the inverse scale space flow. A detailed analysis of this scheme has been given in [126]. An advantage for the numerical realization is the fact that the relaxed inverse scale space method consists of two evolution equations, which can be integrated more efficiently than the original inverse scale space flow.

The application of Bregman iterations and inverse scale space methods to wavelet denoising has been investigated in [200]. It is well-known that the variational problem analogous to the ROF functional (cf. [62])

$$\frac{\lambda}{2} \int_\Omega (\sum c_j \psi_j - f)^2 + \sum |c_j| \to \min_{(c_j)}, \qquad (104)$$

with $\{\psi_j\}$ denoting the wavelet basis, results into soft-thresholding (also called soft shrinkage, cf. [79]). The inverse scale space method instead yields a hard-thresholding formula (with thresholding parameter related to $\frac{1}{t}$) and Bregman iterations yields firm thresholding (cf. [93]), an intermediate between hard and soft thresholding.

6.4 Linearized Bregman Methods

The realization of the Bregman iteration still needs the solution of variational problems of similar structure as the total variation regularized least-squares problem. Since one expects small steps anyway, it seems reasonable to approximate the fidelity. The simplest approximation is a linearization of the first term, which yields the linearized Bregman iteration

$$u_{k+1} = \arg\min_{u \in BV(\Omega)} \left[H(u_k, f) + \partial_u H(u_k, f)(u - u_k) + D_J^{p_k}(u, u_k) \right] \qquad (105)$$

with $p_k \in \partial J(u_k)$. In the case of a quadratic fidelity, the linearization yields the optimality condition

$$\lambda A^*(Au_k - f) + p_{k+1} = p_k. \qquad (106)$$

One observes that the operator A and its adjoint only need to be evaluated in order to carry out the linearized Bregman iterations, which makes the scheme particularly attractive for applications where A is nonlocal, e.g. a convolution operator or even a more complicated image formation model.

A possible disadvantage, which indeed is observed in all numerical tests, is that (106) is not solvable if J is not strictly convex, since in this case the functional to be minimized in (105) is not bounded below. Therefore a multiple of a squared norm is usually added to the regularization in such cases in order to make the scheme work, e.g. in the case of total variation

$$J(u) = \frac{\kappa}{2}\|u\|_{L^2}^2 + |u|_{TV}. \qquad (107)$$

Ideally one tries to choose κ small in order to approximate pure total variation regularization, which can indeed be achieved. Thus a similar behaviour of reconstructions as in the original Bregman iteration is obtained, with a significant improvement of computational efficiency if A can be evaluated easily. A key observation for the realization of efficient computational methods is that the subproblem in each step of such a linearized Bregman iteration can be rewritten as the minimizer of

$$J^k(u) = \frac{\kappa}{2}\|u - f^k\|_{L^2}^2 + |u|_{TV}, \qquad (108)$$

with f^k precomputed from the previous iteration. Thus one just solves a TV-denoising problem via the ROF-model in each step. Recently the linearized Bregman method and variations were applied with some success to compressive sampling problems (cf. [55, 56]), with a regularization term of the form

$$J(u) = \frac{\kappa}{2}\|u\|_{L^2}^2 + |\langle u, \varphi_k \rangle|_{\ell^1}, \qquad (109)$$

where (ψ_k) is an orthonormal basis of some Hilbert space. In that case the subproblem in each step can be computed explicitely by thresholding of coefficients. The convergence analysis of linearized Bregman iterations can be found in [20, 21] in the total variation case, in [56] in the case of ℓ^1 regularization, and in [170] for regularization with powers of strictly convex norms in Banach spaces.

6.5 Total Variation Flow

Total variation flow (or TV flow) is the scale space version of total variation in the denoising case (cf. [38, 40, 41, 194]). It is obtained if the ROF-model (2) is considered as an implicit time discretization of a flow with time step $\tau = \frac{1}{\lambda}$ and iterated to compute $u(t + \tau)$ as a solution of

$$\frac{\lambda}{2} \int_\Omega (u - u(t))^2 \, dx + TV(u). \qquad (110)$$

From the optimality condition one can derive an evolution law in the form of the differential inclusion

$$\partial_t u(t) = -p(t), \qquad p(t) \in \partial TV(u(t)) \qquad (111)$$

with initial value

$$u(0) = f. \tag{112}$$

An analysis of total variation flow can be given in two ways: In [7–9, 25] the evolution is interpreted as a degenerate parabolic problem and appropriate entropy solution techniques are constructed. The approach in [87] rather stays with the variational interpretation of (111) as a gradient flow for total variation and defines weak solutions via a variational inequality. The existence of the flow as well as numerical schemes are then analyzed using a smooth approximation of total variation. Uniqueness and large time-behaviour (cf. [87]), as well as stability estimates (cf. [50]), can be infered directly from the gradient flow interpretation. Note that for two solutions u_1 and u_2 of (111) with subgradients p_1 respectively p_2 we obtain using the duality product with $u_1 - u_2$

$$\frac{1}{2}\frac{d}{dt}\int_\Omega (u_1 - u_2)^2 \, dx = \langle \partial_t u_1 - \partial_t u_2, u_1 - u_2 \rangle = -\langle p_1 - p_2, u_1 - u_2 \rangle.$$

The right-hand side equals a negative Bregman distance and is hence not positive, which directly yields that the flow is non-expansive in the L^2-norm. Further inspections of the behaviour of the dual variable p can then be used to derive error estimates also in the Bregman distance (cf. [50]).

Visually solutions obtained with TV flow behave similarly than those obtained from minimizing the ROF-model, in particular they suffer from the same systematic errors. A surprising result in [39] shows that they are actually the same in the case $d = 1$. In particular for small noise levels (for which one uses only a small final time in the flow), the solutions of (111) can be integrated very efficiently by explicit methods and hence outperform the variational and inverse scale space model with respect to computational speed. However, due to the systematic errors and the difficulties to choose a stopping time in a robust automatic way, some care is needed. For problems different from denoising the total variation flow and similar scale space methods cannot be used, since the construction would rely on the inversion of the operator A, which is neither stable nor computationally efficient in most typical imaging tasks.

6.6 Other Multiscale Approaches

A multiscale approach with similar appearance as Bregman iterations was introduced by Tadmor et al. [181], who constructed hierarchical decompositions based on the ROF-functional. The starting point is the usual decomposition

$$f = u_0 + v_0, \quad u_0 = \arg\min_{u \in BV(\Omega)} \left[\frac{\lambda_0}{2} \|u - f\|_{L^2}^2 + TV(u) \right], \tag{113}$$

typically with λ_0 small in order to obtain really a coarse scale in u_0. The residual v_0 respectively a further sequence of residuals v_j can now be further decomposed into

$$v_{j-1} = u_j + v_j, \quad u_j = \arg\min_{u \in BV(\Omega)} \left(\frac{\lambda_j}{2} \|u - v_{j-1}\|_{L^2}^2 + TV(u) \right). \tag{114}$$

In order to guarantee a decrease of scales λ_j needs to be decreased, usually via the dyadic sequence

$$\lambda_j = 2\lambda_{j-1}. \tag{115}$$

From this scheme one obtains a decomposition in the form

$$f = \sum_{j=0}^{\infty} u_j, \tag{116}$$

where increasing index marks decreasing scales.

Variants of this approach have been investigated in [182], where rather general norms have been investigated, and in [185], where the fidelity term has been changed to a squared H^{-1}-norm. A time-continuous version analogous to inverse scale space flows has been investigated by Athavale and Tadmor [11].

7 Nonquadratic Fidelity

So far we have laid our attention on quadratic fidelity terms, i.e. Gaussian noise models in the Bayesian framework, since they allow an introduction to the analysis without too many technical issues. In practice one sometimes encounters nonquadratic fidelities instead, due to two reasons: non-Gaussianity of the noise model or / and nonlinearity of the image formation model. We will briefly give an overview on recent developments for those two cases, which are both still subject of very active research.

7.1 Non-Gaussian Noise

In several applications we find noise models different from the simple additive Gaussian assumptions. Such cases can still be treated in the MAP framework, since only $p(f|u)$ is to be changed. In the following we discuss some interesting cases and provide useful references.

Laplace Noise

Motivated by noise models with additive Laplacian distributed errors, various authors have considered models of the form

$$E(u) = \lambda \int_\Sigma |f - Au| \, dy + TV(u), \qquad (117)$$

with $A : BV(\Omega) \to L^1(\Sigma)$ (cf. [70, 101]).

The existence of minimizers is analogous to the quadratic case considered above, but uniqueness remains unclear since the fidelity term is not strictly convex even for A having trivial nullspace. An interesting property of L^1 fidelities is the possibility of exact reconstruction. If the data f satisfy a source condition, then the reconstruction is exact for finite λ, i.e. for λ sufficiently large a minimizer of (117) satisfies $Au = f$ (cf. [28]). Bregman iterations have been generalized to the case of L^1-fidelity in [104], it turned out that well-posedness and convergence can only be achieved for λ changing during the iteration.

For data being indicator functions, the L^1-fidelity model has also interesting geometrical properties, first of all there is an indicator function solution again (cf. [70]) and the structure of the resulting edge set can be analyzed (cf. [188, 189]).

Poisson Statistics

Poisson-distributed data are the natural case for many image formation models, where one registers counts of photons (e.g. CCD-cameras, cf. [173, 174]) or positrons (e.g. PET, cf. [183, 197]). In the case of high count rates the Poisson distribution can be well approximated by Gaussians, but for lower count rates it becomes more appropriate to use variational models asymptotically derived directly from the Poisson statistics. In the case of denoising this yields variational problems of the form

$$E(u) = \lambda \int_\Omega \left(f \log \frac{f}{u} - f + u \right) dx + TV(u) \qquad (118)$$

to be minimized subject to nonnegativity. The extension to image reconstruction problems is given by

$$E(u) = \lambda \int_\Sigma \left(f \log \frac{f}{Au} - f + Au \right) dy + TV(u) \qquad (119)$$

with Σ denoting the measurement domain.

Such models have been proposed and tested already in the late nineties (cf. [113, 150]), motivated in particular by the bad noise statistics in PET. Due to the algorithms used at that time, the application for large regularization parameters and effective implementation was not able at this time, which prevented significant practical impact despite the enormous potential in such applications. With improvements in the computational approaches such as splitting methods these nonlinear models recently again received strong attention and now seem to find their practice in microscopy and PET (cf. [23, 42, 44, 77, 123, 165]).

An interesting aspect of nonlinear fidelities motivated by the Poisson case (but actually more general) has been investigated in [44], namely the appropriate construction of Bregman iterations. It turns out that Bregman iterations can be constructed in a primal as well as in a dual fashion, which is equivalent in the case of quadratic fidelities, but produces different methods in the case of nonquadratic ones.

Multiplicative Noise

Another interesting class of noise models are multiplicative ones, i.e. $f = (A\hat{u})\,\eta$, which is found in several applications.

A Gaussian statistic for the noise has been assumed in [164], which leads to the nonconvex variational problem

$$\frac{\lambda}{2} \int_\Omega (\frac{f}{u} - 1)^2 \, dx + TV(u) \to \min_u, \quad \int_\Omega \left(\frac{f}{u} - 1\right) dx = 0. \qquad (120)$$

Aubert and Aujol [13] modeled the noise as Gamma distributed instead, leading to

$$\lambda \int_\Omega (\log \frac{Au}{f} + \frac{f}{Au}) \, dx + TV(u) \to \min_u, \qquad (121)$$

which is closer to the Poisson noise case and at least convex in a reasonable subset of BV functions. In the case of denoising a substitution to a new variable $v = \ln u$ is possible, which leads to a convex variational model if the total variation of v is used for regularization instead of the total variation of e^v (cf. [108]).

Other Fidelities

Several other fidelity terms have been investigated recently, with different motivations. Bonesky et al. [34] have investigated powers of Banach space norm, to which also inverse scale space methods and their analysis were generalized in [51].

Motivated by the examples mentioned above and the modeling of textures, Meyer [134] proposed to use the dual norm of BV as a fidelity, i.e. the variational problem

$$\lambda \|u - f\|_* + TV(u) \to \min_{u \in BV(\Omega)}, \qquad (122)$$

for which also interesting theoretical results have been obtained. Due to difficulties in the numerical solution, several authors have considered other dual Sobolev norms instead (cf. [147, 185–187]).

Other data fidelities (and also a constraint to probability densities) have been considered in applications to density estimation (cf. [135, 144]). In [54] the Wasserstein metric has been used as a fidelity, and the arising regularized optimal transport has been analyzed.

7.2 Nonlinear Image Formation Models

Another motivation for considering nonquadratic fidelity terms are nonlinear image formation models as appearing e.g. in ultrasound imaging, optical tomography, or other inverse problems in partial differential equations. The main theoretical change in this case is to replace the linear forward operator A by a nonlinear one, which we denote by F in the following. The natural extension of the variational model is the minimization

$$\frac{\lambda}{2} \|F(u) - f\|^2 + J(u) \to \min_u. \tag{123}$$

Such an approach has been proposed and tested e.g. by Luce and Perez [131] in an application to parameter identification. Under standard conditions on the operator F, the existence analysis can be carried out as in the linear case (cf. [2]). A major theoretical difference to the previously discussed cases is that the fidelity can now become non-convex, hence there is no uniqueness in general and there might be undesired local minima.

Error estimates in the nonlinear case can be derived similar to the linear case [107, 161], however there are differences related to the conditions needed on the nonlinearity of the operator. Depending on the degree of linearization used, several variants of Bregman iterations can be constructed for nonlinear models (cf. [20, 21, 170]).

8 Related Regularization Techniques

In the following we discuss some extensions and variants of regularization with total variation, which can be used for several purposes in image processing and analysis.

8.1 Higher-Order Total Variation

In several applications it is interesting to obtain piecewise linear instead of piecewise constant structures. This can be achieved by the higher-order total variation

$$TV_2(u) := \sup_{\substack{g \in C_0^\infty(\Omega;\mathbb{R}^{d \times d}) \\ \|g\|_\infty \leq 1}} \int_\Omega u \, (\nabla \cdot)^2 g \, dx, \qquad (124)$$

or, alternatively by the variation defined via the Laplacian

$$TV_\Delta(u) := \sup_{\substack{g \in C_0^\infty(\Omega) \\ \|g\|_\infty \leq 1}} \int_\Omega u \, \Delta g \, dx. \qquad (125)$$

The second order total variation has been investigated and applied in [68, 105, 171, 176], duality properties have been investigated in [175]. The extension to higher orders is obvious, but so far hardly investigated, in particular in applications, probably also due to the difficulty of treating higher-order functionals.

The higher-order total variation has received some attention recently in combination with the total variation in inf-convolution functionals as proposed in [61],

$$ICTV(u) = \inf_{w \in BV_2(\Omega)} (TV(u - w) + TV_2(w)). \qquad (126)$$

The idea of regularization with the inf-convolution is an optimal decomposition of the solution into a component in $BV(\Omega)$ and another one in

$$BV_2(\Omega) = \{ u \in BV(\Omega) \mid TV_2(u) < \infty \}.$$

In this way staircasing can be avoided (at least to some extent) since in smoothly varying regions the higher order total variation dominates. Clearly the staircasing then appears in the derivative or in other words the variational approach favours a combination of piecewise constant and piecewise linear structures. This can further be improved by considering inf-convolutions with even higher order total variations, of course at the expense of higher computational effort. A slight modification of the inf-convolution approach that further improves the quality of results has been proposed recently (cf. [29, 36, 171, 172]), a detailed analysis of the schemes in terms of underlying function spaces and detailed structure of solutions (e.g. eigenvalue problems as in the TV case above) remains an interesting open problem.

8.2 Total Variation of Vector Fields

So far we have only discussed methods for scalar-valued functions of bounded variation. In several cases one would however prefer to work with vector fields, e.g.

for color or spectral images (cf. [33, 69]) or flow fields (cf. [195]), or even tensors such as in DT-MRI techniques (cf. [73, 177]). The extension of total variation to vector valued functions is rather obvious, namely by taking the total variation of each component. However, there are some subtleties, in particular in the way the components are combined. Formally we may think of

$$TV(\mathbf{u}) = \int_\Omega \|(|\nabla u_1|, |\nabla u_2|, \ldots, |\nabla u_m|)\|_p \, dx \qquad (127)$$

for each vector valued function $\mathbf{u} = (u_1, \ldots, u_m) : \Omega \to \mathbb{R}^m$.

Different choices of p have similar effects as in sparsity regularization. For $p = 1$ the norm favours the vector $(|\nabla u_1|, \ldots, |\nabla u_m|)$ to be locally sparse, which means in particular that edge sets for the different components u_i will usually not coincide. This can be reasonable in some applications if the u_i represent some kind of complementary variables, but it is e.g. not desirable for color images, where most edges are expected to be at the same places for the different color components. In such cases $p \geq 2$ is more appropriate. For tensor images an analogous discussion applies to the choice of the suitable matrix norm.

8.3 Smoothed Total variation

In order to avoid difficulties with the non-differentiability of the total variation, i.e. of the Euclidean norm at zero, smoothed versions are frequently used (cf. [78, 190, 191]).

The most frequently used approximation of the total variation is of the form

$$TV^\epsilon(u) = \int_\Omega \sqrt{|\nabla u|^2 + \epsilon^2} \, dx. \qquad (128)$$

The Huber norm is given by

$$H_\epsilon(u) = \int_\Omega h_\epsilon(|\nabla u|), \qquad (129)$$

where h_ϵ is a locally quadratic approximation of the identity, i.e.

$$h_\epsilon(t) = \begin{cases} t & t > \epsilon \\ \frac{t^2}{\epsilon} & t \leq \epsilon \end{cases} \qquad (130)$$

The Huber norm can also be obtained as a Moreau-Yosida regularization (cf. [125] of the total variation

$$h_\epsilon(\nabla u) = \min_s \left(|s| + \frac{1}{2\epsilon} |s - \nabla u|^2 \right). \qquad (131)$$

A disadvantage of a smoothed approximation is the structure of solutions. Due to the differentiability at zero, the sharp edges can become smeared out. For very small ϵ that is used to obtain an appropriate approximation of the total variation, the parameter dependence can lead to significant slow down in computational methods for differentiable problems.

8.4 Nonlocal Regularization

Nonlocal regularization techniques have emerged recently from a proposed linear filter by Buades et al. [45] of the form

$$u(x) = \frac{1}{C_f(x)} \int_\Omega w_f(x,y) f(y)\, dy \qquad (132)$$

with weight function

$$w_f(x,y) = \exp\left(-\frac{1}{h^2} \int_{\mathbb{R}^d} G_a(t) |f(x+t) - f(y+t)|^2\, dt\right) \qquad (133)$$

with Gaussian G_a. This nonlocal filter generalizes neighbourhood filters in the pixel domain (like Yaroslavsky and SUSAN filters). It allows to take advantage of patches appearing in similar form throughout the image domain, which in particular applies to textures. The significant visual improvement in this respect led to an enormous increase of research in this direction.

The relation to total variation techniques was established first by Kindermann et al. [116] with the observation that the nonlocal filter can be realized equivalently by solving a quadratic variational problem of the form

$$\frac{\lambda}{2} \int_\Omega (u-f)^2\, dx + \frac{1}{2} \int_\Omega \int_\Omega w_f(x,y)(u(x) - u(y))^2\, dx\, dy. \qquad (134)$$

Modifications of this quadratic problem towards a nonlocal version of the total variation (respectively the ROF functional) are possible, but not completely unique. In particular one can define a nonlocal total variation as

$$TV_{NL1}(u) = \int_\Omega \int_\Omega \sqrt{w_f(x,y)} |u(x) - u(y)|\, dx\, dy \qquad (135)$$

and

$$TV_{NL2}(u) = \int_\Omega \sqrt{\int_\Omega w_f(x,y)(u(x)-u(y))^2\, dx}\, dy. \qquad (136)$$

This variational problem can be analyzed in appropriate function spaces and generalized for other imaging and inversion tasks. An additional difficulty for problems where the forward operator A is deviating strongly from the identity is that the weights cannot be computed directly from the data. Two approaches have been proposed, a computation from a rough solution by linear inversion (cf. [128]) or an updating of the weights from the current solution in an iterative process (cf. [154, 204]).

As mentioned above, nonlocal techniques can lead to high quality texture reconstructions in images with repeated structures. A major drawback of nonlocal methods is the computational effort, since effectively the dimension of the problem is doubled (i.e. the number of degrees of freedoms in numerical computations is squared). If no spatial restriction is introduced already the construction of the weights can lead to overwhelming effort for large size 2D and for 3D images.

9 Numerical Methods

Over the last decades, a variety of numerical methods for TV-minimization has been proposed, we only provide a short overview here focusing on state-of-the art approaches.

The basic setup we consider in this section is the minimization of functionals $J : X \to \mathbb{R} \cup \{+\infty\}$ of the form

$$J(u) = \frac{\lambda}{2}\|Au - f\|^2 + \sup_{p \in K}\langle u, Dp\rangle, \qquad (137)$$

where $A : X \to Y$ and $D : Z \to X$ are linear operators, usually A being bounded and compact, while D might be unbounded or with high norm. We assume that X and Z are Banach spaces with a convex bounded subset $K \subset Z$, and that Y is a Hilbert space. This covers the continuous (with D being the divergence) as well as the discretized setting.

The regularization functional

$$R(u) = \sup_{p \in K}\langle u, Dp\rangle \qquad (138)$$

is the convex conjugate of the characteristic function of K and hence convex and positively one-homogeneous. We consider the typical case of K representing a point-wise constraint, which is the case in all common applications we have in mind. If X is a space of (generalized) functions on Ω (either a continuous or discrete set equipped with a positive measure), then the constraint set K is of the form

$$K = \{\, p \in Z \mid p(x) \in M \text{ for almost every } x \in \Omega \,\} \qquad (139)$$

A Guide to the TV Zoo

with $M \subset \mathbb{R}^m$ being a bounded convex set. With the convex conjugate of the charactistic function of M, which we denote by $F : \mathbb{R}^m \to \mathbb{R} \cup \{+\infty\}$,

$$F(s) = \sup_{q \in M} q \cdot s \qquad (140)$$

the regularization functional can (at least formally) be written as

$$R(u) = \int F(D^* u) \, d\mu, \qquad (141)$$

with μ an appropriate measure (see below for standard examples) and D^* being the adjoint of D in the L^2-scalar product with this measure.

The driving example is of course total variation regularization. In this case Z is the space of continuous vector fields vanishing at the boundary of a domain $\Omega \subset \mathbb{R}^d$, $D = \nabla \cdot$ is the divergence operator, X an L^p-space on Ω, and K is the unit ball

$$K = \{ \, p \in Z \mid \|p\|_\infty \leq 1 \, \}. \qquad (142)$$

Note that K can indeed be defined by the pointwise constraint $|p(x)| \leq 1$ for all $x \in \Omega$ and hence (141) holds for weakly differentiable functions u, where $D^* = -\nabla$ and $F(p) = |p|$, i.e., the total variation as defined in the first part.

A second important case are so-called *sparse* or *compressive* models, which are based on penalizing the coefficients in some orthonormal basis (or more generally in a frame) of a separable Hilbert space by a weighted ℓ^1-norm. By a suitable redefinition of the operator A as the effective operator acting on the coefficients, such algorithms can be rewritten in the above form with $X = \ell^2(\mathbb{N})$, $Z = \ell^\infty(\mathbb{N})$, D is the multiplication operator

$$(u_n)_{n \in \mathbb{N}} \mapsto (w_n u_n)_{n \in \mathbb{N}},$$

where (w_n) is a suitable sequence of positive weights, often identical to one, and

$$K = \{ \, p \in Z \mid |p_n| \leq 1, \forall \, n \in \mathbb{N} \, \}. \qquad (143)$$

The resulting regularization functional is then the weighted ℓ^1-norm

$$R(u) = \sum_{n \in \mathbb{N}} w_n |u_n|. \qquad (144)$$

This is a special case of (141) with μ being the sum of weighted discrete point measures.

9.1 Optimality Conditions

In the following we provide a unifying formulation of optimality conditions for (137), which gives a common viewpoint on many existing methods. For this sake we introduce a gradient variable v and the corresponding constraint

$$D^*u + v = 0. \tag{145}$$

Under the main assumption that the regularization functional has the representation (141), the unconstrained variational problem (137) can then be rephrased as the constrained problem

$$\frac{\lambda}{2}\|Au - f\|^2 + \int F(v)\,d\mu \to \min \quad \text{subject to} \quad D^*u + v = 0. \tag{146}$$

The solution of the constrained problem is a saddle-point of the Lagrangian (minimum with respect to u and v, maximum with respect to w) given by

$$L(u, v, w) = \frac{\lambda}{2}\|Au - f\|^2 + \int F(v)\,d\mu + \int (D^*u + v)w\,d\mu. \tag{147}$$

The Lagrangian is differentiable with respect to u and w, hence for an optimal solution the variations with respect to this variable need to vanish. Moreover, the Lagrangian is convex with respect to v, hence zero needs to be an element of the subgradient. Note that the minimization problem with respect to v for fixed u and w is equivalent to the minimization of

$$\int (F(v) + vw)\,d\mu, \tag{148}$$

which is achieved by the pointwise minimization of $F(v) + vw$ for $v(x)$ μ-almost everywhere.

As a consequence of the above considerations we obtain the following optimality system for the constrained problem

$$\lambda A^* A u + D w = \lambda A^* f \tag{149}$$

$$\partial F(v) + w \ni 0 \tag{150}$$

$$D^*u + v = 0 \tag{151}$$

As we shall see below, most of the methods indeed work without the variable v, which is the special case of an iteration method that always keeps (151) exactly satisfied. The above formulation has several advantages however. First of all it is quite general and allows to interpret many schemes as iterations on these optimality conditions, in particular also the recently proposed split Bregman method. Secondly,

A Guide to the TV Zoo

the difficulties due to the gradient operator D and the nondifferentiability of F become separated. This allows to immediately generalize the schemes below to quite arbitrary operators D, and allows to investigate (150) as a pointwise relation. Several approximations of this relation will indeed distinguish different schemes, they can be more easily understood when considered in the pointwise sense.

9.2 Primal Approaches

We start with primal iteration approaches, corresponding to iterations on (150) or smoothed versions thereof, usually with additional damping in (149).

Most of these approaches are based on simplifying the subgradient relation as an approximation by smoothing, i.e., F is approximated by a family of parameter-dependent differentiable functionals F_δ such that $F_\delta \to F$ in a suitable sense (e.g. uniformly) as $\delta \to 0$. Several special approximations have been proposed depending on the specific choice of F. In the case of F being the Euclidean norm in $\mathbb{R}^m$, the C^∞-approximation

$$F_\delta(s) = \sqrt{|s|^2 + \delta^2} \tag{152}$$

received most attention (cf. [78, 163, 190–192]) due to its simplicity and high regularity. A quite general way to obtain first-order regular (differentiable with Lipschitz-continuous derivative) approximations is the *Moreau-Yosida regularization* (cf. [81]) defined by

$$F_\delta(s) = \inf_\sigma [F(\sigma) + \frac{1}{2\delta}|s - \sigma|^2]. \tag{153}$$

Even if it cannot be computed in an explicit form it is useful for computation, since one can just minimize the whole functional with respect to the new variable as well. In the case of F being the Euclidean norm in $\mathbb{R}^m$, the Moreau-Yosida regularization can be explicitely calculated as the so-called *Huber norm* (cf. [10, 109])

$$F_\delta(s) = \begin{cases} \frac{1}{2\delta}|s|^2 & \text{if } |s| \le \epsilon \\ |s| - \epsilon & \text{if } |s| > \epsilon. \end{cases} \tag{154}$$

A potential disadvantage of such approaches is the slow growth of the regularized functional F_δ at zero. Hence, sparsity is not promoted, since replacing small values by zero has only little gain in the regularization term (which is then completely balanced by losses in fidelity terms).

The first scheme proposed in [163] was a gradient descent method, which iterates (150) explicitly as

$$w^{k+1} = F'_\delta(v^k) \tag{155}$$

and then inserts w^{k+1} in (149). In order to obtain convergence, damping (with rather small damping parameter τ depending on the image size) needs to be used to compute

$$u^{k+1} = u^k - \tau(\lambda A^*(Au^k - f) + Dw^{k+1}). \qquad (156)$$

Finally, (151) is enforced exactly, i.e. one computes

$$v^{k+1} = D^*u^{k+1}. \qquad (157)$$

The main disadvantages of the gradient descent approach are the dependence on δ and in particular on the grid size (respectively number of pixels). In total variation, the role of τ corresponds to a time step in the explicit time discretization of a parabolic partial differential equation and thus needs to be very small (of the order of the grid size squared) to maintain stability and convergence to the minimizer.

A popular approach to reduce the severe restriction on τ is to use some kind of semi-implicit time stepping, which still yields linear problems in each time step. In this way (156) and (157) are solved as a linear system coupled with

$$L(v^k)v^{k+1} + w^{k+1} = L(v^k)v^k - F_\delta(v^k), \qquad (158)$$

where $L(v^k)$ is a (discretized) differential operator such that $L(v)v$ approximates $F'_\delta(v)$. A quite popular method of this kind is the lagged diffusivity approximation (cf. [191, 192]) using (152) and

$$L(v)\varphi = \frac{\varphi}{\sqrt{|v|^2 + \delta^2}}, \qquad (159)$$

which yields $L(v)v = F'_\delta(v)$. In the denoising case $A = I$ also the Newton method on the smoothed optimality system can be put in this form if $L = F''_\delta$ and $\tau = \frac{1}{\lambda}$. Besides the dependence of δ, a major disadvantage is that the matrix for the linear system to be solved has to be changed in each iteration step (an approximation L independent of v^k usually does not provide reasonable results) and may be badly conditioned for small δ.

9.3 Dual Approaches

Instead of using (150), the dual approach rather iterates the equivalent dual subgradient relation

$$v + \partial F^*(w) \ni 0. \qquad (160)$$

A Guide to the TV Zoo

Note that the convex conjugate $F^*(w)$ is again the characteristic function of M. Thus, (160) can be interpreted as the optimality condition for minimizing the linear functional $\langle v, w \rangle$ with respect to $w \in M$.

As in the primal case one can approximate the dual subgradient relation, respectively the inclusion $q \in M$, via smoothing. Again one can construct the Moreau-Yosida regularization

$$F_\delta^*(q) = \inf_r [F^*(r) + \frac{1}{2\delta}|r-q|^2] = \inf_{r \in M} \frac{1}{2\delta}|r-q|^2.$$

It is straight-forward to show that the minimum is attained for $r = P_M(q)$, where P_M is the projection onto M. Hence, we have

$$F_\delta^*(q) = \frac{1}{2\delta}|q - P_M(q)|^2 = \frac{1}{2\delta}\text{dist}(q, M)^2.$$

This smoothing is a penalty method approximating F^* from below, i.e. we have $F^*(q) \geq F_\delta^*(q)$ for all q. If F is the Euclidean norm, i.e. M the Euclidean unit ball in $\mathbb{R}^m$, we have $P_M(q) = \frac{q}{|q|}$ if $q \notin M$. Hence

$$2\delta F_\delta^*(q) = |q - \frac{q}{|q|}|^2 = |q|^2 - 2|q| + 1 = (|q|-1)^2 \quad \text{for } q \notin M.$$

A compact form for F_δ^* is the quadratic penalty

$$F_\delta^*(q) = \frac{1}{2\delta}\max\{|q|-1, 0\}^2. \tag{161}$$

A standard alternative to penalty schemes are barrier approaches, which amounts to approximating F^* from above, i.e. by a family of functionals F_δ^* differentiable on M such that $F_\delta^*(q) \geq F^*(q)$ for all q at least close to the boundary of M. The derivation of specific barrier functionals is usually based on a particular perturbation related to the complementarity condition for the constraint $q \in M$. Since the complementarity condition needs to be written for the specific inequalities defining M, the barrier functional has to be derived separately in each case. If F is the Euclidean norm respectively M is the unit ball in $\mathbb{R}^m$, then the resulting standard approximation in interior point methods is based on the logarithmic barrier

$$F_\delta^*(q) = -\delta \log(1 - |q|). \tag{162}$$

In [139] Newton-type methods for barrier as well as penalty methods are constructed and also parallelization techniques are discussed. It is shown that in the case of penalty methods some slope is introduced in the usually flat parts of the reconstruction but the edges are sharp, while barrier methods keep the flat regions with introducing some smoothing of the edges. However, if the penalty parameter is sufficiently small compared to the grid size, this effect disappears.

In dual iteration approaches, (149) and (151) are satisfied exactly, i.e.

$$\lambda A^* A u^{k+1} + D w^{k+1} = \lambda A^* f \qquad (163)$$

$$D^* u^{k+1} + v^{k+1} = 0, \qquad (164)$$

while iterative approximation is carried out on (160) in order to determine w^{k+1}. Note that A^*A needs to be regular in order to solve (163) and (164) for u^{k+1} and v^{k+1}, which is not an issue for denoising, but prevents the use of dual methods for ill-posed problems such as deblurring or inpainting.

An obvious approach is to use a fixed point iteration on (160), i.e.,

$$w^{k+1} + \tau \partial F^*(w^{k+1}) \ni w^k - \tau v^k. \qquad (165)$$

It is easy to see that this is the optimality condition for minimizing

$$J_k(w) = \frac{1}{2} \|w - w^k + \tau v^k\|^2 + \tau F(w),$$

and since F it is the characteristic function of M it is further equivalent to minimize the least-squares term subject to $w \in M$, which yields

$$w^{k+1} = P_M(w^k - \tau v^k),$$

where P_M is the projection onto M. Since v^k is the gradient of the dual objective with respect to w, this iteration is just the dual projected gradient method proposed in [205]. An older and more popular approach in the same spirit is due to Chambolle [60] in the case of M being the unit ball (in the Euclidean norm, but extensions to other norms are obvious). Chambolle did not use exact projection, but

$$w^{k+1} = \frac{w^k - \tau v^k}{1 + \tau |v^k|},$$

which is still consistent. The behaviour of the iteration is very similar to the projected gradient method. Again a major disadvantage is that τ needs to be chosen very small in order to obtain a stable iteration (note that the dual gradient is a differential operator of the form $v \mapsto \nabla \cdot (a \nabla v)$).

9.4 Augmented Lagrangian Methods

Recently the most popular methods are of Augmented Lagrangian type, although the seminal paper formulated them differently motivated from the Bregman iteration above, see [96] for this Split Bregman method in detail. The major idea in our setup

A Guide to the TV Zoo

of (149)–(151) is to use an augmented Lagrangian approach, i.e. add multiples of the (151) to (149) and (150), which clearly yields an equivalent system that however can be solved uniquely for u and v. Then an iterative update of w in (151) is performed.

The augmented optimality system, with parameter $\mu > 0$ consists of (151) and

$$\lambda(A^*A + \mu DD^*)u + \mu Dv + Dw = \lambda A^* f \qquad (166)$$

$$\partial F(v) + \mu v + \mu D^* u + w \ni 0. \qquad (167)$$

In the case of total variation DD^* is the Laplacian, which can be inverted if the mean value or boundary values are fixed. Moreover, $\mu v + F(v)$ is maximally monotone and thus has a single-valued inverse, which can be computed explicitly by shrinkage formulas for the typical choices of M.

In its standard form the iteration first solve (166) and (167) for u and v, i.e.,

$$\lambda(A^*A + \mu DD^*)u^{k+1} + \mu Dv^{k+1} = \lambda A^* f - Dw^k \qquad (168)$$

$$\partial F(v^{k+1}) + \mu v^{k+1} + \mu D^* u^{k+1} \ni -w^k, \qquad (169)$$

and then perform a gradient ascent step for the dual variable w, i.e.,

$$w^{k+1} = w^k + \mu(D^* u^{k+1} + v^{k+1}). \qquad (170)$$

A difficulty in the realization of this iteration is the fact that (168) and (169) are fully coupled, which is usually solved via an inner iteration alternating between these two. This idea can also be directly incorporated in the outer iteration, replacing (168) by

$$\lambda(A^*A + \mu DD^*)u^{k+1} = \lambda A^* f - \mu Dv^k - Dw^k, \qquad (171)$$

yielding an approach equivalent to the alternating direction of multipliers method (ADMM, cf. [35, 158]). Variants can be found in [14, 85, 203].

10 Segmentation via TV Relaxation

While all above approaches were based on computing an image intensity as a function with values in $\mathbb{R}$ (or a closed subinterval), region-based segmentations rather look for a decomposition of Ω into subsets with different characteristics. The relation to functions can be made by two approaches, either via characteristic functions

$$\chi(x) = \begin{cases} 1 & x \in \Sigma \subset \Omega \\ 0 & \text{else} \end{cases} \qquad (172)$$

or via level set functions (cf. [145])

$$\Sigma = \{ x \in \Omega \mid \phi(x) < 0 \} \qquad (173)$$

with ϕ continuous. Modern region-based segmentation schemes are based on minimizing an energy depending on the image characteristics in the two subregions (some volume integrals of f) penalized by the perimeter of $\partial \Sigma$ in order to avoid a classification of noisy parts as small regions to be segmented. The perimeter provides the main link to total variation methods, since the coarea formula yields

$$TV(u) = \int_{\mathbb{R}} \operatorname{Per}(\partial\{u > \alpha\}) d\alpha, \qquad (174)$$

where $\operatorname{Per}(M)$ denotes the perimeter of a set M (the $d-1$-dimensional Hausdorff-measure). For $u = \chi$ or $u = H(\phi)$ with χ and ϕ as above and H the Heaviside function we obtain $\{u > \alpha\} = \emptyset$ for $\alpha \geq 1$ and $\{u > \alpha\} = \Omega$ for $\alpha < 0$. In both cases the perimeter of $\partial\{u > \alpha\}$ is zero. Moreover, for $\alpha \in (0, 1)$ we have

$$\{u > \alpha\} = \Sigma.$$

Hence, the coarea formula implies

$$\operatorname{Per}(\partial \Sigma) = \int_0^1 \operatorname{Per}(\partial \Sigma) d\alpha = TV(u).$$

Thus, the minimization of shape functionals penalized by perimeter as encountered in image segmentation can be rewritten into problems with total variation minimization.

A very interesting connection has been established in a series of papers by Chan et al. [70, 71], who analyzed total variation problems with one-homogeneous fitting

$$\lambda \int_\Omega |u - f| \, dx + TV(u) \to \min_u, \qquad (175)$$

as well as the regularized linear functional

$$\int_\Omega g \, u \, dx + TV(u) \to \min_u, \qquad (176)$$

subject to bound constraints $0 \leq u \leq 1$. Based on the co-area formula and a layer-cake formula for the first integral a connection to shape optimization problems can be derived. We review this analysis in the case of (176). First of all, the existence of a minimizer of (176) with either the bound constraint

$$0 \leq u \leq 1 \qquad \text{a.e. in } \Omega \qquad (177)$$

or the discrete constraint

$$u \in \{0, 1\} \quad \text{a.e. in } \Omega \tag{178}$$

can be shown:

Lemma 10.1. *Let* $g \in L^1(\Omega)$. *Then there exists a minimizer of the constrained minimization problems* (176) *subject to* (177) *as well as of* (176) *subject to* (178).

Proof. The admissible set is a subset of the unit ball in $L^\infty(\Omega)$ and on sublevel sets of the objective functional

$$E(u) = \int_\Omega g\, u\, dx + TV(u)$$

we have

$$TV(u) \leq C - \int g\, u\, dx \leq C + \|g\|_{L^1}.$$

Hence, the sublevel sets of J intersected with the admissible set are bounded in $L^\infty(\Omega) \cap BV(\Omega)$ and consequently compact in the weak* topologies of both spaces.

As we have shown before the total variation is lower semicontinuous in the weak* topology of $BV(\Omega)$ and the first integral is a bounded linear functional (of g) on $L^1(\Omega)$ hence weak* continuous in $L^\infty(\Omega) = L^1(\Omega)^*$. Thus, the functional J is lower semicontinuous with compact sublevel sets, which implies the existence of a minimizer. □

The major result is the following exact relaxation of the problem for characteristic functions:

Theorem 10.2. *Let u be a minimizer of* (176) *subject to* (177). *Then for almost all* $\alpha \in (0, 1)$, *the thresholded function*

$$\chi^\alpha(x) = \begin{cases} 1 & \text{if } u(x) > \alpha \\ 0 & \text{else} \end{cases} \tag{179}$$

is a solution of (176) *subject to* (178).

Proof. We first of all derive a layer-cake representation of the linear functional via

$$\int_\Omega g\, u\, dx = \int_\Omega g(x) \int_0^{u(x)} d\alpha\, dx$$

$$= \int_\Omega g(x) \int_0^1 \chi^\alpha(x)\, d\alpha\, dx$$

$$= \int_0^1 \left[\int_\Omega g\, \chi^\alpha\, dx \right] d\alpha.$$

For the total variation we have from the co-area formula

$$TV(u) = \int_0^1 \text{Per}(\partial\{u > \alpha\}) d\alpha = \int_0^1 TV(\chi^\alpha) \, dx.$$

Together we find that for u being a minimizer of (176) subject to (177)

$$E(u) = \int_0^1 E(\chi^\alpha) \, d\alpha \geq \int_0^1 E(u) \, d\alpha = E(u),$$

where we have used $E(\chi^\alpha) \geq E(u)$ for all α, since χ^α is admissible for the minimization problem (176) subject to (177) and u is the minimizer. The above relation is only possible if $E(\chi^\alpha) = E(u)$ for almost all $\alpha \in (0, 1)$. Hence, for almost every α the characteristic function χ^α is a minimizer of (176) subject to (177). On the other hand (177) is a relaxation of (178), thus χ^α needs to be a minimizer of (176) subject to (178). □

The exact relaxation result in Theorem 10.2 has two important consequences: First of all, the nonconvex $0 - 1$ minimization can be solved equivalently by minimizing a convex problem, which resolves issues with minimizers. Secondly, techniques from the analysis of total variation minimization, e.g. existence, asymptotic behaviour, and error estimates can be carried over.

A particularly interesting example is the relaxation of the Chan–Vese model [66], which can be written as the minimization of

$$J_{CV}(\chi, c_1, c_2) = \frac{\lambda}{2} \int_\Omega [\chi(c_1 - f)^2 + (1 - \chi)(c_2 - f)^2] \, dx + TV(\chi) \qquad (180)$$

subject to (178). The minimization can be realized in an alternating fashion, with respect to χ and with respect to the constants c_1 and c_2. For given χ, the minimizers c_1 and c_2 can be computed explicitly as mean values of f, namely

$$c_1 = \frac{\int_\Omega \chi f \, dx}{\int_\Omega \chi \, dx}, \qquad c_2 = \frac{\int_\Omega (1 - \chi) f \, dx}{\int_\Omega (1 - \chi) \, dx}. \qquad (181)$$

For the minimization with respect to χ Theorem (10.2) can be applied with

$$g = \frac{\lambda}{2}[\chi(c_1 - f)^2 + (1 - \chi)(c_2 - f)^2]. \qquad (182)$$

Hence, the minimization of J_{CV} subject to (178) can be relaxed exactly to the minimization with (177).

Extensions of the relaxation to more general problems in topology optimization are considered in [46], and a region-based version of the Mumford-Shah model, which allows exact relaxation is introduced in [168, 198]. Cremers et al. [156] introduce a lifting method, which allows to treat some nonconvex problems in imaging by convex relaxation after considering the graph and the associated 0-1 function in a higher dimension. Extensions to multi-label problems are considered in [124, 155], those relaxations are not exact however.

11 Applications

So far we have focused on the methodology of total variation whose development was heavily motivated by practical applications in the past. In this section and the following we finally take an opposite viewpoint and give a (certainly uncomplete) overview of the use of total variation methods in practices. We start in this section with applications in classical imaging tasks from computer vision and image analysis, before we proceed to applications in natural and life sciences or disciplines like the arts.

The most obvious application part is image analysis such as denoising, which was the original motivation for the ROF-model. Recent approaches rather use modified versions of total variations and in particular nonlocal versions (cf. e.g. [116, 129, 130, 166]), which are able to deal better with textures (assumed to be self-repeating patterns in the image). Another obvious application field of total variation techniques is segmentation as we have seen from the previous chapter. Other image analysis tasks where total variation methods have been applied successfully are zooming and superresolution (cf. e.g. [4, 199, 201]), colour enhancement (cf. e.g. [33, 74, 133, 142]), fusion (cf. e.g. [22, 157, 193]), flow estimation and video treatment (cf. e.g. [43, 76, 195, 196], inpainting (cf. [52, 64, 65, 84]), and image decomposition into a cartoon and texture part (cf. e.g. [12, 16, 17, 30, 40, 83, 134, 147, 186, 187]). In particular the latter two have led to an investigation of a variety of novel data fidelity terms, since standard L^2-terms are not appropriate. In particular the work by Meyer [134], who used the dual norm of the TV-norm as data fidelity, was quite influential and motivated several studies of the dual and predual space of BV. We also refer to [63] for total variation methods in image analysis.

We have already discussed deblurring and deconvolution above and here mention further work in blind deconvolution. The extension from deconvolution to blind deconvolution is rather straight-forward, one usually looks for the point-spread function in a parametric or non-parametric form as well as for the image. Since the point-spread function is usually smooth, the majority of approaches restricts the total variation regularization to the image. With alternating minimization approaches the numerical realization is quite straight-forward and analogous to standard deconvolution. An interesting question is whether total variation regularization on the image combined with classical smoothness priors on the point-spread function is suitable to overcome the non-uniqueness in blind deconvolution, which is confirmed by promising computational results, but has hardly been investigated by theoretical analysis (cf. e.g. [19, 31, 48, 67, 102]).

Due their favourable properties, total variation methods have found access to many applied areas of biomedical imaging. Due to the enormous amount of applications in the last two decades, we can only give a selection here and further links to literature:

- *Limited angle CT:* Due to the mild ill-posedness and the low noise levels, regularization is not a big topic in modern standard X-ray tomography. The situation changes in limited angle CT as e.g. appearing in modern C-Arm devices, where the ill-posedness of the reconstruction problem is much more severe. TV regularized reconstructions are usually obtained with the standard quadratic fidelity term and the forward operator being the Radon transform with limited angles, see e.g. [149, 153, 184]
- *Magnetic resonance (MR) imaging:* Image reconstruction in MR is mainly achieved by inverting a Fourier transform, which can be carried out efficiently and robustly if a sufficient number of Fourier coefficients, i.e. measurements, is available, which is usually the case. Regularization and thus also total variation approaches are mainly needed for special applications such as fast MR protocols, which do not allow to measure enough frequencies. This motivation has also triggered a lot of research in compressed sensing, where for this application effectively a discretized total variation is used (cf. e.g. [132]). Improved approaches use a decomposition of total variation with a second functional, e.g. based on wavelets, which can take care of the typical slopes arising between edges in MR images, cf. [32, 103, 115, 118–120].
- *Emission tomography:* Emission tomography techniques (cf. [197]) used in nuclear medicine such as *Positron Emission Tomography* (PET) and *Single Photon Emission Computed Tomography* (SPECT) are a natural target for total variation regularization due to inherent high noise level, more precisely the Poisson statistics of the data. We refer to [23, 113, 150, 165–167].
- *Microscopy:* In modern fluorescence microscopy low photon counts are an issue in particular in live imaging and in high resolution imaging at nanoscopic scales. Here total variation methods with data terms appropriate for Poisson noise have been quite benefitial for denoising and (blind) deconvolution (cf. e.g. [42, 44, 77, 91, 159, 167, 180]).
- *Other modalities:* For certain highly ill-posed image reconstruction tasks with expected sharp edges the use of total variation as a regularization is a quite obvious choice, e.g. in optical tomography (cf. e.g. [1, 80, 152]) or in electron tomography (cf. e.g. [24, 97]). A less obvious application of total variation is the study of EEG/MEG reconstructions, where usually rather ℓ^1 priors are used since brain activity is often sparse. Adde et al. [3] however compared total variation methods to standard density reconstruction techniques and other nonlinear diffusions and found it clearly outperforms them. In ultrasound, denoising and segmentation is of interest, where main attention has to be focused on appropriate modelling of speckle noise in the variational problems (cf. e.g. [112, 168]).

Since imaging is nowadays of high importance, it is not surprising that total variation methods have found their place there as well. Examples are astronomy (cf. e.g. [151] and references therein) or geosciences, where hyperspectral imaging is becoming a standard approach (cf. e.g. [111, 136, 137, 202]). Less expected applications appear in the tracking of sharp fronts in weather forecast (cf. [92]) and in the arts, where frescoes have been reconstructed by mathematical techniques instead of human experience (cf. [18, 89, 90]).

12 Open Questions

We finally provide a discussion of some problems related to the TV zoo, which may be particularly rewarding for future research. Again the choice of questions and viewpoints presented here are heavily motivated by our own research and may be subjective, moreover various yet unexpected questions due to the fast progress in this field the will certainly pop up in the next years:

- *Local choice of regularization parameters:* This issue receives growing interest in many recent investigations, motivated by different scales appearing in images locally as well as statistical models. Whereas the effects of locally large and small regularization parameters seem rather obvious in classical smoothing techniques (e.g. squared Sobolev norms), where small local regularization parameters allow high variation and approximate discontinuities (such as in the Ambrosio–Tortorelli model [5]), the effect in total variation is less clear and remains to be analyzed (cf. e.g. [106]).
- *Nonlinear inversion vs. two-step methods:* As we have discussed above, total variation methods can be incorporated into inversion methods and image reconstruction approaches. However, it is not obvious that the additional difficulties and increased computational effort is indeed necessary. In several cases results of a similar quality are achieved by simply using fast linear reconstruction methods with a subsequent application of the ROF model or variants thereof (cf. [166, 167]). A detailed analysis under which conditions the incorporation of total variation or similar methods into the reconstruction algorithms yields significantly superior results is still open.
- *Improved TV models:* Several improvements of total variation, in particular related to the staircasing effect, are currently a topic receiving increasing attention. A particularly timely topic seems to be the adaptive coupling of several TV-type functionals, which might concern anisotropy (cf. [30, 83]) or the combination of total variation and higher-order total variation. The latter has already been investigated via inf-convolution in [61], recently further improvements have been proposed, which yield promising results but also require a more detailed analysis (cf. [29, 36, 171, 172]).
- *4D regularization:* In many applications high-dimensional image structures appear, e.g. spectral or time-resolved images. In such cases it seems reasonable to apply total variation methods for the two or three spatial dimensions, while it makes less sense to favour piecewise constant structures in the additional directions. The appropriate combination of total variation with other regularization functionals related to the additional dimensions remains a challenging topic for future research.

- *Bayesian modelling:* While we have concentrated on variational techniques such as MAP estimation here, in wide areas of statistics it is much more common to use conditional mean (CM) estimates. In principle a prior probability related to total variation can be defined in discretizations in the form $p(u) \sim e^{-\alpha TV(u)}$ and the limit can be studied (cf. [121, 129]). However, convergence of CM estimates appears only with different scaling than in the case of MAP estimation and the limit is then a Gaussian measures, thus edges are not preserved. Appropriate models for edge preserving priors remain to be developed in a TV context. An interesting result in this direction is obtained in [75], who studies the choice $p(u) \sim e^{-\alpha TV(u)^2}$ instead.

Acknowledgements The work of MB has been supported by the German Science Foundation DFG through the project *Regularization with Singular Energies* and SFB 656 *Molecular Cardiovascular Imaging*, and by the German Ministry of Education and Research (BMBF) through the project *INVERS: Deconvolution problems with sparsity constraints in nanoscopy and mass spectrometry*.

The work of SO has been supported by NSF grants DMS0835863 and DMS0914561, ONR grant N000140910360 and ARO MURI subs from Rice University and the University of North Carolina.

The authors thank Alex Sawatzky (WWU Münster) for careful proofreading and comments to improve the manuscript.

References

1. J.F. Abascal, J. Chamorro-Servent, J. Aguirre, J.J. Vaquero, S. Arridge, T. Correia, J. Ripoll, M. Desco, Fluorescence diffuse optical tomography using the split Bregman method. Med. Phys. **38**, 6275 (2011)
2. R. Acar, C.R. Vogel, Analysis of total variation penalty methods. Inverse Probl. **10**, 1217–1229 (1994)
3. G. Adde, M. Clerc, R. Keriven, Imaging methods for MEG/EEG inverse problem. Int. J. Bioelectromagn. **7**, 111–114 (2005)
4. A. Almansa, V. Caselles, G. Haro B. Rouge, Restoration and zoom of irregularly sampled, blurred and noisy images by accurate total variation minimization with local constraints. Multiscale Mod. Simulat. **5**, 235–272 (2006)
5. L. Ambrosio, V. Tortorelli, Approximation of functionals depending on jumps by elliptic functionals via Γ- convergence. Comm. Pure Appl. Math. **43**, 999–1036 (1990)
6. L. Ambrosio, N. Fusco, D. Pallara, *Functions of Bounded Variation and Free Discontinuity Problems* (Oxford University Press, Oxford, 2000)
7. F. Andreu, C. Ballester, V. Caselles, J.M. Mazon, Minimizing total variation flow. Differ. Integr. Equat. **14**, 321–360 (2001)
8. F. Andreu, C. Ballester, V. Caselles, J.M. Mazon, The Dirichlet problem for the total variation flow. J. Funct. Anal. **180**, 347–403 (2001)
9. F. Andreu, V. Caselles, J.I. Diaz, J.M. Mazon, Some qualitative properties for the total variation flow. J. Funct. Anal. **188**, 516–547 (2002)
10. U. Ascher, E. Haber, *Computational Methods for Large Distributed Parameter Estimation Problems with Possible Discontinuities.* Symp. Inverse Problems, Design and Optimization, Rio, 2004

11. P. Athavale, E. Tadmor, Multiscale image representation using integro-differential equations. Inverse Probl. Imag. **3**, 693–710 (2009)
12. G. Aubert, J.F. Aujol, Modeling very oscillating signals, application to image processing. Appl. Math. Optim. **51**, 163–182 (2005)
13. G. Aubert, J.F. Aujol, A Variational approach to remove multiplicative noise. SIAM J. Appl. Math. **68**, 925–946 (2008)
14. J.F. Aujol, Some first-order algorithms for total variation based image restoration. J. Math. Imag. Vis. **34**, 307–327 (2009)
15. J.F. Aujol, A. Chambolle, Dual norms and image decomposition models. IJCV **63**, 85–104 (2005)
16. J.F. Aujol, G. Aubert, L. Blanc-Feraud, A. Chambolle, Image decomposition into a bounded variation component and an oscillating component. J. Math. Imag. Vis. **22**, 71–88 (2005)
17. J.F. Aujol, G. Gilboa, T. Chan, S. Osher, Structure-texture image decomposition - modeling, algorithms, and parameter selection. Int. J. Comput. Vis. **67**, 111–136 (2006)
18. W. Baatz, M. Fornasier, P. Markowich, C.B. Schönlieb, *Inpainting of ancient Austrian frescoes*. Conference Proceedings of Bridges 2008, Leeuwarden, 2008, pp. 150–156
19. S.D. Babacan, R. Molina, A.K. Katsaggelos, Variational Bayesian blind deconvolution using a total variation prior. IEEE Trans. Image Proc. **18**, 12–26 (2009)
20. M. Bachmayr, *Iterative Total Variation Methods for Nonlinear Inverse Problems*, Master Thesis (Johannes Kepler University, Linz, 2007)
21. M. Bachmayr, M. Burger, Iterative total variation methods for nonlinear inverse problems. Inverse Probl. **25**, 105004 (2009)
22. C. Ballester, V. Caselles, L. Igual, J. Verdera, B. Rouge, A variational model for P+XS image fusion. IJCV **69**, 43–58 (2006)
23. J. Bardsley, A. Luttman, Total variation-penalized Poisson likelihood estimation for ill-posed problems . Adv. Comp. Math. **31**, 35–59 (2009)
24. C. Bazan, *PDE-Based Image and Structure Enhancement for Electron Tomography of Mitochondria*, PhD-Thesis (San Diego State University, San Diego, 2009)
25. G. Bellettini, V. Caselles, The total variation flow in $\mathbf{R}^N$. J. Differ. Equat. **184**, 475–525 (2002)
26. G. Bellettini, V. Caselles, M. Novaga, Explicit solutions of the eigenvalue problem $-\operatorname{div}(Du/|Du|) = u$. SIAM J. Math. Anal. **36**, 1095–1129 (2005)
27. M. Benning, *Singular Regularization of Inverse Problems* (PhD Thesis, WWU Münster, 2011)
28. M. Benning, M. Burger, Error estimation with general fidelities. Electron. Trans. Numer. Anal. **38**, 44–68 (2011)
29. M. Benning, C. Brune, M. Burger, J. Müller, Higher-order TV methods—enhancement via Bregman iteration. J. Sci. Comput. **54**(2–3), 269–310 (2013)
30. B. Berkels, M. Burger, M. Droske, O. Nemitz, M. Rumpf, Cartoon extraction based on anisotropic image classification, in *Vision, Modeling, and Visualization 2006: Proceedings*, ed. by L. Kobbelt, T. Kuhlen, T. Aach, R. Westerman (IOS Press, Aachen, 2006)
31. J.M. Bioucas-Dias, M.A.T. Figueiredo, J.P. Oliveira, Total variation-based image deconvolution: a majorization-minimization approach, in *2006 IEEE International Conference on Acoustics, Speech and Signal Processing, 2006. ICASSP 2006 Proceedings*, vol. 2, pp. II, 14–19 May 2006
32. K.T. Block, M. Uecker, J. Frahm, Undersampled radial MRI with multiple coils. Iterative image reconstruction using a total variation constraint. Magn. Reson. Med. **57**, 1086–1098 (2007)
33. P. Blomgren, T. Chan, Color TV: Total variation methods for restoration of vector valued images. IEEE Trans. Image Proc. **7**, 304–309 (1998)
34. T. Bonesky, K.S. Kazimierski, P. Maass, F. Schöpfer, T. Schuster, Minimization of Tikhonov functionals in Banach spaces. Abstr. Appl. Anal. **19**, 192679 (2008)
35. S. Boyd, N. Parikh, E. Chu, B. Peleato, J. Eckstein, Distributed optimization and statistical learning via the alternating direction method of multipliers. Found. Trends Mach. Learn. **3**, 1–122 (2011)

36. K. Bredies, K. Kunisch, T. Pock, Total generalized variation. SIAM J. Imag. Sci. **3**, 492–526 (2010)
37. L.M. Bregman, The relaxation method for finding the common point of convex sets and its application to the solution of problems in convex programming. USSR Comp. Math. Math. Phys. **7**, 200–217 (1967)
38. M. Breuss, T. Brox, A. Bürgel, T. Sonar, J. Weickert, Numerical aspects of TV flow. Numer. Algorithms **41**, 79–101 (2006)
39. A. Briani, A. Chambolle, M. Novaga, G. Orlandi, On the gradient flow of a one-homogeneous functional, Preprint (SNS, Pisa, 2011)
40. T. Brox, J. Weickert, A TV flow based local scale measure for texture discrimination, in *Computer Vision - ECCV 2004*, ed. by T. Pajdla, J. Matas (Springer, Berlin, 2004), pp. 578–590
41. T. Brox, M. Welk, G. Steidl, J. Weickert, Equivalence results for TV diffusion and TV regularization, in *Scale Space Methods in Computer Vision*, ed. by L.D. Griffin, M. Lillholm (Springer, Berlin, 2003), pp. 86–100
42. C. Brune, A. Sawatzky, M. Burger, Bregman-EM-TV methods with application to optical nanoscopy, in *Proceedings of the 2nd International Conference on Scale Space and Variational Methods in Computer Vision*. LNC, vol. 5567 (Springer, Berlin, 2009), pp. 235–246
43. C. Brune, H. Maurer, M. Wagner, Detection of intensity and motion edges within optical flow via multidimensional control. SIAM J. Imag. Sci. **2**, 1190–1210 (2009)
44. C. Brune, A. Sawatzky, M. Burger, Primal and dual Bregman methods with application to optical nanoscopy. Int. J. Comput. Vis. **92**, 211–229 (2011)
45. A. Buades, B. Coll, J.M. Morel, A review of image denoising algorithms, with a new one. Multiscale Model. Simulat. **4**, 490–530 (2005)
46. M. Burger, M. Hintermüller, Projected gradient flows for BV / level set relaxation . PAMM **5**, 11–14 (2005)
47. M. Burger, S. Osher, Convergence rates of convex variational regularization. Inverse Probl. **20**, 1411–1421 (2004)
48. M. Burger, O. Scherzer, Regularization methods for blind deconvolution and blind source separation problems. Math. Contr. Signals Syst. **14**, 358–383 (2001)
49. M. Burger, G. Gilboa, S. Osher, J. Xu, Nonlinear inverse scale space methods for image restoration. Comm. Math. Sci. **4**, 179–212 (2006)
50. M. Burger, E. Resmerita, L. He, Error estimation for Bregman iterations and inverse scale space methods. Computing **81**, 109–135 (2007)
51. M. Burger, K. Frick, S. Osher, O. Scherzer, Inverse total variation flow. SIAM Multiscale Mod. Simul. **6**, 366–395 (2007)
52. M. Burger, L. He, C.B. Schönlieb, Cahn-Hilliard inpainting and a generalization for grayvalue images. SIAM J. Imag. Sci. **2**, 1129–1167 (2009)
53. M. Burger, M. Möller, M. Benning, S. Osher, An adaptive inverse scale space method for compressed sensing. Math. Comput. (2012, to appear)
54. M. Burger, M. Franek, C.-B. Schönlieb, Regularized regression and density estimation based on optimal transport. Appl. Math. Res. Express **2012**, 209–253 (2012)
55. J. Cai, S. Osher, Z. Shen, Linearized Bregman iterations for compressed sensing. Math. Comput. **78**, 1515–1536 (2008)
56. J. Cai, S. Osher, Z. Shen, Convergence of the linearized Bregman iteration for l_1-norm minimization. Math. Comput. **78**, 2127–2136 (2009)
57. J. Cai, S. Osher, Z. Shen, Linearized Bregman iteration for frame based image deblurring. SIAM J. Imag. Sci. **2**, 226–252 (2009)
58. V. Caselles, A. Chambolle, M. Novaga, The discontinuity set of solutions of the TV denoising problem and some extensions. MMS **6**, 879–894 (2007)
59. V. Caselles, A. Chambolle, M. Novaga, Regularity for solutions of the total variation denoising problem. Rev. Mat. Iberoamericana **27**, 233–252 (2011)
60. A. Chambolle, An algorithm for total variation regularization and denoising. J. Math. Imag. Vis. **20**, 89–97 (2004)

61. A. Chambolle, P.L. Lions, Image recovery via total variational minimization and related problems. Numer. Math. **76**, 167–188 (1997)
62. A. Chambolle, R. DeVore, N.Y. Lee, B. Lucier, Nonlinear wavelet image processing: Variational problems, compression, and noise removal through wavelet shrinkage. IEEE Trans. Image Proc. **7**, 319–335 (1998)
63. A. Chambolle, V. Caselles, D. Cremers, M. Novaga, T. Pock, An Introduction to total variation for image analysis, in *Theoretical Foundations and Numerical Methods for Sparse Recovery*, ed. by M. Fornasier. Radon Series in Applied and Computational Mathematics (De Gruyter, Berlin, 2010)
64. T. Chan, J. Shen, Mathematical models for local non-texture inpainting. SIAM J. Appl. Math. **62**, 1019–1043 (2001)
65. T. Chan, J. Shen, *Image Processing and Analysis* (SIAM, Philadelphia, 2005)
66. T. Chan, L.A. Vese, Active contours without edges. IEEE Trans. Image Process. **10**, 266–277 (2001)
67. T. Chan, C.W. Wong, Total variation blind deconvolution. IEEE Trans. Imag. Proc. **7**, 370–375 (1998)
68. T. Chan, A. Marquina, P. Mulet, Second order differential functionals in total variation-based image restoration. CAM Report 98–35, UCLA, 1998
69. T. Chan, S.H. Kang, J. Shen, Total variation denoising and enhancement of color images based on the CB and HSV color models. J. Vis. Comm. Image Represent. **12**, 422–435 (2001)
70. T.F. Chan, S. Esedoglu, Aspects of total variation regularized L^1 function approximation. SIAM J. Appl. Math. **65**, 1817–1837 (2005)
71. T.F. Chan, S. Esedoglu, M. Nikolova, Algorithms for finding global minimizers of denoising and segmentation models. SIAM J. Appl. Math. **66**, 1632–1648 (2006)
72. G. Chavent, K. Kunisch, Regularization of linear least squares problems by total bounded variation. ESAIM Cont. Optim. Calc. Var. **2**, 359–376 (1997)
73. O. Christiansen, T.-M. Lee, J. Lie, U. Sinha, T.F. Chan, Total variation regularization of matrix-valued images. Int. J. Biomed. Imaging **2007**, Article ID 27432, 11 p. (2007). doi:10.1155/2007/27432
74. L. Cinque, G. Morrone, Retinex vombined with total variation for image illumination normalization, in *Image Analysis and Processing Ű ICIAP 2009*. LNCS, vol. 5716 (Springer, Berlin, 2009), pp. 958–964
75. S. Comelli, A Novel Class of Priors for Edge-Preserving Methods in Bayesian Inversion. Master thesis, University of Milano, 2011
76. R. Deriche, P. Kornprobst, G. Aubert, Optical flow estimation while preserving its discontinuities: A variational approach, in *Proceedings of the Asian Conference on Computer Vision, ACCVŠ95*, Singapore, 1995
77. N. Dey, L. Blanc-Feraud, C. Zimmer, Z. Kam, P. Roux, J.C. Olivo-Marin, J. Zerubia, Richardson-Lucy algorithm with total variation regularization for 3D confocal microscope deconvolution. Microsc. Res. Tech. **69**, 260–266 (2006)
78. D. Dobson, O. Scherzer, Analysis of regularized total variation penalty methods for denoising. Inverse Probl. **12**, 601–617 (1996)
79. D. Donoho, I. Johnstone, Ideal spatial adaptation via wavelet shrinkage. Biometrika **81**, 425–455 (1994)
80. A. Douiri, M. Schweiger, J. Riley, S.R. Arridge, Local diffusion regularisation method for optical tomography reconstruction using robust statistics. Optic. Lett. **30**, 2439–2441 (2005)
81. I. Ekeland, R. Temam, *Convex Analysis and Variational Problems*. Corrected Reprint Edition (SIAM, Philadelphia, 1999)
82. H.W. Engl, M. Hanke, A. Neubauer, *Regularization of Inverse Problems* (Kluwer, Dordrecht, 1996)
83. S. Esedoglu, S.J. Osher, Decomposition of images by the anisotropic Rudin-Osher-Fatemi model. Comm. Pure Appl. Math. **57**, 1609–1626 (2004)
84. E. Esser, Primal Dual Algorithms for Convex Models and Applications to Image Restoration, Registration and Nonlocal Inpainting. PhD thesis, UCLA, 2010

85. E. Esser, X. Zhang, T. Chan, A general framework for a class of first order primal-dual algorithms for tv minimization. SIAM J. Imag. Sci. **3**, 1015–1046 (2010)
86. L.C. Evans, R.F. Gariepy, *Measure Theory and Fine Properties of Functions* (CRC Press, Boca Raton, 1992)
87. X. Feng, A. Prohl, Analysis of total variation flow and its finite element approximations. ESAIM: Math. Mod. Numer. Anal. **37**, 533–556 (2003)
88. J. Flemming, B. Hofmann, Convergence rates in constrained Tikhonov regularization: equivalence of projected source conditions and variational inequalities. Inverse Probl. **27**, 085001 (2011)
89. M. Fornasier, Mathematics enters the picture, in *Mathknow Mathematics*, ed. by M. Emmer, A. Quarteroni. Applied Sciences and Real Life (Springer, Milan, 2009), pp. 217–228
90. M. Fornasier, G. Teschke, R. Ramlau, A comparison of joint sparsity and total variation minimization algorithms in a real-life art restoration problem. Adv. Comput. Math. **31**, 301–329 (2009)
91. M. Freiberger, C. Clason, H. Scharfetter, Total variation regularization for nonlinear fluorescence tomography with an augmented Lagrangian splitting approach. Appl. Optic. **49**, 3741–3747 (2010)
92. M.A. Freitag, N.K. Nichols, C.J. Budd, Resolution of sharp fronts in the presence of model error in variational data assimilation, Preprint (University of Bath, 2010)
93. H.Y. Gao, A.G. Bruce, WaveShrink with firm shrinkage. Statist. Sinica **7**, 855–874 (1997)
94. S. Geman, D. Geman, Stochastic relaxation, Gibbs distribution and the Bayesian restoration of images. IEEE Trans. Pattern Anal. Mach. Intell. **6**, 721–741 (1984)
95. E. Giusti, *Minimal Surfaces and Functions of Bounded Variation* (Birkhäuser, Boston, 1984)
96. T. Goldstein, S. Osher, The split Bregman method for L1 regularized problems. SIAM J. Imag. Sci. **2**, 323–343 (2009)
97. B. Goris, M.W. Van den Broek, K.J. Batenburg, H.H. Mezerji, S. Bals, Electron tomography based on a total variation minimization reconstruction technique. Ultramicroscopy **113**, 120–130 (2012)
98. M. Grasmair, Generalized Bregman distances and convergence rates for non-convex regularization methods. Inverse Probl. **26**, 115014 (2010)
99. M. Grasmair, Linear convergence rates for Tikhonov regularization with positively homogeneous functionals. Inverse Probl. **27**, 075014 (2011)
100. A. Haddad, Y. Meyer, Variational Methods in Image Processing. CAM-Report 04–52, UCLA, 2004
101. A. Haddad, Texture separation in $BV - G$ and $BV - L^1$ models. SIAM Multiscale Model. Simul. **6**, 273–286 (2007)
102. L. He, A. Marquina, S. Osher, Blind deconvolution using TV regularization and Bregman iteration. Int. J. Imag. Syst. Tech. **15**, 74–83 (2005)
103. L. He, T.C. Chung, S. Osher, T. Fang, P. Speier, MR Image Reconstruction by Using the Iterative Refinement Method and Nonlinear Inverse Scale Space Methods. CAM Report 06–35, UCLA, 2005
104. L. He, M. Burger, S. Osher, Iterative total variation regularization with non-quadratic fidelity. J. Math. Imag. Vis. **26**, 167–184 (2006)
105. W. Hinterberger, O. Scherzer, Variational methods on the space of functions of bounded Hessian for convexification and denoising. Computing **76**, 109–133 (2006)
106. M. Hintermüller, M. Monserrat Rincon-Camacho, Expected absolute value estimators for a spatially adapted regularization parameter choice rule in L1-TV-based image restoration. Inverse Probl. **26**, 085005 (2010)
107. B. Hofmann, B. Kaltenbacher, C. Pöschl, O. Scherzer, A Convergence rates result for Tikhonov regularization in Banach spaces with non-smooth operators. Inverse Probl. **23**, 987–1010 (2007)
108. Y.M. Huang, M.K. Ng, Y.W. Wen, A new total variation method for multiplicative noise Removal. SIAM J. Imag. Sci. **2**, 20–40 (2009)
109. P.J. Huber, Robust estimation of a location parameter. Ann. Math. Stat. **35**, 73–101 (1964)

110. J. Idier, *Bayesian Approach to Inverse Problems* (Wiley, New York, 2008)
111. M.D. Iordache, J.M. Bioucas-Dias, A. Plaza, Total variation regulatization in sparse hyperspectral unmixing, in *3rd Workshop on Hyperspectral Image and Signal Processing: Evolution in Remote Sensing* (IEEE, New York, 2011), pp. 1–4
112. Z. Jin, X. Yang, A variational model to remove the multiplicative noise in ultrasound images. J. Math. Imag. Vis. **74**, 39–62 (2011)
113. E. Jonsson, S.C. Huang, T. Chan, Total Variation Regularization in Positron Emission Tomography. CAM Report 98-48, UCLA, 1998
114. J.P. Kaipio, E. Somersalo, *Statistical and Computational Inverse Problems* (Springer, New York, 2004)
115. S. Keeling, C. Clason, M. Hintermüller, F. Knoll, A. Laurain, G. Winckel, An image space approach to Cartesian based parallel MR imaging with total variation regularization. Med. Image Anal. **16**, 189–200 (2012)
116. S. Kindermann, S. Osher, P.W. Jones, Deblurring and denoising of images by nonlocal functionals. Multiscale Model. Simul. **4**, 1091–1115 (2005)
117. S. Kindermann, S. Osher, J. Xu, Denoising by BV-duality. J. Sci. Comp. **28**, 411–444 (2006)
118. F. Knoll, M. Unger, C. Clason, C. Diwoky, T. Pock, R. Stollberger, Fast reduction of undersampling artifacts in radial MR angiography with 3D total variation on graphics hardware. Magn. Reson. Mater. Phy. **23**, 103–114 (2010)
119. F. Knoll, K. Bredies, T. Pock, R. Stollberger, Second order total generalized variation (TGV) for MRI. Magn. Reson. Med. **65**, 480–491 (2011)
120. F. Knoll, C. Clason, K. Bredies, M. Uecker, R. Stollberger, Parallel imaging with nonlinear reconstruction using variational penalties. Magn. Reson. Med. **67**, 34–41 (2012)
121. M. Lassas, S. Siltanen, Can one use total variation prior for edge preserving Bayesian inversion. Inverse Probl. **20**, 1537–1564 (2004)
122. P. Lax, *Functional Analysis* (Wiley, New York, 2002)
123. T. Le, R. Chartrand, T.J. Asaki, A variational approach to reconstructing images corrupted by Poisson noise. J. Math. Imag. Vis. **27**, 257–263 (2007)
124. J. Lellman, J. Kappes, J. Yuan, F. Becker, C. Schnörr, Convex multi-class image labeling by simplex-constrained total variation Technical report (University of Heidelberg, 2008)
125. C. Lemarechal, C. Sagastizabal, Practical aspects of the Moreau-Yosıda regularization: theoretical preliminaries. SIAM J. Optim. **7** 367–385 (1997)
126. J. Lie, J.M. Nordbotten, Inverse scale spaces for nonlinear regularization. J. Math. Imag. Vis. **27**, 41–50 (2007)
127. D.A. Lorenz, Convergence rates and source conditions for Tikhonov regularization with sparsity constraints. J. Inverse Ill-Posed Probl. **16**, 463–478 (2008)
128. Y. Lou, X. Zhang, S. Osher, A. Bertozzi, Image recovery via nonlocal operators. J. Sci. Comput. **42**, 185–197 (2010)
129. C. Louchet, Modeles variationnels et Bayesiens pour le d'ebruitage d'images: de la variation totale vers les moyennes non-locales. PhD thesis, University Paris-Descartes, 2008
130. C. Louchet, L. Moisan, Total Variation as a local filter. SIAM J. Imag. Sci. **4**, 651–694 (2011)
131. R. Luce, S. Perez, Parameter identification for an elliptic partial differential equation with distributed noisy data. Inverse Probl. **15**, 291–307 (1999)
132. M. Lustig, D.L. Donoho, J.M. Pauly, Sparse MRI: The application of compressed sensing for rapid MR imaging. Magn. Reson. Med. **58**, 1182–1195 (2007)
133. W. Ma, S. Osher, A TV Bregman Iterative Model of Retinex Theory. CAM-Report 10–13, UCLA, 2010
134. Y. Meyer, *Oscillating Patterns in Image Processing and Nonlinear Evolution Equations* (AMS, Providence, 2001)
135. G. Mohler, A. Bertozzi, T. Goldstein, S. Osher, Fast TV regularization for 2D maximum penalized likelihood estimation. J. Comput. Graph. Stat. **20**, 479–491 (2011)
136. M. Möller, A Variational Approach for Sharpening High-Dimensional Images. Diploma thesis, WWU Münster, 2009

137. M. Möller, T. Wittman, A. Bertozzi, M. Burger, A variational approach for sharpening high-dimensional images. SIAM J. Imag. Sci. **5**, 150–178 (2012)
138. J.M. Morel, S. Solimini, *Variational Methods for Image Segmentation* (Birkhäuser, Boston, 1995)
139. J. Müller, Parallel Total Variation Minimization. Diploma thesis, WWU Münster, 2008
140. D. Mumford, J. Shah, Optimal approximations by piecewise smooth functions and associated variational problems. Comm. Pure Appl. Math. **42**, 577–685 (1989)
141. F. Natterer, F. Wübbeling, *Mathematical Methods in Image Reconstruction* (SIAM, Philadelphia, 2001)
142. M.K. Ng, W. Wang, A total variation model for retinex. SIAM J. Imag. Sci. **4**, 345–365 (2011)
143. A. Obereder, S. Osher, O. Scherzer, On the use of dual norms in bounded variation type regularization, in *Properties from Incomplete Data*, ed. by R. Klette et al. (Kluwer, Dordrecht, 2005), pp. 373–390
144. A. Obereder, O. Scherzer, A. Kovac, Bivariate density estimation using BV regularisation. Comput. Stat. Data Anal. **51**, 5622–5634 (2007)
145. S. Osher, R. Fedkiw, *Level Set Methods and Dynamic Implicit Surfaces* (Springer, New York, 2002)
146. S. Osher, O. Scherzer, G-norm properties of bounded variation regularization. Comm. Math. Sci. **2**, 237–254 (2004)
147. S. Osher, A. Sole, L. Vese, Image decomposition and restoration using total variation minimization and the H^{-1} norm. SIAM Multiscale Model. Simul. **1**, 349–370 (2003)
148. S. Osher, M. Burger, D. Goldfarb, J. Xu, W. Yin, An iterative regularization method for total variation based image restoration. Multiscale Model. Simul. **4**, 460–489 (2005)
149. Y. Pan, R.T. Whitaker, A. Cheryauka, D. Ferguson, TV-regularized iterative image reconstruction on a mobile C-ARM CT, in *Proceedings of SPIE Medical Imaging 2010*, vol. 7622 (SPIE, San Diego, 2010)
150. V.Y. Panin, G.L. Zeng, G.T. Gullberg, Total variation regulated EM Algorithm. IEEE Trans. Nucl. Sci. **NS-46**, 2202–2010 (1999)
151. E. Pantin, J.L. Starck, F. Murtagh, Deconvolution and blind deconvolution in astronomy, in *Blind Image Deconvolution: Theory and Applications*, ed. by K. Egiazarian, P. Campisi (CRC Press, Boca Raton, 2007), pp. 277–317
152. K.D. Paulsen, H. Jiang, Enhanced frequency-domain optical image reconstruction in tissues through total-variation minimization. Appl. Opt. **35**, 3447–3458 (1996)
153. M. Persson, D. Bone, H. Elmqvist, Total variation norm for three-dimensional iterative reconstruction in limited view angle tomography. Phys. Med. Biol. **46**, 853–866 (2001)
154. G. Peyre, S. Bougleux, L. Cohen, Non-local regularization of inverse problems, in *Proceedings of the 10th European Conference on Computer Vision*. LNCS, vol. 5304 (Springer, Berlin, 2008), pp. 57–68
155. T. Pock, A. Chambolle, H. Bischof, D. Cremers, A convex relaxation approach for computing minimal partitions, in *IEEE Conference on Computer Vision and Pattern Recognition (CVPR)*, 2009
156. T. Pock, D. Cremers, H. Bischof, A. Chambolle, Global solutions of variational models with convex regularization. SIAM J. Imag. Sci. **3**, 1122–1145 (2010)
157. T. Pock, A. Chambolle, D. Cremers, H. Bischof, A convex relaxation approach for computing minimal partitions, in *IEEE Conference on Computer Vision and Pattern Recognition, 2009. CVPR 2009*, 20–25 June 2009 (2009), pp. 810–817. doi:10.1109/CVPR.2009.5206604
158. Z. Qin, D. Goldfarb, S. Ma, An Alternating Direction Method for Total Variation Denoising, Preprint (Columbia University, 2011)
159. S. Remmele, M. Seeland, J. Hesser, Fluorescence microscopy deconvolution based on Bregman iteration and Richardson-Lucy algorithm with TV regularization, in *Proceedings of the Workshop: Bildverarbeitung fi£¡ur die Medizin 2008* (Springer, Berlin, 2008), pp. 72–76
160. E. Resmerita, Regularization of ill-posed problems in Banach spaces: convergence rates. Inverse Probl. **21**, 1303–1314 (2005)

161. E. Resmerita, O. Scherzer, Error estimates for non-quadratic regularization and the relation to enhancing. Inverse Probl. **22**, 801–814 (2006)
162. W. Ring, Structural properties of solutions to total variation regularization problems. Math. Model. Numer. Anal. **34**, 799–810 (2000)
163. L.I. Rudin, S.J. Osher, E. Fatemi, Nonlinear total variation based noise removal algorithms. Physica D **60**, 259–268 (1992)
164. L. Rudin, P.L. Lions, S. Osher, Multiplicative denoising and debluring: theory and algorithms, in *Geometric Level Sets in Imaging, Vision, and Graphics*, ed. by S. Osher, N. Paragios (Springer, New York, 2003), pp. 103–119
165. A. Sawatzky, C. Brune, F. Wübbeling, T. Kösters, K. Schäfers, M. Burger, Accurate EM-TV algorithm in PET with low SNR, 2008. IEEE Nuclear Science Symposium Conference Record. doi: 10.1109/NSSMIC.2008.4774392
166. A. Sawatzky, (Nonlocal) Total Variation in Medical Imaging. PhD thesis, WWU Münster, 2011
167. A. Sawatzky, C. Brune, J. Müller, M. Burger, Total variation processing of images with Poisson statistics, in *Proceedings of the 13th International Conference on Computer Analysis of Images and Patterns*, LNCS, vol. 5702 (Springer, Berlin, 2009), pp. 533–540
168. A. Sawatzky, D. Tenbrinck, X. Jiang, M. Burger, A Variational Framework for Region-Based Segmentation Incorporating Physical Noise Models. CAM Report 11–81, UCLA, 2011
169. O. Scherzer, C. Groetsch, Inverse scale space theory for inverse problems, in *Scale-Space and Morphology in Computer Vision. Proceedings of the third International Conference Scale-space 2001*, ed. by M. Kerckhove (Springer, Berlin, 2001), pp. 317–325
170. F. Schoepfer, A.K. Louis, T. Schuster, Nonlinear iterative methods for linear ill-posed problems in Banach spaces. Inverse Probl. **22**, 311–329 (2006)
171. S. Setzer, G. Steidl, Variational methods with higher order derivatives in image processing, in *Approximation XII*, ed. by M. Neamtu, L.L.Schumaker (Nashboro Press, Brentwood, 2008), pp. 360–386
172. S. Setzer, G. Steidl, T. Teuber, Infimal convolution regularizations with discrete l1-type functionals. Comm. Math. Sci. **9**, 797–872 (2011)
173. D.L. Snyder, A.M. Hammoud, R.L. White, Image recovery from data acquired with a charge-coupled-device camera. J. Opt. Soc. Am. A **10**, 1014–1023 (1993)
174. D.L. Snyder, C.W. Helstrom, A.D. Lanterman, M. Faisal, R.L. White, Compensation for readout noise in CCD images. J. Opt. Soc. Am. A **12**, 272–283 (1995)
175. G. Steidl, A note on the dual treatment of higher order regularization functionals. Computing **76**, 135–148 (2006)
176. G. Steidl, S. Didas, J. Neumann, Relations between higher order TV regularization and support vector regression, in *Scale-Space and PDE Methods in Computer Vision*, ed. by R. Kimmel, N. Sochen, J. Weickert (Springer, Berlin, 2005), pp. 515–527
177. G. Steidl, S. Setzer, B. Popilka, B. Burgeth, Restoration of matrix fields by second order cone programming. Computing **81**, 161–178 (2007)
178. D. Strong, T. Chan, Exact Solutions to Total Variation Regularization Problems. CAM-Report 96-46, UCLA, 1996
179. D.M. Strong, J.F. Aujol, T. Chan, Scale recognition, regularization parameter selection, and Meyer's G norm in total variation regularization. SIAM J. Multiscale Model. Simul. **5**, 273–303 (2006)
180. R. Stück, M. Burger, T. Hohage, The iteratively regularized GaussŰNewton method with convex constraints and applications in 4Pi microscopy. Inverse Probl. **28**, 015012 (2012)
181. E. Tadmor, S. Nezzar, L. Vese, A multiscale image representation using hierarchical (BV;L2) decompositions. Multiscale Model. Simul. **2**, 554–579 (2004)
182. E. Tadmor, S. Nezzar, L. Vese, Multiscale hierarchical decomposition of images with applications to deblurring, denoising and segmentation. Comm. Math. Sci. **6**, 281–307 (2008)
183. Y. Vardi, L.A. Shepp, L. Kaufman, A statistical model for Positron emission tomography. JASA **80**, 8–37 (1985)

184. J. Velikina, S. Leng, G.H. Chen, Limited view angle tomographic image reconstruction via total variation minimization. Proc. SPIE v6510 i1
185. L. Vese, L. Lieu, Image restoration and decomposition via bounded total variation and negative Hilbert-Sobolev spaces. Appl. Math. Optim. **58**, 167–193 (2008)
186. L. Vese, S. Osher, Modeling textures with total variation minimization and oscillating patterns in image processing. J. Sci. Comput. **19**, 533–572 (2003)
187. J. Velikina, S. Leng, G.-H. Chen, Limited view angle tomographic image reconstruction via total variation minimization, in *Proceedings of the SPIE 6510, Medical Imaging 2007: Physics of Medical Imaging, 651020*, 14 March, 2007. doi:10.1117/12.713750; http://dx.doi.org/10.1117/12.713750
188. K.R. Vixie, Some properties of minimizers for the Chan-Esedoglu L1TV functional. ArXiv 0710.3980 (2007)
189. K.R. Vixie, S.P. Morgan, L1TV computes the flat norm for boundaries. Abstr. Appl. Anal. (2007), Article ID 45153
190. C.R. Vogel, Nonsmooth Regularization, in *Inverse Problems in Geophysical Applications*, ed. by H.W. Engl et al. (SIAM, Philadelphia, 1997), pp. 1–11
191. C.R. Vogel, *Computational Methods for Inverse Problems* (SIAM, Philadelphia, 2002)
192. C.R. Vogel, M.E. Oman, Iterative methods for total variation denoising. SIAM J. Sci. Comput. **17**, 227–238 (1996)
193. W.W. Wang, P.L. Shui, X.C. Feng, Variational models for fusion and denoising of multifocus images. IEEE Signal Process. Lett. **15**, 65–68 (2008)
194. J. Weickert, *Anisotropic Diffusion in Image Processing* (Teubner, Stuttgart, 1998)
195. J. Weickert, C. Schnörr, A theoretical framework for convex regularizers in PDE-based computation of image motion. Int. J. Comp. Vis. **45**, 245–264 (2001)
196. M. Werlberger, T. Pock, M. Unger, H. Bischof, Optical flow guided TV-L1 video interpolation and restoration, in *Energy Minimization Methods in Computer Vision and Pattern Recognition*, ed. by Y. Boykov, F. Kahl, V. Lempitsky, F.R. Schmidt. Lecture Notes in Computer Science , vol. 6819 (2011), pp. 273–286
197. M.N. Wernick, J.N. Aarsvold (eds.), *Emission Tomography: The Fundamentals of PET and SPECT* (Academic, San Diego, 2004)
198. D. Wirtz, SEGMEDIX: Development and Application of a Medical Image Segmentation Framework. Diploma thesis, WWU Münster, 2009
199. T. Wittman, Variational Approaches to Digital Image Zooming. PhD thesis, University of Minnesota, 2006
200. J. Xu, S. Osher, Iterative regularization and nonlinear inverse scale space applied to wavelet based denoising. IEEE Trans. Image Proc. **16**, 534–544 (2007)
201. Y.F. Yang, T.T. Wu, Z.F. Pang, Image-zooming technique based on Bregmanized nonlocal total variation regularization. Opt. Eng. **50**, 097008-097008-10 (2011)
202. Q. Zhang, X. Huang, L. Zhang, An energy-driven total variation model for segmentation and object-based classification of high spatial resolution remote-sensing imagery. IEEE Geosci. Rem. Sens. Lett. **10**, 125–129 (2013)
203. X. Zhang, M. Burger, S. Osher, A unified primal-dual algorithm framework based on Bregman iteration. J. Sci. Comp. **3** (2010). doi: 10.1007/s10915-010-9408-8
204. X. Zhang, M. Burger, X. Bresson, S. Osher, Bregmanized nonlocal regularization for deconvolution and sparse reconstruction. SIAM J. Imag. Sci. **3**, 253–276 (2010)
205. M. Zhu, S.J. Wright, T.F. Chan, Duality-based algorithms for total variation image restoration. Comp. Optim. Appl. **47**, 377–400 (2010)

EM-TV Methods for Inverse Problems with Poisson Noise

Alex Sawatzky, Christoph Brune, Thomas Kösters, Frank Wübbeling, and Martin Burger

Abstract We address the task of reconstructing images corrupted by Poisson noise, which is important in various applications such as fluorescence microscopy (Dey et al., 3D microscopy deconvolution using Richardson-Lucy algorithm with total variation regularization, 2004), positron emission tomography (PET; Vardi et al., J Am Stat Assoc 80:8–20, 1985), or astronomical imaging (Lantéri and Theys, EURASIP J Appl Signal Processing 15:2500–2513, 2005). Here we focus on reconstruction strategies combining the expectation-maximization (EM) algorithm and total variation (TV) based regularization, and present a detailed analysis as well as numerical results.

Recently extensions of the well known EM/Richardson-Lucy algorithm received increasing attention for inverse problems with Poisson data (Dey et al., 3D microscopy deconvolution using Richardson-Lucy algorithm with total variation regularization, 2004; Jonsson et al., Total variation regularization in positron emission tomography, 1998; Panin et al., IEEE Trans Nucl Sci 46(6):2202–2210, 1999). However, most of these algorithms for regularizations like TV lead to convergence problems for large regularization parameters, cannot guarantee positivity, and rely on additional approximations (like smoothed TV).

A. Sawatzky (✉) · F. Wübbeling · M. Burger
Institute for Computational and Applied Mathematics, University of Münster, Orléans-Ring 10, 48149 Münster, Germany
e-mail: alex.sawatzky@wwu.de; frank.wuebbeling@wwu.de; martin.burger@wwu.de

C. Brune
Department of Mathematics, University of California, Los Angeles, CA 90095-1555, USA
e-mail: brune@math.ucla.edu

T. Kösters
European Institute for Molecular Imaging, University of Münster, Mendelstr. 11, 48149 Münster, Germany
e-mail: thomas.koesters@wwu.de

The goal of this lecture is to provide accurate, robust and fast EM-TV based methods for computing cartoon reconstructions facilitating post-segmentation and providing a basis for quantification techniques. We illustrate also the performance of the proposed algorithms and confirm the analytical concepts by 2D and 3D synthetic and real-world results in optical nanoscopy and PET.

1 Introduction

Image reconstruction is a fundamental problem in several areas of applied sciences, such as medical imaging, optical microscopy, or astronomy. A prominent example of such problems is positron emission tomography (PET), a biomedical imaging technique in nuclear medicine that generates images of living organisms by visualization of weakly radioactively marked pharmaceuticals, so-called tracers. Due to the possibility of measuring temporal tracer uptake, this modality is particularly suitable for investigating physiological and biochemical processes. Another application of image reconstruction is fluorescence microscopy. It represents an important technique for investigating biological living cells down to nanoscales. In this type of applications image reconstruction arises in terms of deconvolution problems, where undesired blurring effects are caused by diffraction of light.

1.1 Mathematical Motivation

Image reconstruction can be formulated mathematically as a linear inverse and ill-posed problem. Typically, one deals with Fredholm integral equations of the first kind, or more general

$$\bar{f} = \bar{K}\bar{u}$$

with a compact linear operator $\bar{K}$, exact data $\bar{f}$ and the desired exact image $\bar{u}$. However only noisy versions f and K of $\bar{f}$ and $\bar{K}$ are available in practice and an approximate solution u of $\bar{u}$ from

$$f = Ku \qquad (1)$$

is wanted. The computation of u by a direct inversion of K is not reasonable since (1) is ill-posed. In this case regularization techniques are required to enforce stability during the inversion process and to compute useful reconstructions.

A commonly used idea to realize regularization techniques with statistical motivation is the Bayesian model (cf. Chap. 1, A Guide to the TV Zoo, Sect. 2.1), using the posterior probability density $p(u|f)$, given according to Bayes formula by

$$p(u|f) = \frac{p(f|u)\, p(u)}{p(f)}. \qquad (2)$$

The computationally interesting Bayesian approach is the maximum a-posteriori probability (MAP) estimation, which consists of computing an estimate u of the unknown object by maximizing the a-posteriori probability density $p(u|f)$. If the measurements f are available, the density $p(u|f)$ is denoted as the a-posteriori likelihood function, just depending on u. The Bayesian approach (2) has the advantage that it allows for incorporating additional prior information on u via the a-priori probability density $p(u)$ into the reconstruction process. The most frequently used a-priori densities are Gibbs functions [58, 59] of the form

$$p(u) \sim e^{-\alpha R(u)}, \tag{3}$$

where α denotes a positive parameter and R a convex energy functional. On the other hand, typical examples for probability densities $p(f|u)$ in (2) are raw data f drawn from an exponential family distribution. In the canonical case of additive white Gaussian noise with expectation 0 and variance σ^2 one finds

$$p(f|u) \sim e^{-\frac{1}{2\sigma^2} \|Ku - f\|^2_{L^2(\Sigma)}},$$

and the minimization of the negative log-likelihood function leads to the classical Tikhonov regularization methods [20], based on minimizing a functional of the form

$$\min_u \frac{1}{2} \|Ku - f\|^2_{L^2(\Sigma)} + \alpha R(u). \tag{4}$$

The first, so-called data fidelity, term penalizes the deviation from (1), while $R(u)$ is a regularization term penalizing deviations from a certain ideal structure (smoothness) of the solution.

In applications mentioned at the beginning, the measured data f are stochastic due to the radioactive decay of tracers in PET and due to laser scanning techniques in fluorescence microscopy. The random variables of the measured data in those applications are not Gaussian but Poisson distributed [120] with expected values given by $(Ku)_i$, i.e. it holds

$$p(f|u) = \prod_i \frac{(Ku)_i^{f_i}}{f_i!} e^{-(Ku)_i}.$$

In the following we concentrate on inverse problems with Poisson distributed data, where the MAP estimation via the negative log-likelihood function (2) asymptotically leads to the following variational problem [20],

$$\min_{u \geq 0} \int_\Sigma (Ku - f \log Ku) \, d\mu + \alpha R(u). \tag{5}$$

Up to additive terms independent of u, the data fidelity term here is the so-called Kullback-Leibler divergence (also known as cross entropy or I-divergence) between the probability measures f and Ku. A particular complication of (5) compared to (4) is the strong nonlinearity in the data fidelity term and resulting issues in the computation of minimizers. In addition, a nonnegativity constraint in (5) is essential in applications with Poisson distributed data since the unknown function u usually represents densities or intensity information.

With regard to the a-priori probability density in (3), the specific choice of the regularization functional R in (5) is crucial for the way a-priori information about the expected solution are incorporated into the reconstruction process. Smooth, in particular quadratic, regularizations have attracted most attention in the past, mainly due to simplicity in analysis and computation. However, such regularization approaches always lead to blurring of reconstructions, in particular they cannot yield reconstructions with sharp edges (cf. Chap. 1, Sect. 2.3). Hence, singular regularization energies, especially those of ℓ^1- or L^1-type, have attracted strong attention in variational problems, but are difficult to handle due to their non differentiable nature. Following the overview in chap. 1 we will focus our attention on the total variation (TV) regularization functional. The exact definition of TV [1] is

$$R(u) := |u|_{BV(\Omega)} = \sup_{\substack{g \in C_0^\infty(\Omega, \mathbb{R}^d) \\ \|g\|_\infty \leq 1}} \int_\Omega u \operatorname{div} g \, dx, \qquad (6)$$

which is formally (true if u is sufficiently regular)

$$|u|_{BV(\Omega)} = \int_\Omega |\nabla u| \, dx. \qquad (7)$$

The space of functions with bounded total variation is denoted by $BV(\Omega)$. For further properties and details of BV functions we refer to [1, 3, 52]. The motivation for using TV is the effective suppression of noise and the realization of almost homogeneous regions with sharp edges. These features are attractive for PET and nanoscopic imaging if the goal is to identify object shapes that are separated by sharp edges and shall be analyzed quantitatively afterwards.

Finally, in the literature there are in general two classes of methods that are used to solve reconstruction problems. On the one hand analytical (direct) methods and on the other hand algebraic (iterative) strategies [99]. A classical representative for a direct method is the Fourier-based filtered backprojection (FBP). Although FBP is well understood and can be computed efficiently, iterative strategies receive more and more attention in practice. The major reason is the high noise level, i.e. low signal-to-noise ratio (SNR), and the special type of statistics found in measurements of various applications, such as PET or fluorescence microscopy, which cannot be taken into account by direct methods. Thus, we focus on extensions of the expectation-maximization (EM) or Richardson-Lucy algorithm [42, 90, 106], which is a popular iterative reconstruction method to compute (5) in the absence of an explicit regularization (i.e. with the use of $R \equiv 0$) [114]. However, the EM algorithm is difficult to be generalized to regularized cases and we discuss robust and accurate solutions of this problem for appropriate models of R (which allow 3D real-life investigations) and its analysis.

1.2 Related Methods

In the past various reconstruction methods have been suggested for the regularized Poisson likelihood estimation problem (5), e.g. using Tikhonov regularization [15], diffusion regularization [16], or L^1 regularization functionals [91]. However, most works deal with regularization strategies based on TV functionals (6) or (7), applying e.g. to PET [5, 78, 101], deconvolution problems [17, 43, 53, 113], or denoising problems with K being the identity operator [86]. In the following we attempt to give a short overview of different reconstruction approaches dealing with regularization techniques in the case of Poisson perturbed inverse problems. Obviously, the overview is biased by the major lines of the authors research and we focus on the TV regularized variational problem (5). Roughly speaking, the iterative algorithms for the minimization of this problem can be separated in the following categories:

Steepest Descent Based Methods

The first class of methods based on the idea of steepest descent direction [79]. In [11], Bardsley proposed a computational method based on a projected quasi-Newton iteration [122, 123], where the nonnegativity constraint is guaranteed via a simple projection on the feasible set. On the other hand, the schemes suggested in [43, 78, 101] are implemented as elementary modifications of the EM algorithm with a fully explicit or semi-implicit treatment of TV in the iteration. It is simple to recognize that these algorithms can be interpreted as a scaled steepest descent method [79] with a special scaling operator and a constant step length (cf. [21, 24]). A major disadvantage of these EM based approaches is that the scaling is not necessarily positive and hence the regularization parameter α needs to be chosen sufficiently small, since otherwise the nonnegativity of solutions is not guaranteed and the algorithm cannot be continued. A possibility to overcome the last mentioned difficulty is to use a split-gradient method [84, 85] based on a decomposition of the gradient of the regularization functional in a positive and negative part. Using this approach, the EM based schemes mentioned above can be rewritten as a scaled gradient method, but now with a scaling that is nonnegative (cf. e.g. [21]). In the case of TV regularization, a scaled gradient projection (SGP) method has been applied successfully for the denoising of 2D Poisson data in [128].

However, the main limitation of steepest descent based methods is the demand for continuous differentiability of regularization functionals, which is certainly not satisfied in the case of TV regularization. To overcome the non differentiability of the TV functional, this class of methods uses an approximation of (7) by differentiable functionals of the form

$$|u|_{BV(\Omega)}^{\varepsilon} = \int_{\Omega} \sqrt{|\nabla u|^2 + \varepsilon}\, dx, \qquad \varepsilon > 0, \tag{8}$$

and thus creates blurring effects in reconstructed images (cf. Chap. 1, Sect. 8.3). In addition, due to the parameter dependence on ε in (8), all these algorithms are less robust. For example, in [22, 128], the authors point out that for small values of ε, the SGP method suffers from a very slow convergence and undesired oscillations arise in restored results. However, a stable iterative method has been proposed recently for all $\varepsilon \geq 0$ (i.e. also for the nondifferentiable TV functional (7)) in [23], based on the primal-dual (or saddle-point) equivalent formulation of (5) with TV regularization and an alternating extragradient method.

Operator Splitting Based Methods

The other class of methods based on the idea of operator splitting strategies [88, 112, 119], where the sum of operators will be decomposed into subproblems corresponding to only one of the summands. The advantage compared to steepest descent based methods is that the continuous differentiability of (5) is no longer required such that the total variation defined in (6) or (7) can be used, i.e. do not use the differentiable approximation (8).

In [54] and [113], the authors propose two numerical schemes, called PIDAL and PIDSplit+, using an instance of the alternating direction method of multipliers (ADMM) [45, 56, 57] and the alternating split Bregman method [62], respectively. Both schemes belong to the family of augmented Lagrangian algorithms [51, 61] and can be reinterpreted as Douglas-Rachford splitting strategy applied to the dual problem of (5) with TV regularization [112]. A numerical drawback of the Douglas-Rachford splitting approach is that the PIDAL algorithm requires an inversion of the operator $2I + K^*K$, respectively $I + K^*K - \Delta$ in case of PIDSplit+, where I is the identity operator, K^* the adjoint operator of K, and Δ is the Laplace operator. Hence, these methods are efficient if K^*K is diagonalizable and can be inverted efficiently, as e.g. a convolution operator via fast Fourier transform or discrete cosine transform. In all other cases an exact solution of the corresponding linear system is no longer possible. However, iterative solvers can be applied to compute an approximated solution, e.g. using a (preconditioned) conjugate gradient (PCG) method [79] due to the symmetry and positive definiteness of $2I + K^*K$ and $I + K^*K - \Delta$.

An alternative to the Douglas-Rachford splitting strategy is the forward-backward splitting approach [88, 102, 112, 119]. In contrast to Douglas-Rachford method, it does not require an inversion of an operator during the iteration, however has the numerical drawback that it is not unconditionally stable. In the case of Poisson distributed data, a forward-backward splitting algorithm was proposed in [4, 5] using total variation and wavelet ℓ^1-regularization. In addition, the authors also suggest a recently introduced first-order primal-dual algorithm for non-smooth convex optimization problems with known saddle-point structure by Chambolle and Pock [35].

There are also nested algorithms combining forward-backward and Douglas-Rachford iterations. In [37], the authors derive two types of such an approach and prove the weak convergence of the proposed algorithms. The effectiveness of algorithms is demonstrated on signal-dependent Gaussian noisy and Poisson noisy data, however only for wavelet-based image deblurring problems and not for total variation based ones.

1.3 Contributions and Organization

In the following we discuss robust algorithms for the TV regularized Poisson likelihood estimation problem (5) without approximation of TV as in (8), i.e. TV functionals (6) and (7) will be used. This enables to achieve cartoon reconstructions with sharp edges. For this purpose, a forward-backward (FB) splitting approach will be used that can be implemented by alternating a classical EM reconstruction step and solving a modified version of the Rudin-Osher-Fatemi (ROF) problem [107]. This modified ROF model can be solved in an analogous way to the projected gradient descent algorithm of Chambolle in [33] or the alternating split Bregman algorithm of Goldstein and Osher in [62]. The advantage of the described approach is that the EM algorithm does not need an inversion of the operator K and hence is applicable for an arbitrary operator, e.g. the Radon or X-ray transform [99]. If the implementation of the EM algorithm already exists [81], one can use it without any modifications. Due to the decoupling of reconstruction and smoothing done by the FB splitting approach, this strategy enables a high flexibility, can be performed equally well for large regularization parameters and is also favourably applicable for problems with a low SNR. First versions of the described method have been proposed by the authors in [28, 109] in applied settings related to PET and nanoscopy, and have already found interest as reference method (cf. [14, 38, 113, 127]).

The lecture below is organized as follows. In Sect. 2 we recall a mathematical model for inverse problems with Poisson distributed data. Starting from a statistical view of the image reconstruction problem in form of MAP estimation based on [20], we will proceed to a continuous representation in terms of multidimensional variational problems. An important point in this context is the incorporation of a-priori knowledge via regularization functionals. As a special case, we will derive the well known EM or Richardson-Lucy algorithm with nonnegativity constraints. In Sect. 3, the EM algorithm will be combined with total variation regularization.

We discuss a robust FB-EM-TV algorithm, which is implemented as a two step iteration scheme, provide suitable stopping criterions, and reinterpret the algorithm as a modified forward-backward splitting approach. Subsequently, we present the Bregman-FB-EM-TV method, whereby a loss of contrast in FB-EM-TV will be enhanced by iterative regularization with Bregman distances. In Sect. 4 we will study the FB-EM-TV model from an analytical point of view. After proving the well-posedness of the minimization problem in terms of existence, uniqueness and stability, we will provide a convergence analysis and positivity preserving properties of the FB-EM-TV algorithm. For this purpose, damping conditions will be offered to guarantee convergence of the forward-backward splitting algorithm. The numerical realization of ROF related problems, appearing in the second half step of the splitting strategy, is discussed in Sect. 5. Finally, the performance of presented techniques will be illustrated by synthetical and real-world 2D and 3D reconstructions in high-resolution fluorescence microscopy and PET in medical imaging.

2 Mathematical Modeling and EM Algorithm

2.1 Mathematical Model for Data Acquisition

This section provides an overview of mathematical modelling essential for a reasonable formulation of inverse problems with Poisson noise. In the following we concentrate on the relevant aspects of the model construction and refer to the work of Bertero et al. [20] for a detailed discussion. An imaging system consists in general of two structural elements:

- A collection of different physical components that generate signals containing useful information of spatial properties of an object, as e.g. collimators, mirrors, or lenses.
- A detector system that provides measurements of occurring signals, which causes the undesirable sampling and noise effects in many cases.

Hence, we assume that the raw data have the following properties:

- The data are discrete and the discretization is specified by the physical configuration of detectors. In addition, we assume that the data are given in form of a vector $f \in \mathbb{R}^N$.
- The data are realizations of random variables, since the noise is a random process caused by the detector system or the forward process of an imaging device. Hence, we consider the detected value f_i as a realization of a random variable F_i.

Additionally, a modeling of the imaging apparatus is necessary, which describes the generation and expansion of signals during the data acquisition process. Mathematically, the aim is to find a transformation that maps the spatial distribution of an object to the signals arriving at the detectors. In the following we concentrate on problems where the transformation is given by a linear operator and the data acquisition process can be described by a linear operator equation of the form

$$\bar{f} = \bar{K}\bar{u}. \tag{9}$$

Here, $\bar{K} : U(\Omega) \to V(\Sigma)$ is a linear and compact operator (thus with a nonclosed range), where $U(\Omega)$ and $V(\Sigma)$ are Banach spaces of functions on open and bounded sets Ω and Σ, respectively. In addition, we suppose that $\bar{K}$ preserves nonnegativity so that $\bar{K}u \geq 0$ for all $u \geq 0$. A typical example of (9) is a Fredholm integral equation of the first kind,

$$(\bar{K}u)(x) = \int_\Omega \bar{k}(x, y) u(y) \, dy, \qquad x \in \Sigma = \Omega,$$

where $\bar{k}$ is a nonnegative kernel of the operator $\bar{K}$. In (9), the function $\bar{u}$ describes the sought after exact properties of the object and $\bar{f}$ denotes the exact signals before detection. Problem statements of the type above can be found in numerous real-life applications, such as medical imaging [99, 114, 125], fluorescence microscopy [21, 43, 67], astronomy [20, 83], or radar imaging [72, 99].

The modeling of the data acquisition, in the manner described above, transfers the problem of object reconstruction into the solution of a linear inverse problem of the form (9). However in practice only noisy (and discrete) versions f and K of the exact data $\bar{f}$ and the operator $\bar{K}$ are available, such that only an approximate solution u of $\bar{u}$ can be computed from the equation

$$f = Ku. \tag{10}$$

In contrast to $\bar{K}$ in (9), the operator $K : U(\Omega) \to \mathbb{R}^N$ is a semi-discrete operator based on $\bar{K}$, which transforms the desired properties of the object to discrete raw data. In order to complete the modeling of the image reconstruction problem, we additionally have to take the noise in the measurements f into account. Thus a model for the probability density of the noise is needed and we concentrate on Poisson noise in this lecture. In this case, every F_i corresponds to a Poisson random variable with an expectation value given by $(Ku)_i$, i.e.

$$F_i \text{ is Poisson distributed with parameter } (Ku)_i. \tag{11}$$

Assuming that the random variables F_i are pairwise independent [89, 120], we can write the conditional probability density $p(f|u)$ of data f for a given image u as

$$p(f|u) = \prod_{i=1}^{N} p(f_i|u).$$

Combined with (11), this property leads to the following conditional probability density

$$p(f|u) = \prod_{i=1}^{N} \frac{(Ku)_i^{f_i}}{f_i!} e^{-(Ku)_i} . \qquad (12)$$

Hence, a complete model for the process of data generation and data acquisition is available, if the operator K and the conditional probability density $p(f|u)$ are known.

2.2 Statistical Problem Formulation of Image Reconstruction

Due to the compactness of the operator $\bar{K}$, (9) is an ill-posed problem [49, 64]. Note that the problem (10) is not ill-posed in a strong sense, because the operator K has a finite range. Nonetheless, the problem is highly ill-conditioned since K approximates $\bar{K}$, hence still some type of regularization is required to enforce stability during the inversion process and to compute useful reconstructions. A frequently used class of regularization techniques are variational methods based on the minimization of functionals of the form

$$\frac{1}{s} \|Ku - f\|^s_{L^s(\Sigma)} + \alpha R(u), \qquad \alpha > 0, \qquad s \in [1, \infty) . \qquad (13)$$

From the viewpoint of statistical modeling, the functionals in (13) are inappropriate for problems with Poisson distributed data, since they result from the assumption of raw data $f = K\bar{u} + \eta$, where η is a vector valued random variable with statistically independent and identically distributed components drawn from an exponential family distribution. Typical examples are that η is Laplace distributed ($s = 1$) or Gaussian distributed ($s = 2$) [20].

In the following we consider the statistical problem formulation of image reconstruction problems in the case of Poisson distributed raw data. An often used idea to achieve regularization techniques with statistical motivation is the Bayesian model obtaining the a-posteriori probability density of u for a given value f from Bayes formula

$$p(u|f) = \frac{p(f|u)\, p(u)}{p(f)} . \qquad (14)$$

Inserting the given measurements f, the density $p(u|f)$ is denoted as the a-posteriori likelihood function, which depends on u only. In order to determine an approximation to the unknown object $\bar{u}$, we use the MAP estimator which maximizes the likelihood function, i.e.

$$u_{MAP} \in \arg\max_{u \geq 0\, a.e.} p(u|f) . \qquad (15)$$

The nonnegativity constraint on the solution is needed, since in typical applications the functions represent densities or intensity information.

For a detailed specification of the likelihood function in (15), we assume that a model for the process of data acquisition and data generation, in the manner of Sect. 2.1, is available. For this reason, we plug the probability density for Poisson noise (12) and the Gibbs a-priori density (3) in the definition of the likelihood function (14) and obtain the following negative log-likelihood function

$$-\log p(u|f) = \sum_{i=1}^{N} \left((Ku)_i - f_i \log(Ku)_i \right) + \alpha R(u) + \text{const}(f),$$

in which the additive terms independent of u are negligible. At this point we pass over from a discrete to a continuous representation of data, which corresponds to the way events on detectors are measured. We assume that the data f in the discrete data space $\mathbb{R}^N$ can be interpreted as a sampling of a function in $V(\Sigma)$, which we denote for the sake of convenience with f again. Then, with the indicator function

$$\chi_{M_i}(x) = \begin{cases} 1, & \text{if } x \in M_i, \\ 0, & \text{else}, \end{cases}$$

where M_i is the region of the i-th detector, we can interpret the discrete data as mean values,

$$f_i = \int_{M_i} f \, dx = \int_{\Sigma} \chi_{M_i} f \, dx.$$

Thus the MAP estimate in (15) can be rewritten as the following continuous variational problem,

$$u_{MAP} \in \arg\min_{u \geq 0 \, a.e.} \int_{\Sigma} (Ku - f \log Ku) \, d\mu + \alpha R(u) \tag{16}$$

with $d\mu = \sum_{i=1}^{N} \chi_{M_i} \, d\lambda$, where λ denotes the Lebesgue measure.

2.3 EM Algorithm

In this section we give a review of a popular reconstruction algorithm for problems with Poisson data, the so-called expectation-maximization (EM) algorithm [42, 99, 114], which finds numerous applications, e.g. in medical imaging, microscopy or astronomy. In the two latter ones, the algorithm is better known as Richardson-Lucy algorithm [90, 106]. The EM algorithm is an iterative procedure to maximize the likelihood function $p(u|f)$ and will form a basis for algorithms discussed later. Here we assume no a-priori probability (i.e. $R \equiv 0$) and the problem in (16) reduces to the following variational problem with a nonnegativity constraint,

$$\min_{u \geq 0 \, a.e.} \int_\Sigma (Ku - f \log Ku) \, d\mu \, . \tag{17}$$

In order to derive the algorithm, we consider the first order optimality condition of the constrained minimization problem (17). Formally, the Karush-Kuhn-Tucker (KKT) conditions [74, Sect. VII, Theorem 2.1.4] provide the existence of a Lagrange multiplier function $\lambda \geq 0$, such that the stationary points of (17) satisfy equations

$$0 = K^* \mathbf{1}_\Sigma - K^* \left(\frac{f}{Ku} \right) - \lambda \, , \tag{18a}$$

$$0 = \lambda u \, , \tag{18b}$$

where K^* is the adjoint operator of K and $\mathbf{1}_\Sigma$ is the characteristic function on Σ. Since the optimization problem (17) is convex (cf. (20) and (21)), every function u satisfying $u \geq 0 \, a.e.$ and (18) is a global minimizer of (17). Multiplying (18a) by u, the Lagrange multiplier function λ can be eliminated by (18b) and division by $K^* \mathbf{1}_\Sigma$ leads to a simple iteration scheme,

$$u_{k+1} = \frac{u_k}{K^* \mathbf{1}_\Sigma} K^* \left(\frac{f}{Ku_k} \right) , \tag{19}$$

which preserves nonnegativity if the operator K preserves nonnegativity and the initialization u_0 is nonnegative. This iteration scheme is the well known EM algorithm, respectively Richardson-Lucy algorithm. In [114], Shepp and Vardi showed that this iteration is a closed example of the EM algorithm proposed by Dempster et al. in [42], who presented the algorithm in a more general setup.

Several convergence proofs of the EM algorithm to the maximum likelihood estimate (15), i.e. the solution of (17), can be found in literature [73, 77, 82, 96, 99, 105, 120]. Most of these works deal with the fully discrete setting, i.e. K is a matrix and u a vector. In this case the existence of a minimizer can be guaranteed since the smallest singular value is bounded away from zero by a positive value. Hence the iterates are bounded during the iteration and convergence is ensured. On the other hand, the convergence is difficult to prove in the continuous setting, i.e. if K is a general continuous operator. In this case the EM algorithm was investigated in [95–97], where the authors proved that, if the algorithm converges to u^*, then u^* is minimizer of (17). Recently, the authors in [105] gave the first convergence result in the continuous setting. They have shown that, if the equation $f = Ku$ has a solution $u^* \geq 0 \, a.e.$ (i.e. if the data are in the range of K), then the iteration (19) converges to u^* with respect to weak topologies on Lebesgue spaces.

It is known that the speed of convergence of iteration (19) is slow. This property is not surprising since the EM algorithm can be rewritten as a scaled gradient method with a constant step length (cf. e.g. [21]). Hence, different accelerating methods have been proposed in the past, as e.g. line-search techniques [2,75], multiplicative relaxation [77], or scaled gradient projection method [24]. Another possibility to accelerate the EM algorithm is the ordered subset EM (OSEM) algorithm, however the convergence properties of OSEM are difficult to assess (cf. [99]). A further property of the EM iteration is a lack of smoothing, whereby the so-called "checkerboard effect" arises, i.e. single pixels become visible in the iterates (see e.g. Fig. 2c). However, this feature points out that the algorithm can provide reasonable reconstructions in the case of sparse objects. Finally, as described in [105], the EM iterates show the following typical behavior for ill-posed problems. The (metric) distance between the iterates and the exact solution decreases initially before it increases as the noise is amplified during the iteration process. This issue might be controlled by using appropriate stopping rules to obtain reasonable results. In [105], it is shown that certain stopping rules indeed allow stable approximations. Another possibility to considerably improve reconstruction results are regularization techniques, which will be discussed in the following.

3 TV Regularization Methods with Poisson Noise

3.1 FB-EM-TV Algorithm

The EM or Richardson-Lucy algorithm, discussed above is the standard iterative reconstruction method for linear inverse problems (10) with incomplete Poisson data. However, by setting $R \equiv 0$ in (16), no a-priori knowledge about the expected solution are taken into account, i.e. different images have the same a-priori probability. Especially in the case of measurements with low SNR, like lower tracer dose rate or tracer with short radioactive half life in PET examinations, the multiplicative fixed point iteration (19) provides unsatisfactory and noisy results even with early termination (see e.g. Fig. 8).

In order to improve the reconstructions from the EM algorithm, we modify (17) according to (16) using the TV functional $|\cdot|_{BV(\Omega)}$, defined in (6) or (7), as regularization energy,

$$\min_{\substack{u \in BV(\Omega) \\ u \geq 0 \, a.e.}} \int_{\Sigma} (Ku - f \log Ku) \, d\mu + \alpha |u|_{BV(\Omega)}, \qquad \alpha > 0. \qquad (20)$$

From the statistical problem formulation in Sect. 2.2 we use the a-priori probability density $p(u)$ in (3) with $R(u) = |u|_{BV(\Omega)}$. This means that images with smaller total variation (higher prior probability) are preferred in the minimization (20).

The expected reconstructions are cartoon-like images, i.e. they will result in almost uniform (mean) intensities inside the different structures which are separated by sharp edges. It is obvious that such an approach cannot be used for studying strange properties inside the structures of an object (which is anyway unrealistic in case of low SNR), but it is well suited for segmenting different structures and analyzing their averages quantitatively.

In order to design a computational method we consider the first order optimality condition of (20). Due to the total variation, this variational problem is not differentiable in the usual sense but still convex (see Sect. 4). This property can be simply shown by extending the data fidelity term in (20) to the Kullback-Leibler (KL) functional D_{KL} [104] without changing the stationary points,

$$\min_{\substack{u \in BV(\Omega) \\ u \geq 0 \text{ a.e.}}} D_{KL}(f, Ku) + \alpha |u|_{BV(\Omega)} \tag{21}$$

with D_{KL} defined by

$$D_{KL}(v, u) = \int_{\Sigma} \left(v \log \frac{v}{u} - v + u \right) d\mu .$$

In the following we consider for a convex functional $J : X \to (-\infty, +\infty]$, X Banach space, a generalized derivative called the subdifferential [47] at a point u defined by

$$\partial J(u) := \{ p \in X^* : J(\tilde{u}) - J(u) - \langle p, \tilde{u} - u \rangle \geq 0, \quad \forall \tilde{u} \in X \}, \tag{22}$$

where X^* denotes the dual space of X. The single elements of $\partial J(u)$ are called subgradients of J at u (cf. mBsO, Sect. 3.5). Due to the continuity of the KL functional and [47, p. 26, Proposition 5.6], the following identity holds

$$\partial_u \left(D_{KL}(f, Ku) + \alpha |u|_{BV(\Omega)} \right) = \partial_u D_{KL}(f, Ku) + \alpha \partial |u|_{BV(\Omega)} .$$

Since the subdifferential of $D_{KL}(f, K \cdot)$ is a singleton, the first optimality condition of (21) for a solution $u \geq 0$ a.e. is given by

$$K^* \mathbf{1}_\Sigma - K^* \left(\frac{f}{Ku} \right) + \alpha p = 0, \qquad p \in \partial |u|_{BV(\Omega)} , \tag{23}$$

where K^* denotes the adjoint of K. Formally, this condition is a nonlinear integrodifferential equation using the definition of TV in (7),

$$K^* \mathbf{1}_\Sigma - K^* \left(\frac{f}{Ku} \right) - \alpha \operatorname{div} \left(\frac{\nabla u}{|\nabla u|} \right) = 0 .$$

The simplest iteration scheme to compute a solution of the variational problem (20) is a gradient-type method, which however is not robust in the case of TV and severe step size restrictions are needed since the subgradient p of TV is treated explicitly. A better idea is to use an iteration scheme that evaluates the nonlocal term (including the operator K) in (23) at the previous iterate u_k and the local term (including the subgradient of TV) at the new iterate u_{k+1}, i.e.

$$\mathbf{1}_\Omega - \frac{1}{K^*\mathbf{1}_\Sigma} K^*\left(\frac{f}{Ku_k}\right) + \alpha \frac{1}{K^*\mathbf{1}_\Sigma} p_{k+1} = 0, \quad p_{k+1} \in \partial|u_{k+1}|_{BV(\Omega)}, \tag{24}$$

where we additionally divide (23) by $K^*\mathbf{1}_\Sigma$. Here, the new iterate u_{k+1} appears only as a point of reference for the subdifferential of total variation. This is a considerable drawback since u_{k+1} cannot be determined from (24) due to the lack of a one-to-one relation between subgradients (dual variables) and primal variable u. In addition, such iteration schemes cannot guarantee preservation of nonnegativity. An improved method can be obtained by approximating the constant term $\mathbf{1}_\Omega$ in (24) by $\frac{u_{k+1}}{u_k}$ such that u_{k+1} appears directly, i.e.

$$u_{k+1} - \frac{u_k}{K^*\mathbf{1}_\Sigma} K^*\left(\frac{f}{Ku_k}\right) + \alpha \frac{u_k}{K^*\mathbf{1}_\Sigma} p_{k+1} = 0, \quad p_{k+1} \in \partial|u_{k+1}|_{BV(\Omega)}. \tag{25}$$

In order to verify that the iteration scheme (25) actually preserves nonnegativity, we can proceed in an analogous way to the EM algorithm in Sect. 2.3. Accordingly, the KKT conditions formally provide the existence of a Lagrange multiplier function $\lambda \geq 0$, such that the stationary points of (21) satisfy

$$0 \in K^*\mathbf{1}_\Sigma - K^*\left(\frac{f}{Ku}\right) + \alpha \partial|u|_{BV(\Omega)} - \lambda, \tag{26a}$$

$$0 = \lambda u. \tag{26b}$$

Multiplying (26a) by u, the multiplier function λ can be eliminated by (26b) and the subsequent division by $K^*\mathbf{1}_\Sigma$ leads to a fixed point equation of the form

$$u - \frac{u}{K^*\mathbf{1}_\Sigma} K^*\left(\frac{f}{Ku}\right) + \alpha \frac{u}{K^*\mathbf{1}_\Sigma} p = 0, \quad p \in \partial|u|_{BV(\Omega)}, \tag{27}$$

which is just the optimality condition (23) multiplied by u, i.e. this multiplication corresponds to the nonnegativity constraint in (20). We see that the iteration (25) is just a semi-implicit approach to the fixed point equation (27). In Sect. 4.3 we will show that this iteration method actually preserves positivity if the operator K preserves positivity and the initialization u_0 is positive.

We observe that the second term in the iteration (25) is just a single EM step in (19). Consequently, the method (25) solving the variational problem (20) can be implemented as a nested two step iteration

$$u_{k+\frac{1}{2}} = \frac{u_k}{K^*1_\Sigma} K^* \left(\frac{f}{Ku_k} \right),$$ (EM step)

(28a)

$$u_{k+1} = u_{k+\frac{1}{2}} - \alpha \frac{u_k}{K^*1_\Sigma} p_{k+1}, \quad p_{k+1} \in \partial |u_{k+1}|_{BV(\Omega)}.$$ (TV step)

(28b)

Thus the algorithm alternates an EM reconstruction step with a TV correction step to compute a solution of (20). In Sect. 3.3 we will see that this iteration scheme can be interpreted as a modified forward-backward (FB) splitting strategy and denote it thus as FB-EM-TV algorithm. The complex half step (28b) can be realized by solving the convex variational problem

$$u_{k+1} \in \arg\min_{u \in BV(\Omega)} \left\{ \frac{1}{2} \int_\Omega \frac{K^*1_\Sigma \left(u - u_{k+\frac{1}{2}}\right)^2}{u_k} dx + \alpha |u|_{BV(\Omega)} \right\}.$$ (29)

Inspecting the first order optimality condition confirms the equivalence of this minimization with the TV correction step (28b). Note that (29) is just a modified version of the Rudin-Osher-Fatemi (ROF) model (cf. Chap. 1), with the difference of a weight $\frac{K^*1_\Sigma}{u_k}$ in the fidelity term. This analogy creates the opportunity to carry over efficient numerical schemes known for the ROF model. In the numerical realization in Sect. 5 we offer two algorithms, which are similar to the projected gradient descent algorithm of Chambolle in [33] and the alternating split Bregman algorithm of Goldstein and Osher in [62]. In this way, the weighted ROF problem (29) can be solved efficiently, obtaining accurate and robust algorithms.

3.2 Damped FB-EM-TV Algorithm

The alternating structure of the iteration (28) has the particular advantage that the interaction between reconstruction and denoising can be controlled via a simple adaption of the TV correction step. A possibility is a damped TV correction step,

$$u_{k+1} = (1-\omega_k) u_k + \omega_k u_{k+\frac{1}{2}} - \omega_k \alpha \frac{u_k}{K^*1_\Sigma} p_{k+1}, \quad \omega_k \in (0, 1],$$ (30)

which relates the current EM iterate $u_{k+\frac{1}{2}}$ with the previous TV denoised iterate u_k via a convex combination by using a damping parameter ω_k. The damped half step (30) can be realized in analogy to (29) by minimizing the variational problem

EM-TV Methods for Inverse Problems with Poisson Noise 87

$$u_{k+1} \in \underset{u \in BV(\Omega)}{\arg\min} \left\{ \frac{1}{2} \int_\Omega \frac{K^* \mathbf{1}_\Sigma \left(u - \tilde{u}_{k+\frac{1}{2}}\right)^2}{u_k} \, dx + \omega_k \, \alpha \, |u|_{BV(\Omega)} \right\} \quad (31)$$

with

$$\tilde{u}_{k+\frac{1}{2}} = \omega_k \, u_{k+\frac{1}{2}} + (1 - \omega_k) \, u_k \, .$$

This damping is not only motivated by numerical results (see Sect. 5), but also required to attain a monotone descent of the objective functional in (20) during the minimization process (see Sect. 4.4). Finally, for $\omega_k = 1$, the iteration (30) simplifies to the original TV denoising step (28b). For small ω_k, the iterations stay close to regularized solutions u_k. For an adequate choice of $\omega_k \in (0, 1]$, the convergence of the two step iteration (28) with respect to the damped regularization step (30) will be proved in Theorem 4. Additionally, an explicit bound on ω_k will be presented for the convergence in the special case of denoising problems (see Corollary 2), i.e. K being the identity operator.

3.3 *(Damped) FB-EM-TV Algorithm in Context of Operator Splitting*

In Sects. 3.1 and 3.2, the FB-EM-TV reconstruction method has been presented as a two step algorithm (28) with an additional damping modification (30). This two step strategy can be interpreted as an operator splitting algorithm. Recently, several works in literature picked up these splitting ideas, providing efficient algorithms in image processing, see e.g. [26, 39, 40, 111]. Most of the papers dealing with convex splitting strategies go back to early works of Douglas and Rachford [44] and other authors in [88] and [119].

The optimality condition (23) of the variational problem (20) can be interpreted as a decomposition problem ($C = A + B$), with respect to the convex Kullback-Leibler functional and the convex TV regularization term (respectively their subdifferentials). Hence we consider the stationary equation

$$0 \in C(u) := \underbrace{K^* \mathbf{1}_\Sigma - K^* \left(\frac{f}{Ku}\right)}_{=: A(u)} + \underbrace{\alpha \, \partial |u|_{BV(\Omega)}}_{=: B(u)} \, . \quad (32)$$

Using these notations, the two step iteration (28) with the modified regularization step (30) and $\omega_k \in (0, 1]$ reads

$$\frac{K^*\mathbf{1}_\Sigma \left(u_{k+\frac{1}{2}} - u_k\right)}{u_k} + A(u_k) = 0$$

$$\frac{K^*\mathbf{1}_\Sigma \left(u_{k+1} - \omega_k u_{k+\frac{1}{2}} - (1-\omega_k)u_k\right)}{u_k} + \omega_k B(u_{k+1}) = 0 \quad (33)$$

and can easily be reformulated as a forward-backward splitting algorithm

$$\frac{K^*\mathbf{1}_\Sigma \left(\tilde{u}_{k+\frac{1}{2}} - u_k\right)}{\omega_k u_k} + A(u_k) = 0 \quad \text{(forward step on } A\text{)}$$

$$\frac{K^*\mathbf{1}_\Sigma \left(u_{k+1} - \tilde{u}_{k+\frac{1}{2}}\right)}{\omega_k u_k} + B(u_{k+1}) = 0 \quad \text{(backward step on } B\text{)}$$

with

$$\tilde{u}_{k+\frac{1}{2}} = \omega_k u_{k+\frac{1}{2}} + (1-\omega_k)u_k.$$

Compared to the undamped FB-EM-TV strategy (28), in the case of the damped iteration scheme, the artificial time step size is not only given by u_k, but can also be controlled via the additional damping parameter ω_k. In a more compact form, the whole iteration can be formulated as

$$\begin{aligned} u_{k+1} &= \left(I + \frac{\omega_k u_k}{K^*\mathbf{1}_\Sigma} B\right)^{-1} \left(I - \frac{\omega_k u_k}{K^*\mathbf{1}_\Sigma} A\right) u_k \\ &= (L_k + B)^{-1}(L_k - A)u_k \end{aligned} \quad (34)$$

with a multiplication operator L_k defined by $\frac{K^*\mathbf{1}_\Sigma}{\omega_k u_k}$.

The forward-backward splitting approach has been suggested independently by Lions and Mercier [88] and Passty [102]. In the case of Poisson distributed data, this splitting strategy was also suggested in [4, 5], where a straight-forward use of the forward-backward algorithm was proposed. There are alternatives to the forward-backward splitting method such as the Peaceman-Rachford or Douglas-Rachford splitting schemes, see e.g. [88] or [112], which are indeed unconditionally stable. However, these approaches have the numerical drawback that also an additional backward step on the nonlinear operator A has to be performed, which would mean an inversion of an operator including K^*K, see [54, 113] and Sect. 1.2.

3.4 Stopping Rules and Pseudocode for the (Damped) FB-EM-TV Algorithm

In the following we consider appropriate stopping rules, which will guarantee the accuracy of the (damped) FB-EM-TV algorithm. The error in the optimality condition (23) can be taken as a basic stopping criterion in a suitable norm. For this purpose, a weighted norm deduced from the weighted scalar product will be used,

$$\langle u, v \rangle_w := \int_\Omega u\, v\, w\, d\lambda \quad \text{and} \quad \|u\|_{2,w} := \sqrt{\langle u, u \rangle_w}, \quad (35)$$

with a positive weight function w and the standard Lebesgue measure λ on Ω. Then the error in the optimality condition can be measured reasonably in the norm

$$opt_{k+1} := \left\| K^* \mathbf{1}_\Sigma - K^* \left(\frac{f}{K u_{k+1}} \right) + \alpha\, p_{k+1} \right\|_{2, u_{k+1}}^2. \quad (36)$$

Furthermore, we are not only interested in the improvement of the whole optimality condition (23), but also in the convergence of the sequence of primal functions $\{u_k\}$ and the sequence of subgradients $\{p_k\}$ with $p_k \in \partial |u_k|_{BV(\Omega)}$. In order to establish appropriate stopping rules for these iterates, we consider the damped TV correction step (30) with the EM reconstruction step (28a),

$$u_{k+1} - \omega_k \frac{u_k}{K^* \mathbf{1}_\Sigma} K^* \left(\frac{f}{K u_k} \right) - (1 - \omega_k)\, u_k + \omega_k\, \alpha\, \frac{u_k}{K^* \mathbf{1}_\Sigma}\, p_{k+1} = 0\, .$$

By combining this iteration scheme with the optimality condition (27) evaluated at u_k, which must be fulfilled in the case of convergence, we obtain the optimality statement for the sequences $\{p_k\}$ and $\{u_k\}$,

$$\alpha\, (p_{k+1} - p_k) + \frac{K^* \mathbf{1}_\Sigma\, (u_{k+1} - u_k)}{\omega_k\, u_k} = 0\, .$$

With the aid of the weighted norm (35), additional stopping criteria for the FB-EM-TV algorithm can be defined, which guarantee the accuracy of primal functions $\{u_k\}$ and subgradients $\{p_k\}$,

$$u_{opt_{k+1}} := \left\| \frac{K^* \mathbf{1}_\Sigma\, (u_{k+1} - u_k)}{\omega_k\, u_k} \right\|_{2, u_{k+1}}^2 ,$$

$$p_{opt_{k+1}} := \|\alpha\, (p_{k+1} - p_k)\|_{2, u_{k+1}}^2 . \quad (37)$$

Note that the stopping criteria in (36) and (37) are well defined, since we will prove in Lemma 5 that each iterate u_k of the (damped) FB-EM-TV splitting strategy is positive. Thus, based on previous sections, we can use Algorithm 1 to solve the TV regularized Poisson likelihood estimation problem (20).

Remark 1. Selecting a reasonable regularization parameter α in the model (20) is a common problem. In the case of additive Gaussian noise, there exist several works in literature dealing with this problem, see e.g. [87, 118, 122]. Most of them

Algorithm 1 (Damped) FB-EM-TV Algorithm
---
1. **Parameters:** f, $\alpha > 0$, $\omega \in (0, 1]$, $maxEMIts \in \mathbb{N}$, $tol > 0$
2. **Initialization:** $k = 0$, $u_0 := c > 0$
3. **Iteration:**
   while $((k < maxEMIts)$ and
   $(opt_k \geq tol$ **or** $u_{opt_k} \geq tol$ **or** $p_{opt_k} \geq tol))$ **do** $\triangleright$ (36), (37)
   i) Compute $u_{k+\frac{1}{2}}$ via EM step in (28a).
   ii) Set $\omega_k = \omega$.
   iii) Compute u_{k+1} via modified ROF model (31). $\triangleright$ Section 5
   iv) $k \leftarrow k + 1$
   end while
   return u_k
---

are based on the discrepancy principle and Chi-square distributions, generalized cross validation methods or unbiased predictive risk estimates. Finding an "optimal" regularization parameter is, in general, more complicated for non-Gaussian noise models. Some works addressing this issue can be found e.g. in [13, 14, 22] and the references therein.

3.5 Bregman-FB-EM-TV Algorithm

The discussed FB-EM-TV algorithm (28) solves the estimation problem (20) and provides cartoon-like reconstructions with sharp edges. However, the use of total variation regularization has the drawback that the reconstructed images suffer from contrast reduction [94, 100] (cf. also Chap. 1, Sect. 6). In the following we discuss an extension of the TV regularized Poisson likelihood estimation problem (20), and therewith also of the (damped) FB-EM-TV algorithm, by an iterative regularization strategy to a simultaneous contrast correction. More precisely, a contrast enhancement by Bregman distance iterations will be performed. These techniques have been derived by Osher et al. in [100], with a detailed analysis for Gaussian-type problems (4), and have been generalized to time-continuity [31], L^p-norm data fitting terms [32] and nonlinear inverse problems [8]. Following these methods, an iterative refinement in our case is realized by a sequence of modified Poisson likelihood estimation problems based on (20).

The inverse scale space methods follow the concept of iterative regularization by Bregman distance [27]. We recall from [30] that for a convex and proper functional $J : X \to (-\infty, +\infty]$, X Banach space, the generalized Bregman distance of J at $u \in X$ and $p \in \partial J(u)$ is defined by

$$D_J^p(\tilde{u}, u) := J(\tilde{u}) - J(u) - \langle p, \tilde{u} - u \rangle, \qquad \tilde{u} \in X,$$

where $\langle \cdot, \cdot \rangle$ denotes the standard duality product between the dual space X^* and X. Note that the Bregman distance does not represent a distance in the common (metric) sense, since it is not symmetric in general and the triangle inequality does not hold. However, it is a distance measure in the following sense

$$D_J^p(\tilde{u}, u) \geq 0 \quad \text{and} \quad D_J^p(\tilde{u}, u) = 0 \quad \text{for} \quad \tilde{u} = u.$$

The Bregman distance is convex in the first argument because J is convex and it can be interpreted as the difference between $J(\tilde{u})$ and the Taylor linearization of J around u evaluated in $\tilde{u}$ if, in addition, J is continuously differentiable(see e.g. [32]).

Following the idea of inverse scale space methods discussed in Chap. 1 Sect. 6 in the case of Gaussian noisy data, an iterative contrast enhancement is performed by a sequence of modified FB-EM-TV problems based on (20). This means that the method initially starts with a simple FB-EM-TV algorithm, i.e. it consists of computing a minimizer u^1 of (20). Subsequently, the results will be refined step by step by considering variational problems with a shifted TV term,

$$u^{l+1} \in \underset{\substack{u \in BV(\Omega) \\ u \geq 0 \, a.e.}}{\arg\min} \left\{ D_{KL}(f, Ku) + \alpha \big(|u|_{BV(\Omega)} - \langle p^l, u \rangle \big) \right\}, \quad p^l \in \partial |u^l|_{BV(\Omega)}. \tag{38}$$

Using the Bregman distance, this minimization problem is equivalent to

$$u^{l+1} \in \underset{\substack{u \in BV(\Omega) \\ u \geq 0 \, a.e.}}{\arg\min} \left\{ D_{KL}(f, Ku) + \alpha D_{|\cdot|_{BV(\Omega)}}^{p^l}(u, u^l) \right\}, \quad p^l \in \partial |u^l|_{BV(\Omega)}. \tag{39}$$

Note that the first iterate u^1 can also be characterized by the variational problem (38), if the function u^0 is chosen constant and we set $p^0 := 0 \in \partial |u^0|_{BV(\Omega)}$.

From the statistical problem formulation in Sect. 2.2, the Bregman distance regularized variational problem (39) uses an adapted a-priori probability density $p(u)$ in the Bayesian model formulation (14). Instead of a zero centered a-priori probability $R(u) = |u|_{BV(\Omega)}$ as in the case of the FB-EM-TV algorithm, here a new a-priori probability is considered in every Bregman refinement step which is related to a shifted total variation, i.e we use the following Gibbs function (3),

$$p(u) \sim e^{-\alpha D_{|\cdot|_{BV(\Omega)}}^{p^l}(u, u^l)}.$$

This a-priori probability density means that images with smaller total variation and a close distance to the maximum likelihood estimator u^l of the previous FB-EM-TV problem are preferred in the minimization of (39).

In order to design a two step iteration analogous to the FB-EM-TV algorithm, we consider the first order optimality condition for the variational problem (39). Due to the convexity of the Bregman distance in the first argument, we can determine the subdifferential of (39). Analogous to the derivation of the FB-EM-TV iteration, the subdifferential of a sum of functions can be split to a sum of subdifferentials due to the continuity of the Kullback-Leibler functional. Since the subdifferential of the KL functional can be expressed formally by Fréchet derivatives (23) and it holds

$$\partial \left(- \langle p^l, u \rangle \right) = \left\{ -p^l \right\},$$

the first order optimality condition of (39) for a solution $u^{l+1} \geq 0$ a.e. is given by

$$0 \in K^* \mathbf{1}_\Sigma - K^* \left(\frac{f}{K u^{l+1}} \right) + \alpha \left(\partial |u^{l+1}|_{BV(\Omega)} - p^l \right), \quad p^l \in \partial |u^l|_{BV(\Omega)}. \tag{40}$$

For u^0 constant and $p^0 := 0 \in \partial |u^0|_{BV(\Omega)}$, this condition yields a well defined update formula of the iterates p^l,

$$p^{l+1} := p^l - \frac{1}{\alpha} \left(K^* \mathbf{1}_\Sigma - K^* \left(\frac{f}{K u^{l+1}} \right) \right) \in \partial |u^{l+1}|_{BV(\Omega)}. \tag{41}$$

Analogous to the FB-EM-TV algorithm, we can apply the idea of the nested two step iteration (28) in every refinement step $l = 0, 1, \cdots$. For the solution of (39), condition (40) yields a strategy consisting of an EM reconstruction step

$$u^{l+1}_{k+\frac{1}{2}} = \frac{u^{l+1}_k}{K^* \mathbf{1}_\Sigma} K^* \left(\frac{f}{K u^{l+1}_k} \right), \tag{42}$$

followed by solving the adapted weighted ROF problem

$$u^{l+1}_{k+1} \in \arg\min_{u \in BV(\Omega)} \left\{ \frac{1}{2} \int_\Omega \frac{K^* \mathbf{1}_\Sigma \left(u - u^{l+1}_{k+\frac{1}{2}} \right)^2}{u^{l+1}_k} dx + \alpha \left(|u|_{BV(\Omega)} - \langle p^l, u \rangle \right) \right\}. \tag{43}$$

Following [100], we provide an opportunity to transfer the shift term $\langle p^l, u \rangle$ to the data fidelity term. This approach facilitates the implementation of contrast enhancement with the Bregman distance via a slight modification of the FB-EM-TV algorithm. For this purpose, we use the scaling $K^* \mathbf{1}_\Sigma v^l := \alpha p^l$ and obtain from (40) the following update formula for the iterates v^l,

$$v^{l+1} = v^l - \left(\mathbf{1}_\Omega - \frac{1}{K^* \mathbf{1}_\Sigma} K^* \left(\frac{f}{K u^{l+1}} \right) \right), \quad v^0 = 0. \tag{44}$$

Using this scaled update, we can rewrite (43) to

$$u_{k+1}^{l+1} \in \underset{u \in BV(\Omega)}{\arg\min} \left\{ \frac{1}{2} \int_\Omega \frac{K^* \mathbf{1}_\Sigma \left((u - u_{k+\frac{1}{2}}^{l+1})^2 - 2uu_k^{l+1} v^l \right)}{u_k^{l+1}} dx + \alpha |u|_{BV(\Omega)} \right\}.$$

In the equation

$$(u - u_{k+\frac{1}{2}}^{l+1})^2 - 2uu_k^{l+1} v^l$$
$$= \left(u - (u_{k+\frac{1}{2}}^{l+1} + u_k^{l+1} v^l)\right)^2 - 2u_{k+\frac{1}{2}}^{l+1} u_k^{l+1} v^l + (u_k^{l+1})^2 (v^l)^2,$$

the last two terms on the right-hand side are independent of u and hence the variational problem (43) simplifies to

$$u_{k+1}^{l+1} \in \underset{u \in BV(\Omega)}{\arg\min} \left\{ \frac{1}{2} \int_\Omega \frac{K^* \mathbf{1}_\Sigma \left(u - (u_{k+\frac{1}{2}}^{l+1} + u_k^{l+1} v^l) \right)^2}{u_k^{l+1}} dx + \alpha |u|_{BV(\Omega)} \right\}, \tag{45}$$

i.e. the step (43) can be realized by a minor modification of the TV regularization step introduced in (29).

In Sect. 3.2 a damped variant of the FB-EM-TV algorithm has been discussed. This damping strategy can also be realized in each Bregman refinement step by adapting the TV regularization step (45) to

$$u_{k+1}^{l+1} \in \underset{u \in BV(\Omega)}{\arg\min} \left\{ \frac{1}{2} \int_\Omega \frac{K^* \mathbf{1}_\Sigma \left(u - \tilde{u}_{k+\frac{1}{2}}^{l+1} \right)^2}{u_k^{l+1}} dx + \omega_k^{l+1} \alpha |u|_{BV(\Omega)} \right\} \tag{46}$$

with

$$\tilde{u}_{k+\frac{1}{2}}^{l+1} = \omega_k^{l+1} u_{k+\frac{1}{2}}^{l+1} + \omega_k^{l+1} u_k^{l+1} v^l + (1 - \omega_k^{l+1}) u_k^{l+1}.$$

Consequently, we can use Algorithm 2 to solve the stepwise refinement (38) of the TV regularized Poisson likelihood estimation problem (20).

Remark 2.

- The update variable v in (44) can be interpreted as an error function with reference to the optimality condition of the unregularized Poisson log-likelihood functional (17). Since in every refinement step of the Bregman iteration the

Algorithm 2 Bregman-FB-EM-TV Algorithm
―――
1. **Parameters:** f, $\alpha > 0$, $\omega \in (0,1]$, $maxBregIts \in \mathbb{N}$, $\delta > 0$, $\tau > 1$,
   $maxEMIts \in \mathbb{N}$, $tol > 0$
2. **Initialization:** $l = 0$, $u_0^1 = u_0 := c > 0$, $v^0 := 0$
3. **Iteration:**

   while ($D_{KL}(f, Ku_0^{l+1}) \geq \tau\delta$ and $l < maxBregIts$)) do

       a) Set $k = 0$.

       while (($k < maxEMIts$) and

           ($opt_k^{l+1} \geq tol$ or $u_{opt_k^{l+1}} \geq tol$ or $p_{opt_k^{l+1}} \geq tol$)) do   ▷ (36), (37)

         i) Compute $u_{k+\frac{1}{2}}^{l+1}$ via EM step in (42).

         ii) Set $\omega_k^{l+1} = \omega$.

         iii) Compute u_{k+1}^{l+1} via modified ROF model (46).   ▷ Section 5

         iv) $k \leftarrow k + 1$

       end while

       b) Compute update v^{l+1} via (44).

       c) Set $u_0^{l+2} = u_k^{l+1}$.

       d) $l \leftarrow l + 1$

   end while

   return u_0^{l+1}
―――

function v^{l+1} differs from v^l by the current error in the optimality condition of (17),

$$K^* 1_\Sigma - K^* \left(\frac{f}{Ku} \right) = 0.$$

Hence, caused by the TV regularization step (45), we can expect that the iterative regularization based on the Bregman distance leads to a stepwise contrast enhancement. The reason is that instead of fitting the new regularized solution u_{k+1}^{l+1} to the EM result $u_{k+\frac{1}{2}}^{l+1}$ in the weighted squared norm, as it occurs in the case of the FB-EM-TV step (29), the Bregman refinement strategy (45) uses an adapted "noisy" function in the data fidelity term, where the intensities of the EM solution $u_{k+\frac{1}{2}}^{l+1}$ are increased by a weighted error function v^l. Following additionally [94] or [121], the elements of the dual space of $BV(\Omega)$, in the underlying case $p^l = \frac{K^* 1_\Sigma}{\alpha} v^l \in \partial |u^l|_{BV(\Omega)} \subset (BV(\Omega))^*$, can be characterized as textures respectively strongly oscillating patterns. Based on this interpretation, it makes sense to consider v^l as the current error function of the unregularized Poisson log-likelihood functional (17).

- As an alternative to the approach in Sect. 3.5, a dual inverse scale space strategy based on Bregman distance iteration can be used to obtain simultaneous contrast correction, see [29]. However, both inverse scale space methods compute very similar iterates and we could not observe a difference in the performance so far. But in the case of the dual approach, error estimates and convergence rates for exact and noisy data (see [29]) can be provided, which are not possible for the primal approach in Sect. 3.5 so far.

4 Analysis

In this section we carry out a mathematical analysis of the TV regularized Poisson based variational model (20). We prove that the problem is well-posed, that the FB-EM-TV algorithm preserves the positivity of the solution and that the proposed damped FB-EM-TV iteration scheme has a stable convergence behavior. The same results can also be extended to the general case of regularized Poisson likelihood estimation problem (16) with a wide class of convex regularization functionals, including those which are non differentiable in the classical sense (see [108]).

4.1 Assumptions, Definitions and Preliminary Results

At the beginning we repeat the properties of the operator K and introduce some assumptions, which will be used in the following analysis.

As introduced in Sect. 2.1, the forward operator K is a semi discrete operator based on $\bar{K} : U(\Omega) \to V(\Sigma)$, which transforms, in contrast to $\bar{K}$, a function from $U(\Omega)$ to the discrete data space $\mathbb{R}^N$. In order to be able to present a unified theory with respect to the continuous problem formulation (20), we passed over in Sect. 2.2 from a discrete to a continuous representation of the raw data using a point measure μ. We assumed that any element in the discrete data space $\mathbb{R}^N$ can be interpreted as a sampling of a function in $V_\mu(\Sigma)$, where $V_\mu(\Sigma)$ denotes the space of Lebesgue measurable functions with respect to the measure μ. Note that in the case of a fully continuous formulation of the forward operator, the measure μ has to be set to the Lebesgue measure and we have $V_\mu(\Sigma) = V(\Sigma)$. Finally, for assumptions below, note that the operator $\bar{K}$ is linear, compact and additionally obtains nonnegativity. Hence, based on observations above, we make the following assumptions.

Assumption 1.

1. $K : U(\Omega) \to V_\mu(\Sigma)$ is linear and bounded.
2. K preserves nonnegativity, i.e. it satisfies $Ku \geq 0$ a.e. for any $u \geq 0$ a.e. and the equality is satisfied if and only if $u = 0$.
3. If $u \in U(\Omega)$ satisfies $c_1 \leq u \leq c_2$ a.e. for some positive constants c_1, c_2, then there exist $c_3, c_4 > 0$ such that $c_3 \leq Ku \leq c_4$ a.e. on Σ.
4. If $v \in V_\mu(\Sigma)$ satisfies $\tilde{c}_1 \leq v \leq \tilde{c}_2$ a.e. for some positive constants $\tilde{c}_1, \tilde{c}_2$, then there exist $\tilde{c}_3, \tilde{c}_4 > 0$ such that $\tilde{c}_3 \leq K^*v \leq \tilde{c}_4$ a.e. on Ω.

Remark 3.

- At first glance, Assumption 1 (3) is restrictive, but there are many classes of linear ill-posed problems for which the required condition is satisfied. An example are integral equations of the first kind, which have smooth, bounded and bounded away from zero kernels. Such integral equations appear in numerous fields of application, e.g. in geophysics and potential theory or in deconvolution problems as fluorescence microscopy or astronomy. Another example of operators, which satisfy Assumption 1 (3), is the X-ray transform which assigns the integral values along all straight lines to a function [99]. This transform coincides in two dimensions with the Radon transform and is often applied in medical imaging. The Assumption 1 (3) is satisfied in this example, if the length of the lines is bounded and bounded away from zero.
- Assumption 1 (4) is more of theoretical nature and is required to prove the positivity of the FB-EM-TV solution below. But also here, the assumption is not significantly restrictive in practice. For integral equations of the first kind, the claim is satisfied in terms of the same conditions as in Assumption 1 (3). On the other hand, using the X-ray transform, the assertion is valid if for any point in Σ exists at least a bounded and bounded away from zero straight line intersecting this point. A condition which should be naturally satisfied in practice.

Next we give the definition of the Kullback-Leibler functional in order to simplify the analysis of the regularized Poisson likelihood estimation problem (20).

Definition 1 (Kullback-Leibler Functional). The Kullback-Leibler (KL) functional is a function $D_{KL} : L^1(\Sigma) \times L^1(\Sigma) \to [0, +\infty]$ with $\Sigma \subset \mathbb{R}^m$ bounded and measurable, given by

$$D_{KL}(\varphi, \psi) = \int_{\Sigma} \left(\varphi \log \left(\frac{\varphi}{\psi} \right) - \varphi + \psi \right) dv \quad \text{for all} \quad \varphi, \psi \geq 0 \text{ a.e.},$$
(47)

where v is a measure. Note that, using the convention $0 \log 0 = 0$, the integrand in (47) is nonnegative and vanishes if and only if $\varphi = \psi$.

Remark 4. In the literature exist further notations for the KL functional, like cross-entropy, information for discrimination or Kullback's I-divergence, cf. e.g. [41, 46, 104]. The functional (47) generalizes the well known Kullback-Leibler entropy,

$$E_{KL}(\varphi, \psi) = \int_{\Sigma} \varphi \log \left(\frac{\varphi}{\psi} \right) dv ,$$

for functions that are not necessarily probability densities. In the definition above, you get the extension by adding (linear) terms which are chosen so that (47) is a Bregman distance or divergence with respect to the Boltzmann-Shannon entropy [104].

In the following Lemmas we recall a collection of basic results about the KL functional and the total variation $|\cdot|_{BV(\Omega)}$ functional from [104] and [1], respectively, which will be used in the analysis below. For further information to both terms, we refer to [46, 104] and [1, 3, 6, 52, 60], respectively.

Lemma 1 (Properties of KL Functional). *Let K satisfy Assumption 1 (1) and (2). Then the following statements hold:*

1. *The function $(\varphi, \psi) \mapsto D_{KL}(\varphi, \psi)$ is convex and thus, due to the linearity of the operator K, the function $(\varphi, u) \mapsto D_{KL}(\varphi, Ku)$ is also convex.*
2. *For any fixed nonnegative $\varphi \in L^1(\Sigma)$, the function $u \mapsto D_{KL}(\varphi, Ku)$ is lower semicontinuous with respect to the weak topology of $L^1(\Sigma)$.*
3. *For any nonnegative function φ and ψ in $L^1(\Sigma)$, one has*

$$\|\varphi - \psi\|_{L^1(\Sigma)}^2 \leq \left(\frac{2}{3} \|\varphi\|_{L^1(\Sigma)} + \frac{4}{3} \|\psi\|_{L^1(\Sigma)} \right) D_{KL}(\varphi, \psi).$$

Proof. See [104, Lemmas 3.3–3.4]. □

Corollary 1. *If $\{\varphi_n\}$ and $\{\psi_n\}$ are bounded sequences in $L^1(\Sigma)$, then*

$$\lim_{n \to \infty} D_{KL}(\varphi_n, \psi_n) = 0 \quad \Rightarrow \quad \lim_{n \to \infty} \|\varphi_n - \psi_n\|_{L^1(\Sigma)} = 0.$$

Lemma 2 (Properties of TV Functional). *The following statements hold:*

1. $|\cdot|_{BV(\Omega)}$ *is convex on $BV(\Omega)$.*
2. $|\cdot|_{BV(\Omega)}$ *is lower semicontinuous with respect to the weak topology of $L^1(\Omega)$.*
3. *Any uniformly bounded sequence $\{u_n\}$ in $BV(\Omega)$ is relatively compact in $L^1(\Omega)$.*

Proof. See [1, Theorems 2.3–2.5]. □

4.2 Well-Posedness of the Minimization Problem

In the following we verify existence, uniqueness, and stability of the minimization problem (20). In order to use the known properties of the KL functional from Lemma 1 we add the term $f \log f - f$ to the data fidelity term in (20). Since this expression is independent of the desired function u, the stationary points of the minimization problem are not affected (if they exist) and (20) is equivalent to

$$\min_{\substack{u \in BV(\Omega) \\ u \geq 0 \, a.e.}} F(u) := D_{KL}(f, Ku) + \alpha |u|_{BV(\Omega)}, \quad \alpha > 0, \quad (48)$$

where D_{KL} is the Kullback-Leibler functional as in (47). In the context of Assumption 1 we set here $U(\Omega) = L^1(\Omega)$ and $V_\mu(\Sigma) = L^1_\mu(\Sigma)$ due to the property of the BV-space and the definition of the KL functional, respectively. For the following analysis, the precompactness result from Lemma 2 (3) is of fundamental importance. Hence we introduce the following definition in order to use this property.

Definition 2 (BV-Coercivity). A functional F defined on $L^1(\Omega)$ is BV-coercive (cf. [74], Definition IV.3.2.6), if the sub-level sets of F are bounded in the BV-norm, i.e. for all $r \in \mathbb{R}_{\geq 0}$ the set $\{u \in L^1(\Omega) : F(u) \leq r\}$ is uniformly bounded in the BV-norm; or equivalently

$$F(u) \to +\infty \quad \text{whenever} \quad \|u\|_{BV(\Omega)} \to +\infty. \tag{49}$$

Lemma 3 (BV-Coercivity of the Minimization Functional). *Let K satisfy Assumption 1 (1) and (2). Assume that $\alpha > 0$, $f \in L^1_\mu(\Sigma)$ is nonnegative and that the operator K does not annihilate constant functions. Since K is linear, the latter condition is equivalent to*

$$K\mathbf{1}_\Omega \neq 0, \tag{50}$$

where $\mathbf{1}_\Omega$ denotes the characteristic function on Ω. Then the functional F defined in (48) is BV-coercive.

Remark 5. According to the definition of the function space of functions with bounded (total) variation, $BV(\Omega) \subset L^1(\Omega)$ is valid and the admissible solution set of the minimization problem (48) can be extended from $BV(\Omega)$ to $L^1(\Omega)$. To this end, we continue the total variation to a functional on $L^1(\Omega)$ by setting $|u|_{BV(\Omega)} = +\infty$ if $u \in L^1(\Omega) \setminus BV(\Omega)$, where furthermore solutions from $BV(\Omega)$ are preferred during minimization.

Proof (Lemma 3). For the proof of BV-coercivity, we derive an estimate of the form

$$\|u\|_{BV(\Omega)} = \|u\|_{L^1(\Omega)} + |u|_{BV(\Omega)} \leq c_1 \left(F(u)\right)^2 + c_2 F(u) + c_3, \tag{51}$$

with constants $c_1 \geq 0$, $c_2 > 0$ and $c_3 \geq 0$. Then, the desired coercivity property (49) follows directly from the nonnegativity of the functional F for all $u \in L^1(\Omega)$ with $u \geq 0$ a.e.

For (51) we use that any $u \in BV(\Omega)$ has a decomposition of the form

$$u = w + v, \tag{52}$$

where

$$w = \left(\frac{\int_\Omega u \, dx}{|\Omega|}\right) \mathbf{1}_\Omega \quad \text{and} \quad v := u - w \quad \text{with} \quad \int_\Omega v \, dx = 0. \tag{53}$$

First we estimate $|v|_{BV(\Omega)}$ and $\|v\|_{L^1(\Omega)}$. Since constant functions have no variation, the nonnegativity of the KL functional yields

$$\alpha |v|_{BV(\Omega)} \leq \alpha |u|_{BV(\Omega)} \leq F(u) \quad \Rightarrow \quad |v|_{BV(\Omega)} \leq \frac{1}{\alpha} F(u).$$

This together with the Poincaré-Wirtinger inequality (see e.g. [6, Sect. 2.5.1]) yields an estimate of the L^1-norm,

$$\|v\|_{L^1(\Omega)} \leq C_1 |v|_{BV(\Omega)} \leq C_1 \frac{1}{\alpha} F(u), \tag{54}$$

where $C_1 > 0$ is a constant that depends on $\Omega \subset \mathbb{R}^d$ and d only. Using the decomposition (52) and the estimates for $|v|_{BV(\Omega)}$ and $\|v\|_{L^1(\Omega)}$, the problem (51) reduces to the estimation of the L^1-norm of constant functions, since

$$\begin{aligned}\|u\|_{BV(\Omega)} &\leq \|w\|_{L^1(\Omega)} + \|v\|_{L^1(\Omega)} + |v|_{BV(\Omega)} \\ &\leq \|w\|_{L^1(\Omega)} + (C_1 + 1) \frac{1}{\alpha} F(u).\end{aligned} \tag{55}$$

In order to estimate $\|w\|_{L^1(\Omega)}$, we consider the L^1_μ distance between $Ku = Kw + Kv$ and f. With Lemma 1 (3) we obtain an upper bound,

$$\begin{aligned}\|(Kv - f) + Kw\|^2_{L^1_\mu(\Sigma)} &\leq \left(\frac{2}{3}\|f\|_{L^1_\mu(\Sigma)} + \frac{4}{3}\|Kv + Kw\|_{L^1_\mu(\Sigma)}\right) D_{KL}(f, Ku) \\ &\leq \left(\frac{2}{3}\|f\|_{L^1_\mu(\Sigma)} + \frac{4}{3}\|Kv\|_{L^1_\mu(\Sigma)} + \frac{4}{3}\|Kw\|_{L^1_\mu(\Sigma)}\right) F(u),\end{aligned}$$

and as lower bound,

$$\begin{aligned}\|(Kv - f) + Kw\|^2_{L^1_\mu(\Sigma)} &\geq \left(\|Kv - f\|_{L^1_\mu(\Sigma)} - \|Kw\|_{L^1_\mu(\Sigma)}\right)^2 \\ &\geq \|Kw\|_{L^1_\mu(\Sigma)} \left(\|Kw\|_{L^1_\mu(\Sigma)} - 2\|Kv - f\|_{L^1_\mu(\Sigma)}\right).\end{aligned}$$

Combining (54) with both inequalities yields

$$\begin{aligned}\|Kw\|_{L^1_\mu(\Sigma)} &\left(\|Kw\|_{L^1_\mu(\Sigma)} - 2\left(\|K\| C_1 \frac{1}{\alpha} F(u) + \|f\|_{L^1_\mu(\Sigma)}\right)\right) \\ &\leq \left(\frac{2}{3}\|f\|_{L^1_\mu(\Sigma)} + \frac{4}{3}\|K\| C_1 \frac{1}{\alpha} F(u) + \frac{4}{3}\|Kw\|_{L^1_\mu(\Sigma)}\right) F(u).\end{aligned} \tag{56}$$

This expression contains terms describing the function w only in dependence of the operator K. For the estimate of $\|w\|_{L^1(\Omega)}$, we use the assumption (50) on K. Thus, there exists a constant $C_2 > 0$ with

$$C_2 = \frac{\int_\Sigma |K\mathbf{1}_\Omega|\,d\mu}{|\Omega|} \quad \text{and} \quad \|Kw\|_{L^1_\mu(\Sigma)} = C_2 \|w\|_{L^1(\Omega)}. \tag{57}$$

This identity used in the inequality (56) yields

$$C_2 \|w\|_{L^1(\Omega)} \left(C_2 \|w\|_{L^1(\Omega)} - 2\left(\|K\| C_1 \frac{1}{\alpha} F(u) + \|f\|_{L^1_\mu(\Sigma)} \right) - \frac{4}{3} F(u) \right)$$
$$\leq \left(\frac{2}{3} \|f\|_{L^1_\mu(\Sigma)} + \frac{4}{3} \|K\| C_1 \frac{1}{\alpha} F(u) \right) F(u). \tag{58}$$

In order to receive an estimate of the form (51), we distinguish between two cases:

Case 1: If

$$C_2 \|w\|_{L^1(\Omega)} - 2\left(\|K\| C_1 \frac{1}{\alpha} F(u) + \|f\|_{L^1_\mu(\Sigma)} \right) - \frac{4}{3} F(u) \geq 1, \tag{59}$$

then we conclude from (58) that

$$\|w\|_{L^1(\Omega)} \leq \frac{1}{C_2} \left(\frac{2}{3} \|f\|_{L^1_\mu(\Sigma)} + \frac{4}{3} \|K\| C_1 \frac{1}{\alpha} F(u) \right) F(u),$$

and obtain with (55),

$$\|u\|_{BV(\Omega)} \leq \frac{4 C_1 \|K\|}{3 C_2 \alpha} (F(u))^2 + \left(\frac{2}{3 C_2} \|f\|_{L^1_\mu(\Sigma)} + \frac{C_1 + 1}{\alpha} \right) F(u). \tag{60}$$

Case 2: If the condition (59) does not hold, i.e.

$$\|w\|_{L^1(\Omega)} < \frac{1}{C_2} \left(1 + 2\left(\|K\| C_1 \frac{1}{\alpha} F(u) + \|f\|_{L^1_\mu(\Sigma)} \right) + \frac{4}{3} F(u) \right),$$

then we find from (55) that

$$\|u\|_{BV(\Omega)} \leq \left(\frac{2 \|K\| C_1 \frac{1}{\alpha} + \frac{4}{3}}{C_2} + \frac{C_1 + 1}{\alpha} \right) F(u) + \frac{1 + 2 \|f\|_{L^1_\mu(\Sigma)}}{C_2}. \tag{61}$$

With Assumption 1 (1) we have $\|K\| < \infty$. Moreover, since $f \in L^1_\mu(\Sigma)$, we obtain from (60) and (61) the desired coercivity property (51). □

Theorem 1 (Existence of Minimizers). *Let K satisfy Assumption 1 (1) and (2). Assume that $\alpha > 0$, $f \in L^1_\mu(\Sigma)$ is nonnegative and that the operator K satisfies (50). Then the functional F defined in (48) has a minimizer in $BV(\Omega)$.*

Proof. We use the direct method of the calculus of variations, see e.g. [6, Sect. 2.1.2]: Let $\{u_n\} \subset BV(\Omega)$, $u_n \geq 0$ a.e., be a minimizing sequence of the functional F, i.e.

$$\lim_{n \to \infty} F(u_n) = \inf_{\substack{u \in BV(\Omega) \\ u \geq 0 \text{ a.e.}}} F(u) =: F_{min} < \infty.$$

With assumptions on the operator K, Lemma 3 implies that the functional F is BV-coercive and that all elements of the sequence $\{u_n\}$ are uniformly bounded in the BV-norm. As a consequence of the precompactness result from Lemma 2 (3), there exists a subsequence $\{u_{n_j}\}$ which converges to some $\tilde{u} \in L^1(\Omega)$. Actually the function $\tilde{u}$ lies in $BV(\Omega)$, since $|\cdot|_{BV(\Omega)}$ is lower semicontinuous and the sequence $\{u_n\}$ is uniformly bounded in $BV(\Omega)$. Simultaneously, caused by Lemma 1 (2), also the objective functional F is lower semicontinuous and implies the inequality

$$F(\tilde{u}) \leq \liminf_{j \to \infty} F(u_{n_j}) = F_{min},$$

which means that $\tilde{u}$ is a minimizer of F. □

Next we consider the uniqueness of minimizers, for which it suffices to verify the strict convexity of the objective functional F. For this purpose, it is straight-forward to see that the negative logarithm function is strictly convex and consequently also the function $u \mapsto D_{KL}(f, Ku)$, if $f \in L^1_\mu(\Sigma)$ fulfills $\operatorname{ess\,inf}_\Sigma f > 0$ and the operator K is injective, i.e. the null space of K is trivial since K is linear (cf. Assumption 1 (1)). Since the regularization term is assumed convex we can immediately conclude the following result.

Theorem 2 (Uniqueness of Minimizers). *Let K satisfy Assumption 1 (1) and (2). Assume that K is an injective operator and $f \in L^1_\mu(\Sigma)$ fulfills $\operatorname{ess\,inf}_\Sigma f > 0$. Then the function $u \mapsto D_{KL}(f, Ku)$ and also the functional F from (48) is strictly convex. In particular, the minimizer of F is unique in $BV(\Omega)$.*

After existence and uniqueness of minimizers we prove the stability of the TV regularized Poisson estimation problem (48) with respect to a certain kind of data perturbations. In Sect. 2.1 we described that the given measurements are discrete in practice and can be interpreted as averages of a function $f \in L^1(\Sigma)$. The open question certainly is the suitable choice of the function f. Moreover, the physically limited discrete construction of the detectors leads to a natural loss of information, because not all signals can be acquired. Consequently, a stability result is required guaranteeing the convergence of regularized approximations to a solution u, if e.g. the approximated data converge to a preferably smooth function f. Since the measurements are still realizations of Poisson distributed random variables, it is natural to assess the convergence in terms of the KL functional, as shown in (62).

Theorem 3 (Stability with Respect to Perturbations in Measurements). *Let K satisfy Assumption 1. Fix $\alpha > 0$ and assume that the functions $f_n \in L^1_\mu(\Sigma)$, $n \in \mathbb{N}$, are nonnegative approximations of a data function $f \in L^1_\mu(\Sigma)$ in the form*

$$\lim_{n \to \infty} D_{KL}(f_n, f) = 0. \tag{62}$$

Moreover, let

$$u_n \in \underset{\substack{v \in BV(\Omega) \\ v \geq 0 \, a.e.}}{\arg\min} \left\{ F_n(v) := D_{KL}(f_n, Kv) + \alpha |v|_{BV(\Omega)} \right\}, \qquad n \in \mathbb{N}, \tag{63}$$

and let u be a solution of the regularized problem (48) corresponding to the data function f. We assume also that the operator K does not annihilate constant functions and that $\log f$ and $\log Ku$ belong to the function space $L^\infty_\mu(\Sigma)$, i.e. there exist positive constants $c_1, \ldots, c_4$ such that

$$0 < c_1 \leq f \leq c_2 \qquad \text{and} \qquad 0 < c_3 \leq Ku \leq c_4 \qquad a.e. \text{ on } \Sigma. \tag{64}$$

Suppose that the sequence $\{f_n\}$ is uniformly bounded in the $L^1_\mu(\Sigma)$-norm, i.e. it exists a positive constant c_5 such that

$$\|f_n\|_{L^1_\mu(\Sigma)} \leq c_5, \qquad \forall n \in \mathbb{N}. \tag{65}$$

Then the problem (48) is stable with respect to the perturbations in the data, i.e. the sequence $\{u_n\}$ has a convergent subsequence and every convergent subsequence converges in the L^1-norm to a minimizer of the functional F in (48).

Proof. For the existence of a convergent subsequence of $\{u_n\}$, we use the precompactness result from Lemma 2 (3). To this end we have to show the uniform boundedness of the sequence $\{F_n(u_n)\}$ and the uniform BV-coercivity of functionals F_n, see the definition below.

We show the uniform boundedness of the sequence $\{F_n(u_n)\}$ first. Let $\alpha > 0$ be a fixed regularization parameter. For any $n \in \mathbb{N}$, the definition of u_n as a minimizer of the objective functional F_n in (63) implies that

$$F_n(u_n) = D_{KL}(f_n, Ku_n) + \alpha |u_n|_{BV(\Omega)} \leq D_{KL}(f_n, Ku) + \alpha |u|_{BV(\Omega)}. \tag{66}$$

Hence, the sequence $\{F_n(u_n)\}$ is bounded, if the sequence $\{D_{KL}(f_n, Ku)\}$ on the right-hand side of (66) is bounded. To show this, we use the uniform boundedness of sequence $\{f_n\}$ in the $L^1_\mu(\Sigma)$-norm (65). Condition (62) and the result in Corollary 1 yield the strong convergence of $\{f_n\}$ to f in $L^1_\mu(\Sigma)$, i.e. we have

$$\lim_{n \to \infty} \|f - f_n\|_{L^1_\mu(\Sigma)} = 0. \tag{67}$$

Condition (64) implies together with the inequality

$$\left| D_{KL}(f_n, Ku) - D_{KL}(f, Ku) - D_{KL}(f_n, f) \right| = \left| \int_\Sigma (\log Ku - \log f)(f - f_n) \, d\mu \right|$$

$$\leq \underbrace{\|\log Ku - \log f\|_{L^\infty_\mu(\Sigma)}}_{< \infty} \underbrace{\|f - f_n\|_{L^1_\mu(\Sigma)}}_{\overset{(67)}{\to} 0},$$

the convergence

$$\lim_{n \to \infty} D_{KL}(f_n, Ku) = D_{KL}(f, Ku). \tag{68}$$

Since u is a minimizer of the regularized problem (48) corresponding to the data function f, $D_{KL}(f, Ku)$ and $|u|_{BV(\Omega)}$ are bounded, and thus also the sequence $\{D_{KL}(f_n, Ku)\}$ is bounded, since it converges to $D_{KL}(f, Ku)$. Together with (66) this yields the uniform boundedness of the sequence $\{F_n(u_n)\}$.

Next we prove that the functionals F_n are uniform BV-coercive, i.e. for any sequence $\{u_n\}$ in $L^1(\Omega)$ with $u_n \geq 0$ a.e.,

$$F_n(u_n) \to +\infty \quad \text{whenever} \quad \|u_n\|_{BV(\Omega)} \to +\infty.$$

For the proof we put $u_n = w_n + v_n$ as in (52) and (53), and repeat the proof of Lemma 3 with u_n and F_n instead of u and F, respectively. Since the operator K does not annihilate constant functions, we obtain $\|Kw_n\|_{L^1_\mu(\Sigma)} = C_2 \|w_n\|_{L^1(\Omega)}$ with C_2 as in (57) and hence analogous to (58),

$$C_2 \|w_n\|_{L^1(\Omega)} \left(C_2 \|w_n\|_{L^1(\Omega)} - 2 \left(\|K\| C_1 \frac{1}{\alpha} F_n(u_n) + \|f_n\|_{L^1_\mu(\Sigma)} \right) - \frac{4}{3} F_n(u_n) \right)$$

$$\leq \left(\frac{2}{3} \|f_n\|_{L^1_\mu(\Sigma)} + \frac{4}{3} \|K\| C_1 \frac{1}{\alpha} F_n(u_n) \right) F_n(u_n).$$

Since the sequence $\{f_n\}$ converges strongly to f in $L^1_\mu(\Sigma)$ (67), it is also bounded in the $L^1_\mu(\Sigma)$-norm. The upper bound on each $\|f_n\|_{L^1_\mu(\Sigma)}$ and the boundedness of the operator norm of K yield the uniform BV-coercivity of functionals F_n analogous to both cases in the proof of Lemma 3.

The uniform BV-coercivity implies together with the boundedness of the sequence $\{F_n(u_n)\}$ that the sequence $\{u_n\}$ is uniformly bounded in the BV-norm. Then the precompactness result from Lemma 2 (3) ensures the existence of a subsequence $\{u_{n_j}\}$ which converges strongly to some $\tilde{u} \in L^1(\Omega)$. Actually the function $\tilde{u}$ lies in $BV(\Omega)$, since $|\cdot|_{BV(\Omega)}$ is lower semicontinuous with respect to the weak topology of $L^1(\Omega)$ (see Lemma 2 (2)), i.e. we have

$$|\tilde{u}|_{BV(\Omega)} \leq \liminf_{j \to \infty} |u_{n_j}|_{BV(\Omega)} < \infty.$$

Now let $\{u_{n_j}\}$ be an arbitrary subsequence of $\{u_n\}$ which converges to some $\tilde{u} \in L^1(\Omega)$ with respect to the L^1-norm. The boundedness of the operator K (see Assumption 1 (1)) implies the strong convergence of the sequence $\{Ku_{n_j}\}$ to $K\tilde{u}$ in $L^1_\mu(\Sigma)$, as well as the pointwise convergence almost everywhere on Σ of a subsequence of $\{Ku_{n_j}\}$, again denoted by $\{Ku_{n_j}\}$ without loss of generality. A similar behavior holds also for the sequence $\{f_n\}$, which converges strongly to f in $L^1_\mu(\Sigma)$ (67). Since the functions f_n and u_n are nonnegative for all $n \in \mathbb{N}$ and K is an operator preserving nonnegativity (Assumption 1 (2)), we can apply Fatou's Lemma to the sequence $\{f_{n_j} \log(f_{n_j}/Ku_{n_j}) - f_{n_j} + Ku_{n_j}\}$ and obtain

$$D_{KL}(f, K\tilde{u}) \leq \liminf_{j \to \infty} D_{KL}(f_{n_j}, Ku_{n_j}). \tag{69}$$

Due to the lower semicontinuity of the functional $|\cdot|_{BV(\Omega)}$ and due to (66), (68) and (69), we obtain the following inequality,

$$D_{KL}(f, K\tilde{u}) + \alpha |\tilde{u}|_{BV(\Omega)} \overset{(69)}{\leq} \liminf_{j \to \infty} D_{KL}(f_{n_j}, Ku_{n_j}) + \alpha \liminf_{j \to \infty} |u_{n_j}|_{BV(\Omega)}$$

$$\leq \liminf_{j \to \infty} \left(D_{KL}(f_{n_j}, Ku_{n_j}) + \alpha |u_{n_j}|_{BV(\Omega)} \right)$$

$$\leq \limsup_{j \to \infty} \left(D_{KL}(f_{n_j}, Ku_{n_j}) + \alpha |u_{n_j}|_{BV(\Omega)} \right)$$

$$\overset{(66)}{\leq} \limsup_{j \to \infty} \left(D_{KL}(f_{n_j}, Ku) + \alpha |u|_{BV(\Omega)} \right)$$

$$\overset{(68)}{=} D_{KL}(f, Ku) + \alpha |u|_{BV(\Omega)},$$

which means that $\tilde{u}$ is a minimizer of the functional F in (48). □

Remark 6.

- For the proof of stability, condition (64) is required, which assumes that the functions $\log f$ and $\log Ku$ belong to $L^\infty_\mu(\Sigma)$, where u is a regularized solution of the minimization problem (48). In the case of the data function f, this assumption is not significantly restrictive in most practical situations. The boundedness from above is fulfilled naturally due to the finite acquisition time of the data. The almost everywhere boundedness on Σ away from zero is reasonable, when a sufficient amount of measurements has been collected.

In addition, in most practical applications a certain level of background noise is present, which causes the positivity of the data. In the case of the function Ku, condition (64) is not simple to justify and requires a more precise analysis of the variational problem (48). Due to Assumption 1 (3), it suffices to prove that u is bounded and bounded away from zero. E.g. the authors in [104] show that this condition on u is available, if we replace the TV functional in (48) by the KL functional $D_{KL}(\cdot, u^*)$ as regularization energy, where u^* denotes the a-priori estimation of the solution and satisfies the boundedness condition from above and away from zero. Roughly speaking, this is possible because, during the minimization, the linear part of the KL functional in u tries to keep the function bounded and the log part tries to push the function away from zero. However, in the case of total variation regularization we did not yet succeed to prove a similar property. Nevertheless, in Sect. 4.3 we will show at least that the iterate sequence $\{u_k\}$ of the FB-EM-TV splitting algorithm (28) has the boundedness and the boundedness away from zero property, assuming that the data function f belongs to $L_\mu^\infty(\Sigma)$ with ess inf$_\Sigma f > 0$ and the initialization function u_0 is strictly positive.

- Analogous to the reasoning above, the convergence of subsequences $\{u_{n_j}\}$ in Theorem 3 can also be proved in the $L^p(\Omega)$-norm with $1 \leq p < d/(d-1)$, $\Omega \subset \mathbb{R}^d$, since any uniformly bounded sequence $\{u_n\}$ in $BV(\Omega)$ is actually relatively compact in $L^p(\Omega)$ for $1 \leq p < d/(d-1)$, see [1, Theorem 2.5].

- As in Theorem 3 it is also possible to consider perturbations of the operator K. The proof is similar to the one above and only slight modifications are necessary. However, several assumptions on the perturbed operators K_n are needed, like the boundedness of operators for each $n \in \mathbb{N}$ and the pointwise convergence to K (cf. e.g. [1]). Unfortunately, it is also essential that the operators K_n fulfill the assumption (64), i.e. that $K_n u$ is bounded and bounded away from zero for any $n \in \mathbb{N}$, where u is a solution of the minimization problem (48). Therefore this condition is severely restrictive for the possible perturbations of the operator K.

- We finally mention that some stability estimates for the TV regularized Poisson likelihood estimation problem (48) have been also derived in [12], but in a different setting. There, the assumptions on the possible data perturbations are more restrictive (convergence in the supremum norm), while the assumptions on the operator perturbations are relaxed.

4.3 Positivity Preservation of the FB-EM-TV Algorithm

In the following we investigate the positivity preservation of the iteration sequence $\{u_k\}$ obtained by the FB-EM-TV splitting approach (28) and its damped modification (30). Given a strictly positive $u_k \in BV(\Omega)$ for some $k \geq 0$, it is straight-forward to see that the result $u_{k+\frac{1}{2}}$ of the EM reconstruction half step (28a) is well defined and strictly positive due to the form of this iteration step, if the data function f is strictly positive and the operator K fulfills the positivity preservation properties in Assumption 1 (3) and (4). Consequently, an existence and uniqueness proof for the regularization half step (29) and its damped variant (31), analogous to the classical results for the ROF model in Chap. 1, delivers also the existence of $u_{k+1} \in BV(\Omega)$. In order to show inductively the well-definedness of the complete iteration sequence $\{u_k\}$, it remains to verify that u_{k+1} is strictly positive again.

If any u_k is negative during the iteration, the objective functional in the regularization half step (29) is in general not convex anymore. Moreover, the minimization problem becomes a maximization one and the existence and uniqueness of u_{k+1} cannot be guaranteed. Thus, the nonnegativity of the regularization half step solution, and with it also the nonnegativity of the whole iteration sequence, is strongly desired, in particular since in typical applications the functions represent densities or intensity information. The latter aspect is considered explicitly by using the nonnegativity constraint in the Poisson based log-likelihood optimization problem (20).

In order to clarify the positivity preservation of the (damped) FB-EM-TV iteration scheme, we present a maximum principle for the following weighted ROF problem,

$$\min_{u \in BV(\Omega)} J(u) := \frac{1}{2} \int_\Omega \frac{(u-q)^2}{h} \, dx + \beta \, |u|_{BV(\Omega)}, \quad \beta > 0, \quad (70)$$

which represents the more general form of the regularization half step (29) and its damped modification (31) in the forward-backward splitting strategy.

Lemma 4 (Maximum Principle for the Weighted ROF Problem). *Let $\tilde{u} \in BV(\Omega)$ be a minimizer of the variational problem (70), where the function q belongs to $L^\infty(\Omega)$ with $\operatorname{ess\,inf}_\Omega q > 0$ and let the weighting function h be strictly positive. Then the following maximum principle holds*

$$0 < \operatorname*{ess\,inf}_\Omega q \leq \operatorname*{ess\,inf}_\Omega \tilde{u} \leq \operatorname*{ess\,sup}_\Omega \tilde{u} \leq \operatorname*{ess\,sup}_\Omega q. \quad (71)$$

Proof. Let $\tilde{u}$ be a minimizer of the functional J defined in (70). For the proof of the maximum principle, we show that there exists a function v with

$$0 < \operatorname*{ess\,inf}_\Omega q \leq \operatorname*{ess\,inf}_\Omega v \leq \operatorname*{ess\,sup}_\Omega v \leq \operatorname*{ess\,sup}_\Omega q \quad (72)$$

and

$$J(v) \leq J(\tilde{u}). \quad (73)$$

Then the desired boundedness property (71) follows directly from the strict convexity of the functional J in (70), i.e. from the uniqueness of the solution.

We define the function v as a version of $\tilde{u}$ cut off at $\operatorname{ess\,inf}_\Omega q$ and $\operatorname{ess\,sup}_\Omega q$, i.e.

$$v := \min\{\max\{\tilde{u}, \operatorname{ess\,inf}_\Omega q\}, \operatorname{ess\,sup}_\Omega q\}.$$

With this definition, the property (72) is directly guaranteed. To show (73), we use

$$M := \{x \in \Omega : v(x) = \tilde{u}(x)\} \subseteq \Omega$$

and estimate $|v|_{BV(\Omega)}$ by $|\tilde{u}|_{BV(\Omega)}$ first. Since the function v has (due to its definition) no variation on $\Omega \setminus M$, we obtain

$$|v|_{BV(\Omega)} = |v|_{BV(M)} = |\tilde{u}|_{BV(M)} \leq |\tilde{u}|_{BV(\Omega)}. \tag{74}$$

We see also that the data fidelity terms of J in (70) with respect to v and $\tilde{u}$ coincide on M, due to the definition of the function v. In case of $x \in \Omega \setminus M$, we distinguish between two cases:

Case 1: If $\tilde{u}(x) \geq \operatorname{ess\,sup}_\Omega q$ then $v(x) = \operatorname{ess\,sup}_\Omega q$ and

$$0 \leq v(x) - q(x) = \operatorname{ess\,sup}_\Omega q - q(x) \leq \tilde{u}(x) - q(x)$$
$$\Rightarrow \quad (v(x) - q(x))^2 \leq (\tilde{u}(x) - q(x))^2.$$

Case 2: If $\tilde{u}(x) \leq \operatorname{ess\,inf}_\Omega q$ then $v(x) = \operatorname{ess\,inf}_\Omega q$ and

$$0 \leq -v(x) + q(x) = -\operatorname{ess\,inf}_\Omega q + q(x) \leq -\tilde{u}(x) + q(x)$$
$$\Rightarrow \quad (v(x) - q(x))^2 \leq (\tilde{u}(x) - q(x))^2.$$

Finally, we obtain

$$(v - q)^2 \leq (\tilde{u} - q)^2, \qquad \forall x \in \Omega,$$

and property (73) is fulfilled due to the strict positivity of the weighting function h and (74). □

Lemma 5 (Positivity of the (Damped) FB-EM-TV Algorithm). *Let $\{\omega_k\}$ be a given sequence of damping parameters with $\omega_k \in (0, 1]$ for all $k \geq 0$ and let $\operatorname{ess\,inf}_\Omega u_0 > 0$. Assume that $f \in L^\infty_\mu(\Sigma)$ with $\operatorname{ess\,inf}_\Sigma f > 0$ and that the operator K satisfies the positivity preservation properties in Assumption 1 (3) and (4). Then each half step of the (damped) FB-EM-TV splitting method and therewith also the solution is strictly positive.*

Proof. Since $u_0 > 0$, $f > 0$ and the operator K as well as the adjoint operator K^* does not affect the strict positivity, the first EM reconstruction step $u_{\frac{1}{2}}$ in (28) is strictly positive. Because the TV correction step (28b) can be realized via the weighted ROF problem (29), the maximum principle in Lemma 4 using $q := u_{\frac{1}{2}} > 0$ and $h := \frac{u_0}{K^*1_\Sigma} > 0$ yields $u_1 > 0$. With the same argument, we also obtain $u_1 > 0$, if we take the damped regularization step (30) via the variational problem (31), using the maximum principle with $q := \omega_0 u_{\frac{1}{2}} + (1 - \omega_0) u_0 > 0$ for $\omega_0 \in (0, 1]$. Inductively, the strict positivity of the whole iteration sequence $\{u_k\}$ and with it the strict positivity of the solution is obtained by the same arguments using Lemma 4. □

Finally we consider the positivity preservation of the Poisson denoising strategy, which can be obtained from the (damped) FB-EM-TV algorithm by using the identity operator K. In this case the FB-EM-TV splitting strategy (28) with the damped modification (30) results in the following iteration scheme (note that the EM reconstruction step vanishes in the denoising case, i.e. it holds $u_{k+\frac{1}{2}} = f$),

$$u_{k+1} = (1 - \omega_k) u_k + \omega_k f - \omega_k \alpha u_k p_{k+1}, \qquad p_{k+1} \in \partial |u_{k+1}|_{BV(\Omega)}, \tag{75}$$

with $\omega_k \in (0, 1]$, in order to denoise an image corrupted by Poisson noise. Analogous to the FB-EM-TV algorithm, this iteration step can be realized by solving a weighted ROF problem of the form (cf. (29) and (31)),

$$u_{k+1} \in \underset{u \in BV(\Omega)}{\arg\min} \left\{ \frac{1}{2} \int_\Omega \frac{(u - (\omega_k f + (1 - \omega_k) u_k))^2}{u_k} dx + \omega_k \alpha |u|_{BV(\Omega)} \right\}. \tag{76}$$

Although this denoising iteration is a special case of the damped FB-EM-TV algorithm, we study its properties here explicitly, because it will later simplify the convergence criteria of the Poisson denoising method.

Lemma 6 (Maximum Principle and Positivity of the Poisson Denoising Scheme). *Let $\{\omega_k\}$ be a sequence of damping parameters with $\omega_k \in (0, 1]$ for all $k \geq 0$, the data function f lies in $L^\infty_\mu(\Omega)$ with $\operatorname{ess\,inf}_\Omega f > 0$ and the initialization function u_0 fulfills*

$$0 < \operatorname*{ess\,inf}_\Omega f \leq \operatorname*{ess\,inf}_\Omega u_0 \leq \operatorname*{ess\,sup}_\Omega u_0 \leq \operatorname*{ess\,sup}_\Omega f. \tag{77}$$

Moreover, let $\{u_k\}$ be a sequence of iterates generated by the damped Poisson denoising scheme (76). Then the following maximum principle holds,

$$0 < \operatorname*{ess\,inf}_\Omega f \leq \operatorname*{ess\,inf}_\Omega u_k \leq \operatorname*{ess\,sup}_\Omega u_k \leq \operatorname*{ess\,sup}_\Omega f, \qquad \forall k \geq 0. \tag{78}$$

EM-TV Methods for Inverse Problems with Poisson Noise   109

This result guarantees that each step of the damped Poisson denoising method (75) and with it also the solution is strictly positive.

Proof. For $k = 0$, the condition (78) is satisfied due to (77). For $k \geq 0$, Lemma 4 offers a maximum principle for the Poisson denosing model (76) using $q := \omega_k f + (1 - \omega_k) u_k$ and $h := u_k$, i.e. we have

$$0 < \operatorname*{ess\,inf}_{\Omega} \{ \omega_k f + (1 - \omega_k) u_k \}$$

$$\leq \operatorname*{ess\,inf}_{\Omega} u_{k+1} \leq \operatorname*{ess\,sup}_{\Omega} u_{k+1} \leq \operatorname*{ess\,sup}_{\Omega} \{ \omega_k f + (1 - \omega_k) u_k \} . \tag{79}$$

Due to $\omega_k \in (0, 1]$ for all $k \geq 0$ and the inequalities

$$\operatorname*{ess\,inf}_{\Omega} \{ \omega_k f + (1 - \omega_k) u_k \} \geq \omega_k \operatorname*{ess\,inf}_{\Omega} f + (1 - \omega_k) \operatorname*{ess\,inf}_{\Omega} u_k$$

and

$$\operatorname*{ess\,sup}_{\Omega} \{ \omega_k f + (1 - \omega_k) u_k \} \leq \omega_k \operatorname*{ess\,sup}_{\Omega} f + (1 - \omega_k) \operatorname*{ess\,sup}_{\Omega} u_k ,$$

we obtain from (79) and the induction hypothesis the desired maximum principle (78). □

Remark 7. The assumption (77) on the initialization function u_0 is satisfied in general, since u_0 will be usually chosen as a positive and constant function or as the given noisy image f itself.

4.4 Convergence of the Damped FB-EM-TV Algorithm

In Sect. 3.3 we interpreted the (damped) FB-EM-TV reconstruction method as a forward-backward operator splitting algorithm. In the past, several works in convex analysis have been proposed dealing with the convergence of such splitting strategies for solving decomposition problems, see e.g. Tseng [119] and Gabay [56]. For the discussed algorithm (34),

$$u_{k+1} = \left(I + \frac{\omega_k u_k}{K^* \mathbf{1}_\Sigma} B \right)^{-1} \left(I - \frac{\omega_k u_k}{K^* \mathbf{1}_\Sigma} A \right) u_k , \tag{80}$$

Gabay provided in [56] a proof of weak convergence of the forward-backward splitting approach under the assumption of a fixed damping parameter ω strictly less than twice the modulus of A^{-1}. On the other hand, Tseng gave later in [119] a convergence proof, where in the case of (80), the damping values $\frac{\omega_k u_k}{K^*1_\Sigma}$ need to be bounded in the following way,

$$\varepsilon \leq \frac{\omega_k u_k}{K^*1_\Sigma} \leq 4m - \varepsilon, \qquad \varepsilon \in (0, 2m],$$

where the Kullback-Leibler data fidelity functional needs to be strictly convex with modulus m. Unfortunately, the results above cannot be used in the case of (80), since the modulus assumption on the data fidelity cannot be verified and in particular, we cannot provide the upper bounds for the iterates u_k.

In the following we prove the necessity of a damping strategy manually, in order to guarantee a monotone descent of the objective functional F in (20) with respect to the iterates u_k of the FB-EM-TV algorithm. In the theorem below we establish the convergence of the damped FB-EM-TV splitting algorithm under appropriate assumptions on the damping parameters ω_k.

Theorem 4 (Convergence of the Damped FB-EM-TV Algorithm). *Let K satisfy Assumption 1 and does not annihilate constant functions. Moreover, let $\{u_k\}$ be a sequence of iterates obtained by the damped FB-EM-TV algorithm (33) and let the data function $f \in L^\infty_\mu(\Sigma)$ satisfy $\operatorname{ess\,inf}_\Sigma f > 0$. If there exists a sequence of corresponding damping parameters $\{\omega_k\}$, $\omega_k \in (0, 1]$, satisfying the inequality*

$$\omega_k \leq \frac{\int_\Omega \frac{K^*1_\Sigma (u_{k+1} - u_k)^2}{u_k} dx}{\sup_{v \in [u_k, u_{k+1}]} \frac{1}{2} \int_\Sigma \frac{f(Ku_{k+1} - Ku_k)^2}{(Kv)^2} d\mu} (1 - \varepsilon), \qquad \varepsilon \in (0, 1),$$

(81)

then the objective functional F defined in (48) is decreasing during the iteration. If, in addition, the function K^*1_Σ, the damping parameters and the iterates are bounded away from zero by positive constants c_1, c_2 and c_3 such that for all $k \geq 0$,

$$0 < c_1 \leq K^*1_\Sigma, \qquad 0 < c_2 \leq \omega_k, \qquad 0 < c_3 \leq u_k, \qquad (82)$$

then the sequence of iterates $\{u_k\}$ has a convergent subsequence in the weak topology on $BV(\Omega)$ and in the strong topology on $L^1(\Omega)$ and every such convergent subsequence converges to a minimizer of the functional F.*

Proof. This proof is divided into several steps. First, we show the monotone descent of the objective functional F. In the following steps, we prove the existence of convergent subsequences of the primal iterates $\{u_k\}$ and the subgradients $\{p_k\}$ corresponding to $\{u_k\}$. Subsequently, we verify that the limit p of the dual iterates $\{p_k\}$ is a subgradient of the TV regularization functional at the limit u of the primal

EM-TV Methods for Inverse Problems with Poisson Noise   111

iterates $\{u_k\}$, i.e. $p \in \partial |u|_{BV(\Omega)}$. In the last step, we show that the limit u is a minimizer of the objective functional F.

First step: Monotone descent of the objective functional

To get a descent of the objective functional F using an adequate damping strategy, we look for a condition on the damping parameters $\{\omega_k\}$, which guarantees for all $k \geq 0$ a descent of the form

$$F(u_{k+1}) + \underbrace{\frac{\varepsilon}{\omega_k} \int_\Omega \frac{K^* \mathbf{1}_\Sigma \, (u_{k+1} - u_k)^2}{u_k} \, dx}_{\geq 0} \leq F(u_k), \qquad \varepsilon > 0. \qquad (83)$$

This condition ensures a descent of F, since the second term on the left-hand side of (83) is positive due to $\omega_k > 0$ and due to the strict positivity of the iterates u_k (see Lemma 5). In order to show (83), we start with the damped TV regularization step (30), multiply it by $u_{k+1} - u_k$ and integrate the result over the domain Ω. Thus, for $p_{k+1} \in \partial |u_{k+1}|_{BV(\Omega)}$, we obtain

$$0 = \int_\Omega \frac{K^* \mathbf{1}_\Sigma \, (u_{k+1} - \omega_k \, u_{k+\frac{1}{2}} - (1-\omega_k) \, u_k)(u_{k+1} - u_k)}{u_k} \, dx$$
$$\qquad\qquad\qquad\qquad + \omega_k \, \alpha \, \langle p_{k+1}, u_{k+1} - u_k \rangle$$
$$= \int_\Omega \frac{K^* \mathbf{1}_\Sigma \, (u_{k+1} - u_k)^2}{u_k} + \omega_k \, \frac{K^* \mathbf{1}_\Sigma \, (u_k - u_{k+\frac{1}{2}})(u_{k+1} - u_k)}{u_k} \, dx$$
$$\qquad\qquad\qquad\qquad + \omega_k \, \alpha \, \langle p_{k+1}, u_{k+1} - u_k \rangle .$$

Due to the definition of subgradients in (22), we have

$$\langle p_{k+1}, u_{k+1} - u_k \rangle \geq |u_{k+1}|_{BV(\Omega)} - |u_k|_{BV(\Omega)}$$

and thus

$$\alpha \, |u_{k+1}|_{BV(\Omega)} - \alpha \, |u_k|_{BV(\Omega)} + \frac{1}{\omega_k} \int_\Omega \frac{K^* \mathbf{1}_\Sigma \, (u_{k+1} - u_k)^2}{u_k} \, dx$$
$$\leq -\int_\Omega \frac{K^* \mathbf{1}_\Sigma \, (u_k - u_{k+\frac{1}{2}})(u_{k+1} - u_k)}{u_k} \, dx .$$

Adding the difference $D_{KL}(f, Ku_{k+1}) - D_{KL}(f, Ku_k)$ on both sides of this inequality and considering the definitions of the KL functional D_{KL} in (47) and the objective functional F in (48) yields

$$F(u_{k+1}) - F(u_k) + \frac{1}{\omega_k} \int_\Omega \frac{K^* \mathbf{1}_\Sigma (u_{k+1} - u_k)^2}{u_k} dx$$

$$\leq \int_\Sigma \left(f \log\left(\frac{f}{K u_{k+1}}\right) + K u_{k+1} - f \log\left(\frac{f}{K u_k}\right) - K u_k \right) d\mu$$

$$- \int_\Omega \left(K^* \mathbf{1}_\Sigma (u_{k+1} - u_k) - \frac{K^* \mathbf{1}_\Sigma u_{k+\frac{1}{2}}}{u_k} (u_{k+1} - u_k) \right) dx \qquad (84)$$

$$= \int_\Sigma \left(f \log\left(\frac{f}{K u_{k+1}}\right) - f \log\left(\frac{f}{K u_k}\right) \right) d\mu$$

$$+ \int_\Omega \left(K^* \left(\frac{f}{K u_k}\right) (u_{k+1} - u_k) \right) dx.$$

The last equality in (84) holds, since $u_{k+\frac{1}{2}}$ is given by the EM reconstruction step (28a) and K^* is the adjoint operator of K, i.e. we have

$$\int_\Sigma K u \, d\mu = \langle \mathbf{1}_\Sigma, K u \rangle_\Sigma = \langle K^* \mathbf{1}_\Sigma, u \rangle_\Omega = \int_\Omega K^* \mathbf{1}_\Sigma u \, dx.$$

Next we characterize the right-hand side of (84) using an auxiliary functional

$$G(u) := \int_\Sigma f \log\left(\frac{f}{Ku}\right) d\mu$$

and consider the directional derivatives of G. That is, we define for any $w_1 \in L^1(\Omega)$ a function $\phi_{w_1}(t) := G(u + t w_1)$ and see that the directional derivative $G'(u; w_1)$ of G at u in direction w_1 is given by

$$G'(u; w_1) = \phi'_{w_1}(t)\big|_{t=0} = \int_\Sigma \frac{\partial}{\partial t} \left(f \log\left(\frac{f}{Ku + t Kw_1}\right) \right) d\mu \bigg|_{t=0}$$

$$= \left\langle -\frac{f}{Ku}, Kw_1 \right\rangle_\Sigma = \left\langle -K^*\left(\frac{f}{Ku}\right), w_1 \right\rangle_\Omega.$$

Interpreting the right-hand side of inequality (84) formally as a Taylor linearization of G yields

$$F(u_{k+1}) - F(u_k) + \frac{1}{\omega_k} \int_\Omega \frac{K^* \mathbf{1}_\Sigma (u_{k+1} - u_k)^2}{u_k} dx$$

$$\leq G(u_{k+1}) - G(u_k) - G'(u_k; u_{k+1} - u_k)$$

$$= \frac{1}{2} G''(v; u_{k+1} - u_k, u_{k+1} - u_k), \qquad v \in [u_k, u_{k+1}],$$

$$\leq \sup_{v \in [u_k, u_{k+1}]} \frac{1}{2} G''(v; u_{k+1} - u_k, u_{k+1} - u_k).$$

$$(85)$$

We can now compute the second directional derivative $G''(u; w_1, w_2)$ of G using the function $\phi_{w_2}(t) := G'(u + t w_2; w_1)$ for any $w_2 \in L^1(\Omega)$,

$$G''(u; w_1, w_2) = \phi'_{w_2}(t)\big|_{t=0} = -\int_\Sigma \frac{\partial}{\partial t}\left(\frac{f}{Ku + t Kw_2} Kw_1\right) d\mu \bigg|_{t=0}$$

$$= \int_\Sigma \frac{f\, Kw_2\, Kw_1}{(Ku)^2}\, d\mu .$$

Plugging the computed derivative $G''(u; w_1, w_2)$ in (85), we obtain

$$F(u_{k+1}) - F(u_k) + \frac{1}{\omega_k}\int_\Omega \frac{K^* 1_\Sigma\, (u_{k+1} - u_k)^2}{u_k}\, dx \qquad (86)$$
$$\leq \sup_{v \in [u_k, u_{k+1}]} \frac{1}{2}\int_\Sigma \frac{f\, (Ku_{k+1} - Ku_k)^2}{(Kv)^2}\, d\mu .$$

Finally, we split the third term on the left-hand side of (86) with $\varepsilon \in (0, 1)$,

$$F(u_{k+1}) + \frac{\varepsilon}{\omega_k}\int_\Omega \frac{K^* 1_\Sigma\, (u_{k+1} - u_k)^2}{u_k}\, dx + \frac{1-\varepsilon}{\omega_k}\int_\Omega \frac{K^* 1_\Sigma\, (u_{k+1} - u_k)^2}{u_k}\, dx$$
$$\leq \sup_{v \in [u_k, u_{k+1}]} \frac{1}{2}\int_\Sigma \frac{f\, (Ku_{k+1} - Ku_k)^2}{(Kv)^2}\, d\mu + F(u_k),$$

and obtain condition (83), i.e. a descent of the objective functional F, if

$$\sup_{v \in [u_k, u_{k+1}]} \frac{1}{2}\int_\Sigma \frac{f\, (Ku_{k+1} - Ku_k)^2}{(Kv)^2}\, d\mu \leq \frac{1-\varepsilon}{\omega_k}\int_\Omega \frac{K^* 1_\Sigma\, (u_{k+1} - u_k)^2}{u_k}\, dx . \qquad (87)$$

By solving (87) for w_k, we obtain the required condition (81) for the damping parameters $\{\omega_k\}$ in order to have a descent of the objective functional F. Additionally, by a suitable choice of ε in (81), it can be guaranteed that $\omega_k \leq 1$ for all $k \geq 0$.

Second step: Convergence of the primal iterates

In order to show that the iteration method converges to a minimizer of the functional F, we need a convergent subsequence of the primal iterates $\{u_k\}$. Since the operator K does not annihilate constant functions, the functional F is BV-coercive according to Lemma 3 and we obtain from (51),

$$\|u_k\|_{BV(\Omega)} \leq c_4\, (F(u_k))^2 + c_5\, F(u_k) + c_6 \leq c_4\, (F(u_0))^2 + c_5\, F(u_0) + c_6 ,$$

for all $k \geq 0$ with constants $c_4 \geq 0$, $c_5 > 0$ and $c_6 \geq 0$. The latter inequality holds due to the nonnegativity of F and due to the monotone decrease of the sequence $\{F(u_k)\}$ with the corresponding choice of damping parameters $\{\omega_k\}$ in (81). Thus, the sequence $\{u_k\}$ is uniformly bounded in the BV-norm and the Banach-Alaoglu theorem (see e.g. [93, Theorem 2.6.18]) yields the compactness in the weak* topology on $BV(\Omega)$, which implies the existence of a subsequence $\{u_{k_l}\}$ with

$$u_{k_l} \rightharpoonup^* u \quad \text{in } BV(\Omega).$$

The definition of the weak* topology on $BV(\Omega)$ in [3, Definition 3.11] also implies the strong convergence of the subsequence $\{u_{k_l}\}$ in $L^1(\Omega)$,

$$u_{k_l} \to u \quad \text{in } L^1(\Omega).$$

We can also consider the sequence $\{u_{k+1}\}$ and choose with the same argumentation further subsequences, again denoted by k_l, such that

$$u_{k_l+1} \rightharpoonup^* \tilde{u} \quad \text{in } BV(\Omega),$$
$$u_{k_l+1} \to \tilde{u} \quad \text{in } L^1(\Omega).$$

We show now that the limits of the subsequences $\{u_{k_l}\}$ and $\{u_{k_l+1}\}$ coincide, i.e. it holds $u = \tilde{u}$. For this purpose, we apply inequality (83) recursively and obtain the following estimate,

$$F(u_{k+1}) + \varepsilon \sum_{j=0}^{k} \int_{\Omega} \frac{K*\mathbf{1}_{\Sigma}(u_{j+1} - u_j)^2}{\omega_j u_j} dx \leq F(u_0) < \infty, \quad \forall k \geq 0.$$

The series of functional descent values on the left-hand side is summable and the Cauchy criterion for convergence delivers

$$\lim_{k \to \infty} \int_{\Omega} \frac{K*\mathbf{1}_{\Sigma}(u_{k+1} - u_k)^2}{\omega_k u_k} dx = 0. \tag{88}$$

Additionally, the Cauchy-Schwarz inequality yields the following estimate,

$$\|u_{k+1} - u_k\|_{L^1(\Omega)}^2 \leq \underbrace{\int_{\Omega} \frac{\omega_k u_k}{K*\mathbf{1}_{\Sigma}} dx}_{\overset{(82)}{\leq} c_1 \|u_k\|_{L^1(\Omega)}} \underbrace{\int_{\Omega} \frac{K*\mathbf{1}_{\Sigma}(u_{k+1} - u_k)^2}{\omega_k u_k} dx}_{\overset{(88)}{\to} 0}. \tag{89}$$

The first term on the right-hand side of (89) is uniformly bounded for all $k \geq 0$, since $\omega_k \in (0, 1]$, the function $K*\mathbf{1}_{\Sigma}$ is bounded away from zero (82) and the sequence $\{u_k\}$ is uniformly bounded in the BV-norm. Moreover, since the second term on the right-hand side of (89) converges to zero (cf. (88)), we obtain from (89) that

$$u_{k+1} - u_k \to 0 \quad \text{in } L^1(\Omega), \tag{90}$$

and the uniqueness of the limit implies $u = \tilde{u}$.

Third step: Convergence of the dual iterates

In addition to the second step, we need a convergent subsequence of the subgradients $\{p_k\}$ corresponding to the sequence $\{u_k\}$, i.e. $p_k \in \partial |u_k|_{BV(\Omega)}$. To this end, we use the general property that the subdifferentials of a convex one-homogeneous functional $J : X \to (-\infty, +\infty]$, X Banach space, can be characterized by (cf. Chap. 1, Lemma 3.12)

$$\partial J(u) = \{ p \in X^* : \langle p, u \rangle = J(u), \ \langle p, v \rangle \leq J(v) \ \forall v \in X \}.$$

In the case of TV, the dual norm of each subgradient p_k is bounded by

$$\|p_k\| = \sup_{\|v\|_{BV(\Omega)}=1} \langle p_k, v \rangle \leq \sup_{\|v\|_{BV(\Omega)}=1} |v|_{BV(\Omega)} \leq \sup_{\|v\|_{BV(\Omega)}=1} \|v\|_{BV(\Omega)} = 1.$$

Hence, the sequence $\{p_k\}$ is uniformly bounded in the $(BV(\Omega))^*$-norm and the Banach-Alaoglu theorem yields the compactness in the weak* topology on $(BV(\Omega))^*$, which implies the existence of a subsequence, again denoted by k_l, such that

$$p_{k_l+1} \rightharpoonup^* p \quad \text{in } (BV(\Omega))^*.$$

Fourth step: Show that $p \in \partial |u|_{BV(\Omega)}$

From steps above we have the weak* convergence of sequences $\{u_{k_l}\}$ and $\{u_{k_l+1}\}$ in $BV(\Omega)$, as well as the weak* convergence of $\{p_{k_l+1}\}$ in $(BV(\Omega))^*$. Next we show that the limit p of the dual iterates is a subgradient of $|\cdot|_{BV(\Omega)}$ at the limit u of the primal iterates, i.e. $p \in \partial |u|_{BV(\Omega)}$. Hence we have to prove (see the definition of the subgradients in (22)) that

$$|u|_{BV(\Omega)} + \langle p, v - u \rangle \leq |v|_{BV(\Omega)}, \quad \forall v \in BV(\Omega). \tag{91}$$

Let $p_{k_l+1} \in \partial |u_{k_l+1}|_{BV(\Omega)}$, then the definition of the subgradient of $|\cdot|_{BV(\Omega)}$ in (22) yields

$$|u_{k_l+1}|_{BV(\Omega)} + \langle p_{k_l+1}, v - u_{k_l+1} \rangle \leq |v|_{BV(\Omega)}, \quad \forall v \in BV(\Omega). \tag{92}$$

Since $|\cdot|_{BV(\Omega)}$ is lower semicontinuous, we can estimate the BV-seminorm at u from above,

$$|u|_{BV(\Omega)} \leq \liminf_{l \to \infty} |u_{k_l+1}|_{BV(\Omega)} \leq |u_{k_l+1}|_{BV(\Omega)},$$

and (92) delivers

$$|u|_{BV(\Omega)} + \langle p_{k_l+1}, v - u_{k_l+1}\rangle \leq |v|_{BV(\Omega)}, \qquad \forall v \in BV(\Omega). \tag{93}$$

In the third step we also verified the weak* convergence of $\{p_{k_l+1}\}$ in $(BV(\Omega))^*$, i.e. it holds

$$\langle p_{k_l+1}, v\rangle \to \langle p, v\rangle, \qquad \forall v \in BV(\Omega).$$

In order to prove $p \in \partial |u|_{BV(\Omega)}$, it remains to show with respect to (93) and (91) that

$$\langle p_{k_l+1}, u_{k_l+1}\rangle \to \langle p, u\rangle. \tag{94}$$

For this purpose we consider the complete iteration scheme of the damped FB-EM-TV algorithm (30) with $u_{k+\frac{1}{2}}$ in (28a),

$$u_{k_l+1} - (1-\omega_{k_l})u_{k_l} - \omega_{k_l}\left(\frac{u_{k_l}}{K^*1_\Sigma} K^*\left(\frac{f}{Ku_{k_l}}\right)\right) + \omega_{k_l}\,\alpha\,\frac{u_{k_l}}{K^*1_\Sigma}\,p_{k_l+1} = 0, \tag{95}$$

which is equivalent to

$$-\alpha\, p_{k_l+1} = \frac{K^*1_\Sigma\,(u_{k_l+1} - u_{k_l})}{\omega_{k_l}\,u_{k_l}} + K^*1_\Sigma - K^*\left(\frac{f}{Ku_{k_l}}\right).$$

Multiplying this formulation of the iteration scheme by u_{k_l+1} and integrating over the domain Ω yields

$$\begin{aligned}
-\alpha\,\langle p_{k_l+1}, u_{k_l+1}\rangle &= \int_\Omega \frac{K^*1_\Sigma\,(u_{k_l+1}-u_{k_l})\,u_{k_l+1}}{\omega_{k_l}\,u_{k_l}}\,dx + \left\langle 1_\Sigma - \frac{f}{Ku_{k_l}}, Ku_{k_l+1}\right\rangle \\
&= \underbrace{\int_\Omega \frac{K^*1_\Sigma\,(u_{k_l+1}-u_{k_l})^2}{\omega_{k_l}\,u_{k_l}}\,dx}_{\xrightarrow{(88)} 0} + \left\langle 1_\Sigma - \frac{f}{Ku_{k_l}}, Ku_{k_l+1}\right\rangle \\
&\quad + \underbrace{\int_\Omega \frac{K^*1_\Sigma\,(u_{k_l+1}-u_{k_l})\,u_{k_l}}{\omega_{k_l}\,u_{k_l}}\,dx}_{\xrightarrow{(90)} 0}
\end{aligned} \tag{96}$$

The third term on the right-hand side of (96) vanishes in the limit, since the term $\frac{K^*\mathbf{1}_\Sigma}{\omega_{k_l}}$ is uniformly bounded in the supremum norm (caused by the boundedness away from zero of ω_k (82) and the boundedness preservation of K^* in Assumption 1 (4)) and due to the $L^1(\Omega)$-norm convergence (90). Using the boundedness of the operator K for the convergence of $\frac{f}{Ku_{k_l}}$, we obtain

$$\left\langle \mathbf{1}_\Sigma - \frac{f}{Ku_{k_l}}, Ku_{k_l+1} \right\rangle \to \left\langle \mathbf{1}_\Sigma - \frac{f}{Ku}, Ku \right\rangle$$

and thus can deduce from (96) that

$$-\alpha \langle p_{k_l+1}, u_{k_l+1} \rangle \to \int_\Omega \left(K^*\mathbf{1}_\Sigma - K^*\left(\frac{f}{Ku}\right) \right) u\, dx \stackrel{(98)}{=} -\alpha \langle p, u \rangle .$$

Hence, we can conclude (94) and therewith $p \in \partial |u|_{BV(\Omega)}$.

Fifth step: Convergence to a minimizer of the objective functional

Let $\{u_{k_l}\}$ and $\{u_{k_l+1}\}$ be arbitrary convergent subsequences of the primal iteration sequence $\{u_k\}$, which converge to some $u \in BV(\Omega)$ in the weak* topology on $BV(\Omega)$ and in the strong topology on $L^1(\Omega)$. As seen in the third and fourth step, there exists a weak* convergent subsequence $\{p_{k_l+1}\}$ of the dual iteration sequence $\{p_k\}$, which convergence to some $p \in (BV(\Omega))^*$ such that $p \in \partial |u|_{BV(\Omega)}$. To verify the convergence of the damped FB-EM-TV splitting algorithm, it remains to show that u is a minimizer of the functional F. For this purpose, we consider the complete iteration scheme of the damped FB-EM-TV algorithm (95) with reference to the convergent subsequences and show their weak* convergence to the optimality condition (23) of the variational problem (21). Note that it suffices to prove only the convergence to (23) and not to (27), since the function u is positive due to the strict positivity assumption on the iterates u_k for all $k \geq 0$ in (82). An equivalent formulation to (95) reads as follows

$$\frac{u_{k_l+1} - u_{k_l}}{\omega_{k_l} u_{k_l}} + \mathbf{1}_\Omega - \frac{1}{K^*\mathbf{1}_\Sigma} K^*\left(\frac{f}{Ku_{k_l}}\right) + \frac{\alpha}{K^*\mathbf{1}_\Sigma} p_{k_l+1} = 0 . \quad (97)$$

The convergence can be verified in the following way. Due to the boundedness away from zero assumptions in (82), we use (88) in order to deduce the following convergence,

$$c_1 c_2 c_3 \int_\Omega \frac{(u_{k+1} - u_k)^2}{\omega_k^2 u_k^2} dx \leq \int_\Omega \frac{K^*\mathbf{1}_\Sigma (u_{k+1} - u_k)^2}{\omega_k^2 u_k^2} \omega_k u_k\, dx \stackrel{(88)}{\to} 0 .$$

Since the integrand on the left-hand side is nonnegative, we obtain with the uniqueness of the limit,

$$\lim_{l \to \infty} \frac{u_{k_l+1} - u_{k_l}}{\omega_{k_l} u_{k_l}} = 0 .$$

Therefore, if we pass over to the weak* limit of the subsequences in (97) using the boundedness of the operator K in Assumption 1 (1) for the convergence of $\frac{f}{Ku_{k_l}}$, we obtain that both limit functions u and p of sequences $\{u_{k_l}\}$ and $\{p_{k_l+1}\}$ satisfy the optimality condition (23) of the variational problem (21),

$$\mathbf{1}_\Omega - \frac{1}{K^*\mathbf{1}_\Sigma} K^* \left(\frac{f}{Ku} \right) + \frac{\alpha}{K^*\mathbf{1}_\Sigma} p = 0. \tag{98}$$

This means that the subsequence $\{u_{k_l}\}$ converges in the weak* topology on $BV(\Omega)$ and in the strong topology on $L^1(\Omega)$ to a minimizer of the functional F. □

Remark 8.

- Note that inequality (87) in the proof above motivates the mathematical necessity of a damping in the FB-EM-TV splitting strategy. In the undamped case, i.e. $\omega_k = 1$, the term on the right-hand side of (87) is maximal for $\varepsilon \to 0^+$, due to the strict positivity of $K^*\mathbf{1}_\Sigma$ and u_k for all $k \geq 0$ in (82). In general, one cannot say whether this term is greater than the supremum on the left-hand side of (87) or not and with it whether the objective functional F is decreasing during the iteration or not. Hence, a parameter $\omega_k \in (0, 1)$ is needed, which increases the term on the right-hand side of (87) to guarantee a descent of the objective functional F.
- Analogous to the proof above, the strong convergence of the sequence $\{u_k\}$ to a minimizer of the functional F in Theorem 4 can also be proved in the L^p-norm with $1 \leq p < d/(d-1)$, since any uniformly bounded sequence $\{u_k\}$ in $BV(\Omega)$ is actually relatively compact in $L^p(\Omega)$ for $1 \leq p < d/(d-1)$ (see [1, Theorem 2.5]). Since the subsequence $\{u_{k_l}\}$ is furthermore uniformly bounded in the BV norm, there exists a subsequence $\{u_{k_{l_m}}\}$ with

$$u_{k_{l_m}} \to \tilde{u} \quad \text{in } L^p(\Omega) \quad \text{with} \quad 1 \leq p < d/(d-1).$$

With the uniqueness of the limit and the definition of the weak* topology on $BV(\Omega)$, we obtain

$$u_{k_{l_m}} \rightharpoonup^* u \quad \text{in } BV(\Omega),$$

$$u_{k_{l_m}} \to u \quad \text{in } L^1(\Omega).$$

Due to the uniqueness of the limit, i.e. $\tilde{u} = u$, we can pass over in the proof from $\{u_{k_l}\}$ to $\{u_{k_{l_m}}\}$.
- The assumptions on boundedness away from zero in (82) are reasonable from our point of view. In the case of the function $K^*\mathbf{1}_\Sigma$, the assumption is practical since if there exists a point $x \in \Omega$ with $(K^*\mathbf{1}_\Sigma)(x) = 0$ then it is a-priori impossible to reconstruct the information in this point. Moreover, the assertion on the damping parameters ω_k makes sense because a strong damping is certainly undesirable. The boundedness away from zero of the iterates u_k is satisfied due to

the strict positivity of each half step of the (damped) FB-EM-TV splitting method (see Lemma 5).
- Inspired by the relaxed EM reconstruction strategy proposed in [99, Chap. 5.3.2], another possibility of influencing convergence arises in the FB-EM-TV strategy by adding a relaxation parameter $\nu > 0$ to the EM fixed point iteration in the form,

$$u_{k+\frac{1}{2}} = u_k \left(\frac{1}{K^* \mathbf{1}_\Sigma} K^* \left(\frac{f}{K u_k} \right) \right)^\nu \qquad \text{(relaxed EM step)}.$$

Corresponding, one can obtain a reasonable TV denoising step in the FB-EM-TV splitting idea via

$$u_{k+1} = \left(u_{k+\frac{1}{2}}^{\frac{1}{\nu}} - \alpha u_k^{\frac{1}{\nu}} p_{k+1} \right)^\nu, \quad p_{k+1} \in \partial |u_{k+1}|_{BV(\Omega)}, \quad \text{(relaxed TV step)},$$

with the relaxed EM step $u_{k+\frac{1}{2}}$ above. The relaxed terms in the TV step are necessary to fit the basic variational problem (20) and its optimality condition (23). Due to the computational challenge of the relaxed TV denoising step, which would require again novel methods, a comparison of this strategy with the damping strategy proposed in Sect. 3.2 would go beyond the scope of this lecture.

In practice, determining the damping parameters ω_k via the general condition in (81) is not straight-forward and one would be interested in an explicit bound for all damping parameters ω_k. Such an explicit bound is not possible in the case of a general operator K, but can be provided in case of the Poisson denoising strategy (75), i.e. being identity operator K.

Corollary 2 (Convergence of the Damped Poisson Denoising Scheme). *Let $\{u_k\}$ be a sequence of iterates generated by the damped Poisson denoising scheme (76) and let the given noisy function $f \in L^\infty_\mu(\Omega)$ satisfy $\text{ess}\inf_\Omega f > 0$. In order to guarantee the convergence in the case of the identity operator K, the condition (81) in Theorem 4 on the damping parameters simplifies to*

$$\omega_k \leq \frac{2 (\text{ess}\inf_\Omega f)^2}{(\text{ess}\sup_\Omega f)^2} (1 - \varepsilon), \qquad \varepsilon \in (0, 1). \tag{99}$$

Proof. In the special case of the identity operator K, the maximum principle of the damped Poisson denoising scheme from Lemma 6 is the main idea for simplifying the desired condition (81) on the damping parameters. For this sake, we consider the inequality (87), which guarantees a monotone descent of the objective functional if

$$\frac{1}{2} \int_\Omega \frac{f u_k}{v^2} \frac{(u_{k+1} - u_k)^2}{u_k} d\lambda \leq \frac{1 - \varepsilon}{\omega_k} \int_\Omega \frac{(u_{k+1} - u_k)^2}{u_k} d\lambda, \qquad \forall v \in [u_k, u_{k+1}].$$

The goal is to find an estimate for the coefficients $\frac{f u_k}{2 v^2}$. Due to $v \in [u_k, u_{k+1}]$ and since $\{u_k\}$ are iterates generated by the damped Poisson denoising scheme (76), we can use the maximum principle from Lemma 6 and obtain an estimate

$$\frac{f u_k}{2 v^2} \leq \frac{(\operatorname{ess sup}_\Omega f)(\operatorname{ess sup}_\Omega u_k)}{2 (\operatorname{ess inf}_\Omega \{u_k, u_{k+1}\})^2} \leq \frac{(\operatorname{ess sup}_\Omega f)^2}{2 (\operatorname{ess inf}_\Omega f)^2}, \qquad \forall k \geq 0,$$

which should be less or equal $\frac{1-\varepsilon}{\omega_k}$. Thus, choosing ω_k according to the estimate (99) guarantees a monotone descent of the objective functional. □

5 Numerical Realization of the Weighted ROF Problem

In Sects. 3.1 and 3.5 we presented the (Bregman-)FB-EM-TV algorithm as a nested two step iteration strategy in order to solve the TV regularized Poisson likelihood estimation problem (20). However we left open the numerical realization of the TV correction half step (29), (31), (45) and (46) contained in both iteration strategies. Since all these regularization half steps have a similar form, we can present a uniform numerical framework, which is also valid for the image denoising variational problem (76). The most general form of all the schemes above is

$$\min_{u \in BV(\Omega)} \frac{1}{2} \int_\Omega \frac{(u - q)^2}{h} dx + \beta |u|_{BV(\Omega)}, \qquad \beta > 0, \qquad (100)$$

with an appropriate setting of the "noise" function q, the weight function h and the regularization parameter β. The choice of these parameters with respect to the proposed restoration methods is summarized in Table 1.

The variational problem (100) is just a modified version of the well known ROF model (cf. Chap. 1), with an additional weight h in the data fidelity term. This analogy creates the opportunity to carry over the different numerical schemes known for the ROF model, e.g. we refer to [7,34–36] and the references therein, where most of them can be adapted to the weighted modification (100). Here we consider two popular numerical realizations for TV regularized problems in image processing. In the first one, the exact dual TV functional (6) for the minimization of (100) will be used, which does not need any smoothing of the total variation. Then, an approach analogous to the projected gradient descent algorithm of Chambolle in [33] will be used, which characterizes the subgradients of TV as divergences of vector fields with supremum norm less or equal one. The second numerical scheme will be similar to the alternating split Bregman algorithm of Goldstein and Osher in [62], whereby a slightly modified augmented Lagrangian approach of the alternating split Bregman algorithm will be used in order to handle the weight in the data fidelity term better. Using either method, the weighted ROF problem (100) can be solved efficiently, obtaining accurate and robust algorithms. In the following, we only give a short

Table 1 Overview for the setting of functions q, h and parameter β in (100) with respect to the different algorithms proposed in Sect. 3

Algorithm	q	h	β
Poisson denoising (76)	f	u_k	α
Damped Poisson denoising (76)	$\omega_k f + (1 - \omega_k) u_k$	u_k	$\omega_k \alpha$
FB-EM-TV algorithm (29)	$u_{k+\frac{1}{2}}$	$\dfrac{u_k}{K^* 1_\Sigma}$	α
Damped FB-EM-TV algorithm (31)	$\omega_k u_{k+\frac{1}{2}} + (1 - \omega_k) u_k$	$\dfrac{u_k}{K^* 1_\Sigma}$	$\omega_k \alpha$
Bregman-FB-EM-TV algorithm (45)	$u_{k+\frac{1}{2}}^{l+1} + u_k^{l+1} v^l$	$\dfrac{u_k^{l+1}}{K^* 1_\Sigma}$	α
Damped Bregman-FB-EM-TV algorithm (46)	$\omega_k^{l+1} u_{k+\frac{1}{2}}^{l+1} + \omega_k^{l+1} u_k^{l+1} v^l + (1 - \omega_k^{l+1}) u_k^{l+1}$	$\dfrac{u_k^{l+1}}{K^* 1_\Sigma}$	$\omega_k^{l+1} \alpha$

derivation of both numerical schemes and refer to [108, Sect. 6.3] for a detailed explanation.

First, we establish an iterative algorithm to compute the solution of the variational problem (100) using a modified variant of the projected gradient descent algorithm of Chambolle [33]. We compute the primal variable u from the primal optimality condition with a dual variable $\tilde{g}$ to be determined as a minimizer of a dual problem,

$$u = q - \beta h \operatorname{div} \tilde{g}, \quad \tilde{g} = \arg\min_{g \in C_0^\infty(\Omega, \mathbb{R}^d)} \int_\Omega \frac{(\beta \operatorname{div} g - q)^2}{h} \, dx,$$

$$\text{s.t.} \ |g(x)|_{\ell^2}^2 - 1 \leq 0, \quad \forall x \in \Omega, \qquad (101)$$

where $|\cdot|_{\ell^2}$ is the Euclidean vector norm. For the choice of this vector norm, compare the remark at the end of this chapter. The constraint for the dual variable g in (101) is a consequence of the exact (dual) definition of total variation (6). The dual problem in (101) is a (weighted) quadratic optimization problem with a nonlinear inequality constraint and can be solved with a projected gradient descent algorithm similar to [33],

$$g^{n+1}(x) = \frac{g^n(x) + \tau \left(\nabla(\beta h \operatorname{div} g^n - q)(x) \right)}{1 + \tau \left| \nabla(\beta h \operatorname{div} g^n - q)(x) \right|_{\ell^2}}, \quad \forall x \in \Omega. \qquad (102)$$

In a standard discrete setting on pixels with unit step sizes and first derivatives computed by one-sided differences, the convergence result of Chambolle in

[33, Theorem 3.1] can be transferred to the weighted ROF problem (100). The proof is based on the Banach fixed point theorem and required the condition

$$0 < \tau \leq \left(4 d \gamma \, \|h\|_{L^\infty(\Omega)}\right)^{-1}, \tag{103}$$

in order to obtain a contraction constant less one, where $4\,d$ is the upper bound of the discrete divergence operator. Hence, the convergence of (102) to an optimal solution can be guaranteed if the damping parameter τ satisfies (103). Note that the weight function h can be interpreted as an adaptive regularization, since the regularization parameter β is weighted in (102) by h.

Next, we present an efficient numerical scheme to compute the solution of (100) and follow the idea of the (alternating) split Bregman algorithm proposed by Goldstein and Osher in [62]. For this purpose, we use the formal definition of TV in (7) and rewrite (100) as an equivalent constrained optimization problem of the form

$$\min_{u, \tilde{u}, v} \frac{1}{2} \int_\Omega \frac{(\tilde{u} - q)^2}{h} \, dx + \beta \int_\Omega |v|_{\ell^2} \, dx \quad \text{s.t.} \quad \tilde{u} = u \quad \text{and} \quad v = \nabla u,$$

where $|\cdot|_{\ell^2}$ is the Euclidean vector norm. The difference to the split Bregman algorithm is that an additional auxiliary function $\tilde{u}$ is introduced, which will simplify the handling of the weight function h in the numerical scheme below. Following the idea of augmented Lagrangian methods [55, 61, 76] and using the standard Uzawa algorithm (without preconditioning) [48], we obtain an alternating minimization scheme given by

$$u^{n+1} \in \arg\min_u \left\{ \langle \lambda_2^n, \tilde{u}^n - u \rangle + \frac{\mu_2}{2} \|\tilde{u}^n - u\|_{L^2(\Omega)}^2 \right.$$
$$\left. + \langle \lambda_1^n, v^n - \nabla u \rangle + \frac{\mu_1}{2} \|v^n - \nabla u\|_{L^2(\Omega)}^2 \right\}, \tag{104a}$$

$$\tilde{u}^{n+1} \in \arg\min_{\tilde{u}} \left\{ \frac{1}{2} \int_\Omega \frac{(\tilde{u} - q)^2}{h} \, dx + \langle \lambda_2^n, \tilde{u} - u^{n+1} \rangle \right.$$
$$\left. + \frac{\mu_2}{2} \|\tilde{u} - u^{n+1}\|_{L^2(\Omega)}^2 + \chi_{\tilde{u} \geq 0} \right\}, \tag{104b}$$

$$v^{n+1} \in \arg\min_v \left\{ \beta \int_\Omega |v|_{\ell^2} \, dx + \langle \lambda_1^n, v - \nabla u^{n+1} \rangle \right.$$
$$\left. + \frac{\mu_1}{2} \|v - \nabla u^{n+1}\|_{L^2(\Omega)}^2 \right\}, \tag{104c}$$

$$\lambda_1^{n+1} = \lambda_1^n + \mu_1 \left(v^{n+1} - \nabla u^{n+1} \right), \qquad (104d)$$

$$\lambda_2^{n+1} = \lambda_2^n + \mu_2 \left(\tilde{u}^{n+1} - u^{n+1} \right), \qquad (104e)$$

whereas λ_1 and λ_2 are Lagrange multipliers as well as μ_1 and μ_2 are positive relaxation parameters. Note that we additionally add an indicator function $\chi_{\tilde{u} \geq 0}$ in (104b), defined as 0 if $\tilde{u}$ is nonnegative and $+\infty$ else, in order to guarantee the nonnegativity of the solution. The efficiency of this strategy is strongly dependent on the question, how fast we can solve each of the subproblems (104a), (104b) and (104c). Since the minimization problem (104a) is "decoupled" from the L^1-norm, it is differentiable with a Helmholtz-type optimality equation incorporating Neumann boundary conditions,

$$(\mu_2 I - \mu_1 \Delta) u^{n+1} = \lambda_2^n + \mu_2 \tilde{u}^n - \text{div}(\lambda_1^n + \mu_1 v^n),$$

where I is the identity operator and Δ denotes the Laplace operator. In the discrete setting using finite difference discretization on a rectangular domain Ω, this equation can be solved efficiently by the discrete cosine transform (DCT-II), since the operator $\mu_2 I - \mu_1 \Delta$ is diagonalizable in the DCT-transformed space (cf. e.g. [65, 103]),

$$u^{n+1} = DCT^{-1} \left(\frac{DCT \left(\lambda_2^n + \mu_2 \tilde{u}^n - \text{div}(\lambda_1^n + \mu_1 v^n) \right)}{\mu_2 + \mu_1 \hat{k}} \right), \qquad (105)$$

where DCT^{-1} denotes the inverse discrete cosine transform and $\hat{k}$ represents the negative Laplace operator in the discrete cosine space (see [108, Sect. 6.3.4]),

$$\hat{k}_{p_1,\ldots,p_d} = 4 \sum_{k=1}^{d} \left(\frac{\sin\left(\frac{\pi p_k}{2 N_k}\right)}{h_k} \right)^2.$$

Here we assumed a d-dimensional regular grid of $N_1 \times \cdots \times N_d$ points with a stepsize $h_k = \frac{1}{N_k}$, $k = 1, \ldots, d$, of the image grid in the k-th direction and $0 \leq p_k \leq N_k - 1$ is the subindex in the discrete cosine space.

Moreover, the minimization problem (104b) is also differentiable and can be computed by an explicit formula of the form

$$\tilde{u}^{n+1} = \begin{cases} \dfrac{q + h \left(\mu_2 u^{n+1} - \lambda_2^n \right)}{I + \mu_2 h}, & \text{if } \dfrac{q + h \left(\mu_2 u^{n+1} - \lambda_2^n \right)}{I + \mu_2 h} \geq 0, \\ 0, & \text{else}. \end{cases}$$

For the minimization problem with respect to v in (104c), we use a generalized shrinkage formula presented in [124],

$$v_i^{n+1} = \frac{\left(\nabla_{x_i} u^{n+1} - (1/\mu_1)(\lambda_1^n)_i\right)(x)}{\left|\left(\nabla u^{n+1} - (1/\mu_1)\lambda_1^n\right)(x)\right|_{\ell_2}} \max\left(\left|\left(\nabla u^{n+1} - (1/\mu_1)\lambda_1^n\right)(x)\right|_{\ell_2} - (\beta/\mu_1), 0\right)$$

for any $x \in \Omega$ and $1 \leq i \leq d$, where d is the dimension of Ω, ∇_{x_i} denotes the i-th component of the gradient ∇ and the convention $(0/0) \cdot 0 = 0$ is used.

Finally we remark that the speed of convergence of the augmented Lagrangian based method presented above depends in practice crucially on the values of relaxation parameters μ_1 and μ_2. Both parameters are usually chosen fixed in advance such that it can be expensive to find optimal values. In order to make the convergence speed of the algorithm less dependent on the initially chosen μ_1 and μ_2, an efficient approach can be found in [25, 66] and the references therein presenting an iterative adaptation strategy. Recently an interesting alternative to the iterative adaptation idea of relaxation parameters has been proposed in [63]. The authors present there an acceleration strategy based simply on ADMM with a predictor-corrector type acceleration step using a fixed relaxation parameter.

Remark 9. Note that the definition of the total variation (6) is not unique for $d \geq 2$. Depending on the definition of the supremum norm $\|g\|_\infty = \sup_{x \in \Omega} |g(x)|_{\ell^s}$ with respect to different vector norms on $\mathbb{R}^d$ with $1 \leq s \leq \infty$, one obtains equivalent versions of the *BV* seminorm $|\cdot|_{BV(\Omega)}$. More precisely, a family of total variation seminorms is defined by

$$\int_\Omega |Du|_{\ell^r} = \sup\left\{\int_\Omega u \operatorname{div} g \, dx \, : \, g \in C_0^\infty(\Omega, \mathbb{R}^d), \, |g|_{\ell^s} \leq 1 \text{ on } \Omega\right\},$$

for $1 \leq r < \infty$ and the Hölder conjugate index s, i.e. $r^{-1} + s^{-1} = 1$. The most common formulations are the isotropic total variation ($r = 2$) and the anisotropic total variation ($r = 1$). The different definitions of TV have effects on the structure of solutions obtained during the TV minimization. In the case of isotropic TV, corners in the edge set will not be allowed, whereas orthogonal corners are favored by the anisotropic variant. For a detailed analysis, we refer e.g. to [19, 50, 94, 117].

6 Numerical Results in Fluorescence Microscopy

In this section we illustrate the performance of the described numerical schemes and test the theoretical results by 2D reconstructions on synthetic and real data in fluorescence microscopy.

6.1 Fluorescence Microscopy

In recent years revolutionary imaging techniques have been developed in light microscopy with enormous importance for biology, material sciences, and medicine. The technology of light microscopy has been considered to be exhausted for a couple of decades, since the resolution is basically limited by Abbe's law for diffraction of light. By developing stimulated emission depletion (STED) fluorescence microscopy [71] and 4Pi-confocal fluorescence microscopy [69,70] now resolutions are achieved that are way beyond this diffraction barrier [68,80]. STED microscopy takes an interesting laser sampling approach, which in principle would even allow molecular resolutions. In this technique, fluorescent dyes are stimulated by a small laser spot and are directly quenched by an additional interfering laser spot. Since this depletion spot vanishes at one very small point in the middle, fluorescence of the stimulating spot is only detected at this tiny position. Hence, data with previously unknown resolution can be measured. However, by reaching the diffraction limit of light, measurements suffer from blurring effects and in addition suffer from Poisson noise due to laser sampling.

In the case of optical nanoscopy the linear, compact operator $\bar{K}$ is a convolution operator with a kernel $\bar{k} \in C(\Omega \subset \mathbb{R}^d)$,

$$(\bar{K}u)(x) = (\bar{k} * u)(x) := \int_\Omega \bar{k}(x-y)u(y)\,dy. \tag{106}$$

The kernel is often referred to as the point spread function (PSF), whose Fourier transform is the object transfer function. From a computational point of view, it is important to say that the convolution operator in the proposed algorithms can be computed efficiently by FFT following the Fourier convolution theorem,

$$\bar{k} * u = \mathscr{F}^{-1}\left(\mathscr{F}(\bar{k})\,\mathscr{F}(u)\right).$$

To get an impression of images suffering from blurring effects and Poisson noise, we refer to Fig. 1. In this simulation of measurements, we use a special 4Pi convolution kernel which is illustrated in Fig. 1d, e. Compared to standard convolution kernels, e.g. of Gaussian type, the latter one bears an additional challenge since it varies considerably in structure. This leads to side lobe effects in the object structure of the measured data as we can see in Fig. 1b, c. In practice, this type of convolution is found in 4Pi microscopy, since two laser beams interfere in the focus. Under certain circumstances, convolution kernels can also be locally varying, such that blind deconvolution strategies are needed. Here we assume a 4Pi convolution kernel of the form (see e.g. [9])

$$\bar{k}(x,y) \sim \cos^4\left(\left(\frac{2\pi}{\lambda}\right)y\right)e^{-\left(\frac{x}{\sigma_x}\right)^2 - \left(\frac{y}{\sigma_y}\right)^2}, \tag{107}$$

with the standard deviations σ_x and σ_y, and where λ denotes the refractive index characterizing the doubling properties.

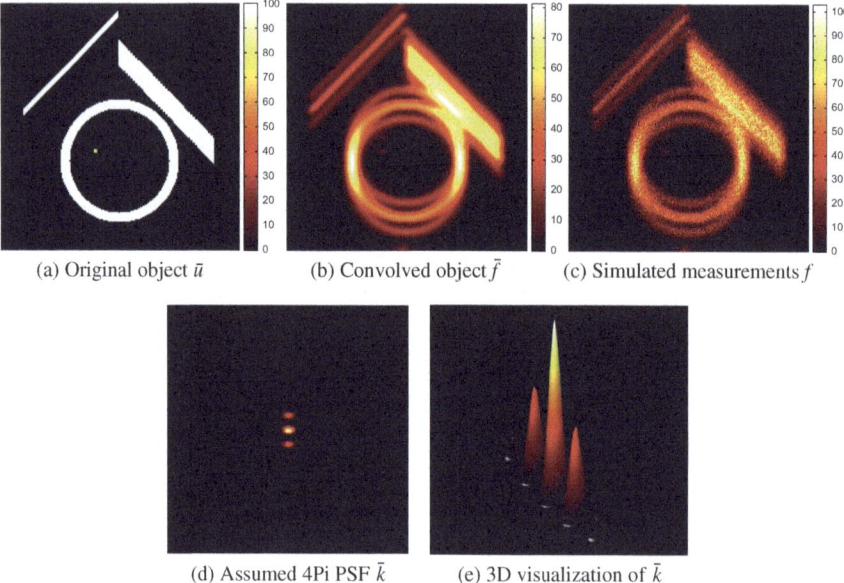

Fig. 1 Synthetic 2D fluorescence microscopy data using 4Pi PSF and Poisson noise. (**a**) Exact object (ground truth) $\bar{u} \in \mathbb{R}^{200 \times 200}$. (**b**) Convolved object $\bar{f} = \bar{K}\bar{u}$ (106) using 4Pi PSF shown in (**d**). (**c**) Simulated measurements f, where $\bar{f}$ in (**b**) is perturbed by Poisson noise. (**d**) Assumed 4Pi microscopy PSF (107) with parameters $\lambda = 0.12$, $\sigma_x = 0.02$, $\sigma_y = 0.07$. (**e**) 3D visualization of the 4Pi PSF in (**d**)

6.2 2D Reconstruction Results on Synthetic Data

We start by illustrating the proposed techniques using synthetic 2D fluorescence microscopy data $f \in \mathbb{R}^{200 \times 200}$ (see Fig. 1c) simulated for a simple object $\bar{u} \in \mathbb{R}^{200 \times 200}$ shown in Fig. 1a. The data are obtained via a 4Pi convolution kernel (107) presented in Fig. 1d, e, where additionally Poisson noise is simulated.

In Fig. 2 we present EM reconstructions for different numbers of iterations following algorithm (19) with data f illustrated in Fig. 1c. We can observe that early stopping in Fig. 2a leads to a natural regularization, however with blurring effects and undesired side lobes in the whole object. A higher number of iterations leads to sharper results, as in Fig. 2b, however the reconstructions suffer more and more from the undesired "checkerboard effect", as in Fig. 2c. In Fig. 2d we additionally display the expected monotone descent of the objective functional in (17) for 100 EM iterations. Finally, we present in Fig. 2e the typical behavior of EM iterates for ill-posed problems as described in [105]. The (metric) distance, here Kullback-Leibler, between the iterates and the exact solution decreases initially before it increases as the noise is amplified during the iteration process. The minimal distance in Fig. 2e is reached approximately after 50 iterations.

In Fig. 3 we illustrate reconstruction results obtained with the FB-EM-TV algorithm using different regularization parameters α. In comparison to EM

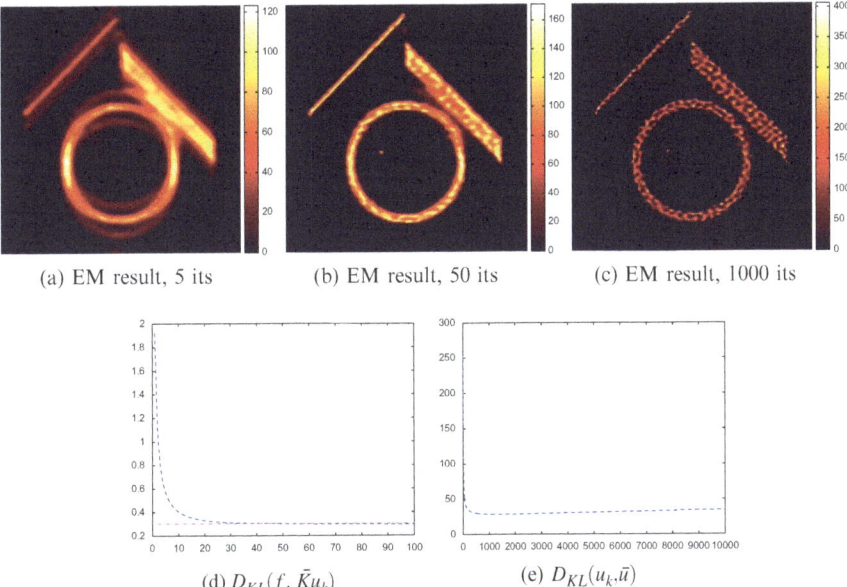

Fig. 2 Synthetic 2D fluorescence microscopy data from Fig. 1: EM reconstructions. (**a–c**) Reconstruction results obtained with the EM algorithm (19) and stopped at different iteration numbers. (**d**) Kullback-Leibler distances D_{KL} between given measurements f and convolved EM iterates $\bar{K} u_k$ for 100 iterations (*blue dashed line*), as well as between f and exact convolved object $\bar{K}\bar{u}$ (*magenta dash-dot line*). (**e**) Kullback-Leibler distance between EM iterates u_k and exact object $\bar{u}$ for 10,000 iterations

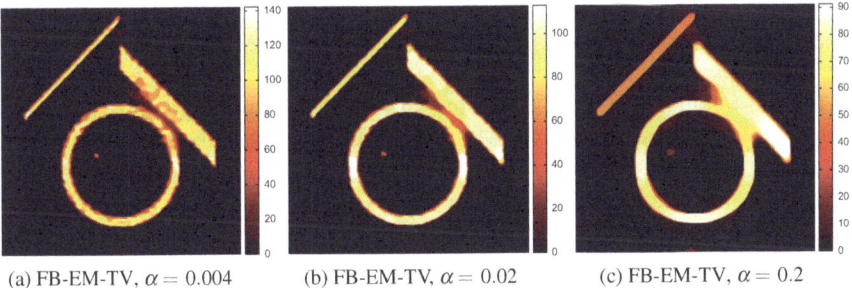

Fig. 3 Synthetic 2D fluorescence microscopy data from Fig. 1: FB-EM-TV reconstructions. (**a–c**) Reconstruction results obtained with the FB-EM-TV splitting algorithm (28) using different regularization parameters α

reconstructions in Fig. 2, the regularized EM algorithm deconvolves the given data without remaining side lobes and reduces noise and oscillations very well. Additionally, the FB-EM-TV algorithm successfully reconstructs the main geometrical configurations of the desired object in Fig. 1a, despite the low SNR of the given data in Fig. 1c. The reconstruction in Fig. 3a is slightly under-smoothed, whereas in Fig. 3c the computed image is over-smoothed. A visually reasonable

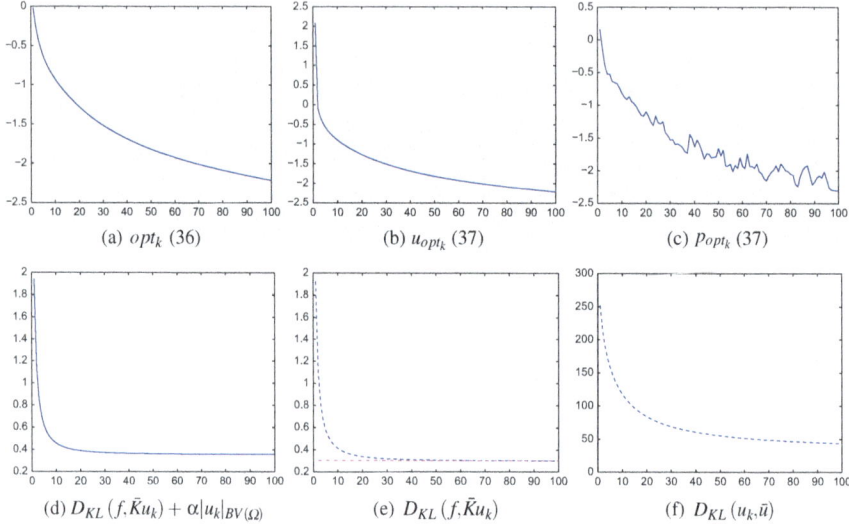

Fig. 4 Synthetic 2D fluorescence microscopy data from Fig. 1: different statistics for the result in Fig. 3b with 100 FB-EM-TV iterations. (**a–c**) Stopping rules proposed in Sect. 3.4. (**d**) Values of the objective functional in (21). (**e**) Kullback-Leibler distances D_{KL} between given measurements f and convolved FB-EM-TV iterates $\tilde{K}u_k$ *(blue dashed line)*, as well as between f and exact convolved object $\tilde{K}\bar{u}$ *(magenta dash-dot line)*. (**f**) Kullback-Leibler distances between FB-EM-TV iterates u_k and exact object $\bar{u}$

reconstruction is illustrated in Fig. 3b. Moreover, different statistical results for the FB-EM-TV reconstruction in Fig. 3b are plotted in Fig. 4. As expected, we can observe a decreasing behavior of the stopping rules proposed in Sect. 3.4, the objective functional values in (21) and Kullback-Leibler distances to the given measurements f and exact image $\bar{u}$.

In Sect. 4.4, in particular in Theorem 4, it was proved that a damping strategy is theoretically required in the FB-EM-TV algorithm in order to attain a monotone descent of the objective functional in (21) during the minimization process and hence to guarantee convergence of the splitting scheme. However, in numerical tests we could observe that the damping strategy is only needed in the case of high values of the regularization parameter α. For example, we compare the behavior of the objective functional values for different damping parameters in the case of $\alpha = 10$ in Fig. 5. Without damping (i.e. $\omega_k = 1$ for all $k \geq 0$) we obtain oscillations in Fig. 5a. These oscillations decrease in the case of a small damping (i.e. in the case of ω_k less one) as plotted in Fig. 5b and vanish if the damping is strong enough, such that a monotone descent in the objective functional can be achieved for $\omega_k = 0.05$ for all $k \geq 0$ in Fig. 5c.

Although the original object in Fig. 1a can be reconstructed quite well with the FB-EM-TV algorithm (see Fig. 3b), we can observe a natural loss of contrast as mentioned in Sect. 3.5. This means that some parts of the test object cannot be separated sufficiently. To overcome this problem we proposed to use inverse scale space methods based on Bregman distance iteration in Sect. 3.5. In Fig. 6 we present

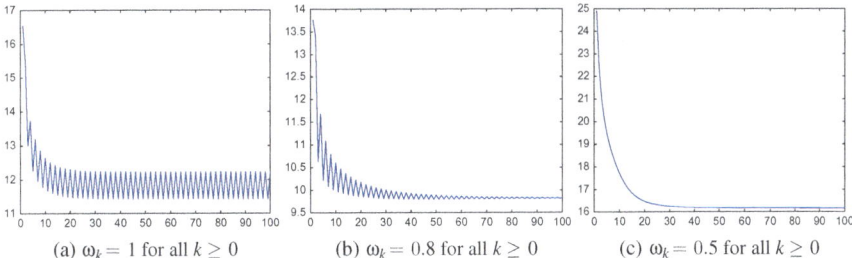

Fig. 5 Synthetic 2D fluorescence microscopy data from Fig. 1: influence of damping parameters in the FB-EM-TV algorithm with respect to the behavior of the objective functional values. (**a–c**) Iterations vs. values of the objective functional in (21) for different damping parameter values ω_k and $\alpha = 10$

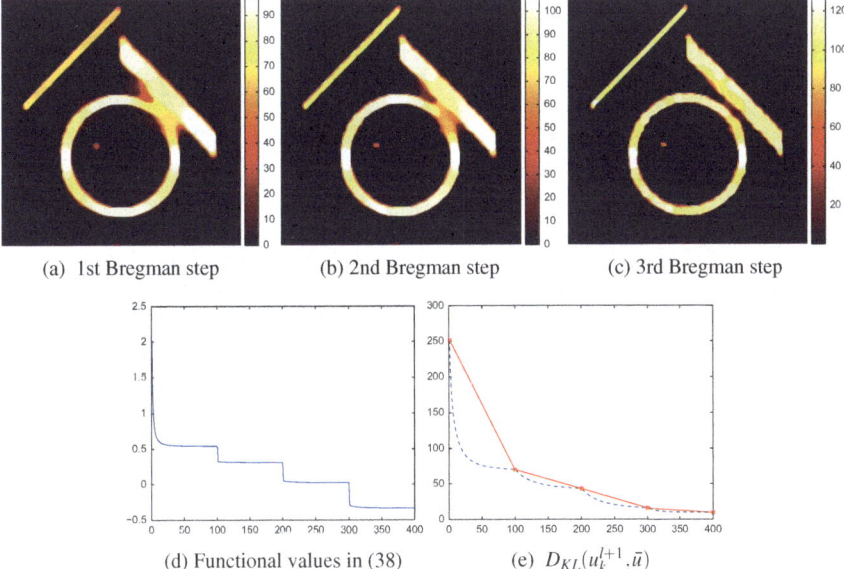

Fig. 6 Synthetic 2D fluorescence microscopy data from Fig. 1: Bregman-FB-EM-TV reconstructions. (**a–c**) Reconstruction results at different contrast enhancement steps of the Bregman-FB-EM-TV algorithm proposed in Sect. 3.5. (**d**) Values of the objective functional in (38) for 4 outer Bregman refinement steps with always 100 inner FB-EM-TV iteration steps. (**e**) Kullback-Leibler distances between Bregman-FB-EM-TV iterates u_k^{l+1} and exact object $\bar{u}$ for 4 outer Bregman refinement steps with always 100 inner FB-EM-TV iteration steps (*blue dashed line*), as well as between the final results of each Bregman refinement step and exact object $\bar{u}$ (*red solid line*)

reconstruction results for different refinement steps of the Bregman-FB-EM-TV algorithm proposed in Sect. 3.5. Corresponding to the characteristic of inverse scale space methods, we observe that the results will be improved with increasing Bregman iteration number with respect to the contrast enhancement, as we can see in the maximal intensity of reconstructions in Fig. 6. In Fig. 6d we also plot the expected monotone descent of the objective functional in (38) for 4 outer Bregman

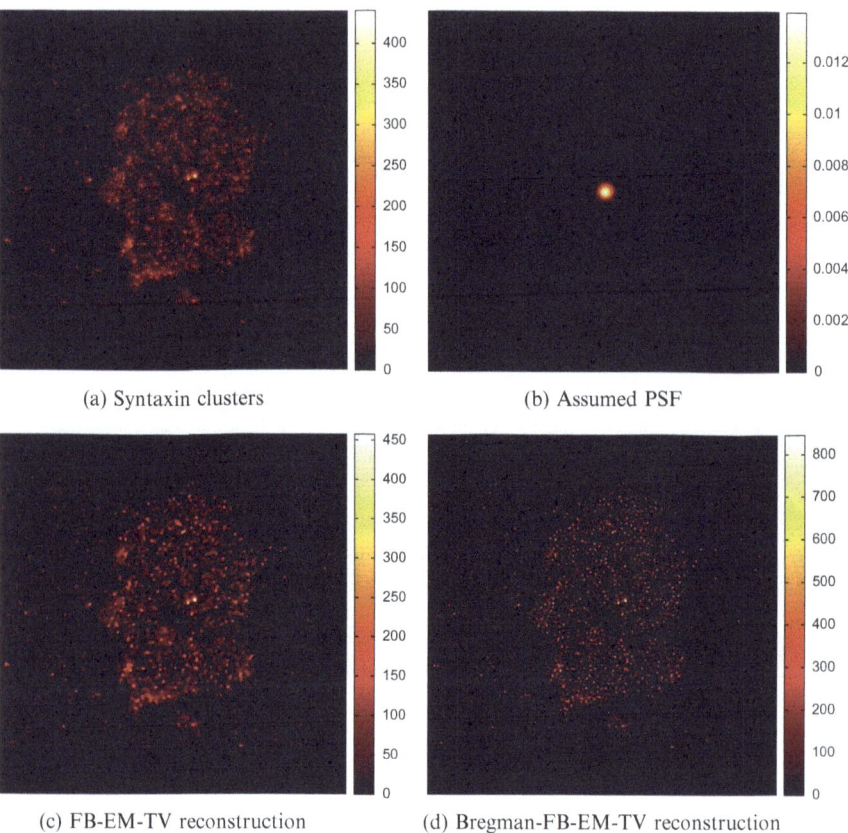

Fig. 7 Immunofluorescence CW-STED microscopy measurements: deconvolution results obtained with the (Bregman-)FB-EM-TV algorithms. (**a**) CW-STED micrograph of protein syntaxin on a membrane sheet of a fixed mammalian (PC12) cell [126], image size 1000 × 1000. (**b**) Assumed point spread function for the reconstruction process. (**c**) Reconstruction with the FB-EM-TV algorithm (28). (**d**) Reconstruction with the Bregman-FB-EM-TV algorithm proposed in Sect. 3.5, here third Bregman refinement step

refinement steps with always 100 inner FB-EM-TV iteration steps. In particular, we can observe the occurring jumps in the functional values which correspond to the contrast refinement effects at each Bregman step. Finally, we present in Fig. 6e the decreasing behavior of the Kullback-Leibler distances to the exact image $\bar{u}$ for the complete Bregman-FB-EM-TV iteration sequence.

6.3 2D Reconstruction Results on Real Data

In this section we perform the proposed algorithms on a 2D real data set in fluorescence microscopy provided by Willig and Schönle (both MPI for Biophysical Chemistry, Göttingen). The data set is illustrated in Fig. 7a and shows syntaxin clusters (a membrane integrated protein participating in exocytosis) in a fixed mammalian (PC12) cell membrane [115]. This micrograph is acquired using STED microscopy with continuous wave (CW) laser beams and has an image size of 1000×1000 pixels. For further information to CW-STED microscopy and technical aspects of the experiment, we refer to [126].

The deconvolution results of the (Bregman-)FB-EM-TV algorithms are presented in Fig. 7c, d using an assumed point spread function illustrated in Fig. 7b. The results show that both methods preserve fine structures in the image well. However, the contrast enhancing property of the Bregman-FB-EM-TV algorithm in Fig. 7d is observable as well, compared to the FB-EM-TV result in Fig. 7c, and the syntaxin clusters can be separated better.

7 Numerical Results in Positron Emission Tomography

In the following we present the performance of the proposed algorithms by 2D and 3D reconstructions results on real data sets in positron emission tomography. The corresponding measurements were acquired in the Department of Nuclear Medicine at the University Hospital of Münster and were provided by Büther and Schäfers (both EIMI, University of Münster).

7.1 Positron Emission Tomography

PET is a biomedical imaging technique, which visualizes biochemical and physiological processes, such as glucose metabolism, blood flow or receptor concentrations (see e.g. [10, 120, 125]). This modality is mainly applied in nuclear medicine and can be used e.g. to detect tumors, to locate areas of the heart affected by coronary artery disease and to identify brain regions influenced by drugs. Therefore, PET is categorized as a functional imaging technique and differs from methods as X-ray computed tomography (CT) that visualize anatomy structures. The data acquisition in PET is based on weak radioactively marked pharmaceuticals (tracers), which are injected into the blood circulation, and bindings dependent on the choice of the tracer to the molecules to be studied. Used markers are radio-isotopes, which decay by emitting a positron, which annihilates almost immediately with an electron. The resulting emission of two photons will then be detected by the tomograph device. Due to the radioactive decay, measured data can be modeled as an inhomogeneous Poisson process with a mean given by the X-ray transform of the spatial tracer distribution [99, Sect. 3.2]. The X-ray transform maps a function on $\mathbb{R}^d$ into the set of its line integrals [99, Sect. 2.2]. More precisely, if $\theta \in S^{d-1}$ and $x \in \theta^\perp$, then the X-ray transform $\tilde{K}$ may be defined by

$$(\tilde{K}u)(\theta, x) = \int_\mathbb{R} u(x + t\theta)\,dt,$$

and corresponds to the integral of u over the straight line through x with direction θ. Up to notation, in the two dimensional case the X-ray transform coincides with the more popular Radon transform, which maps a function on $\mathbb{R}^d$ into the set of its hyperplane integrals [99, Sect. 2.1]. If $\theta \in S^{d-1}$ and $s \in \mathbb{R}$, then the Radon transform can be defined by

$$(\bar{K}u)(\theta, s) = \int_{x \cdot \theta = s} u(x)\, dx = \int_{\theta^\perp} u(s\theta + y)\, dy,$$

and corresponds in the two dimensional case to the integral of u over the straight line represented by a direction $\theta^\perp$ and distance s to the origin.

7.2 2D Reconstruction Results on Real Data

In Fig. 8 we illustrate the performance of the FB-EM-TV algorithm by evaluation of cardiac $H_2\,^{15}O$ PET measurements acquired with a Siemens Ecat Exact scanner [116]. The corresponding sinograms consist of 192 projections × 192 angles and the reconstructed images are given by a grid of 175 × 175 pixels. The EM reconstruction steps were performed using the EMRECON framework [81] and the modified version of the projected gradient descent algorithm of Chambolle described in Sect. 5 was used to solve the weighted ROF problem (100).

The mentioned $H_2\,^{15}O$ tracer will be used in nuclear medicine for the quantification of myocardial blood flow [110]. However, this quantification needs a segmentation of myocardial tissue, left and right ventricle [18, 110], which is extremely difficult to realize due to very low SNR of $H_2\,^{15}O$ data. Hence, to obtain the tracer intensity in the right and left ventricle, we take a fixed 2D layer in two different time frames. The tracer intensity in the right ventricle is illustrated in Fig. 8a, whereby the tracer intensity in the left ventricle is presented in Fig. 8b, using measurements 25 and 45 s after tracer injection in the blood circulation, respectively. To illustrate the SNR problem, we present reconstructions with the classical EM algorithm (19) in Fig. 8 (left). As expected, the results suffer from unsatisfactory quality and are impossible to interpret. Hence, we take EM reconstructions with Gaussian smoothing (Fig. 8 (middle)) as references. The results in Fig. 8 (right) show the reconstructions with the proposed FB-EM-TV algorithm (28). We can see that the results with the FB-EM-TV algorithms are well suited for further use, such as segmentation for quantification of myocardial blood flow, despite the very low SNR of $H_2\,^{15}O$ data [18].

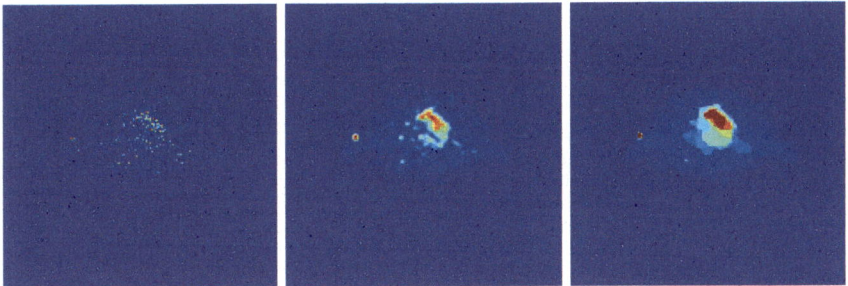

(a) Right ventricle: EM, Gaussian smoothed EM and FB-EM-TV results (from left to right)

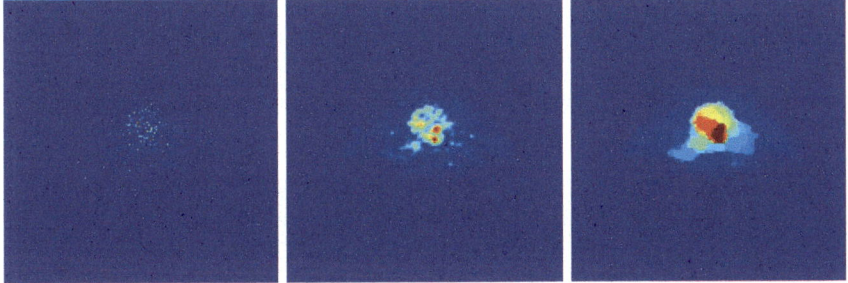

(b) Left ventricle: EM, Gaussian smoothed EM and FB-EM-TV results (from left to right)

Fig. 8 Cardiac $H_2\,^{15}O$ PET measurements: tracer intensity results of different reconstruction methods in two different time frames. (**a**) Tracer intensity in the right ventricle using measurements 25 s after tracer injection in the blood circulation. (**b**) Tracer intensity in the left ventricle using measurements 45 s after tracer injection in the blood circulation

7.3 3D Reconstruction Results on Real Data

In this section we present 3D reconstruction results using cardiac ^{18}F-FDG PET measurements acquired with a Siemens Biograph Sensation 16 scanner [92]. The 47 sinograms consist of 192 projections × 192 angles and the reconstructed images are given by a grid of 175 × 175 × 47 voxels. The corresponding 3D EM reconstruction steps were performed using the EMRECON framework [81] and the augmented Lagrangian method from Sect. 5 was used for occurring TV correction steps. We recall that an inversion of a Laplace operator equation is required in the augmented Lagrangian approach, which can be solved efficiently using the discrete cosine transform (105). To realize this step, we use a MATLAB implementation of the multidimensional (inverse) discrete cosine transform by Myronenko [98].

The ^{18}F-FDG tracer is an important radiopharmaceutical in nuclear medicine and is used for measuring glucose metabolism, e.g. in brain, heart or tumors. In the following, in order to illustrate the 3D data set, we take a fixed transversal, coronal and sagittal slice of reconstructions. In Fig. 9 (left) we display a Gaussian smoothed

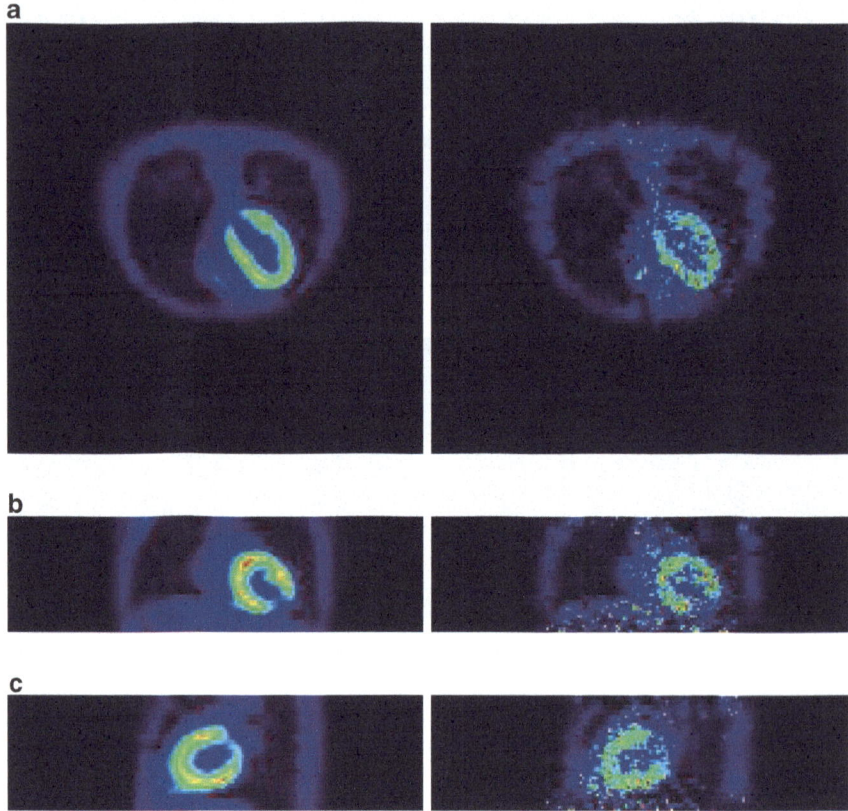

Fig. 9 Cardiac ^{18}F-FDG 3D PET measurements: tracer intensity results obtained with the EM algorithm (19) for different count rates. *Left:* EM reconstruction, 20 iterations, with Gaussian smoothing any 10th step after 20 min data acquisition. *Right:* As left but after 5 s data acquisition. Additionally, the reconstruction is scaled to the maximum intensity of the result on the left-hand side due to the strong presence of noise outside of the region of interest. (**a**) Transversal view: 20 min (*left*) and 5 s (*right*) data acquisition time. (**b**) Coronal view: 20 min (*left*) and 5 s (*right*) data acquisition time. (**c**) Sagittal view: 20 min (*left*) and 5 s (*right*) data acquisition time

EM reconstruction after a data acquisition of 20 min as a ground truth for very high count rates. To simulate low count rates, we take the measurements after the first 5 s only. The corresponding Gaussian smoothed EM reconstruction is illustrated in Fig. 9 (right).

In Fig. 10 we show reconstruction results obtained from the FB-EM-TV algorithm (left) and its extension via Bregman distance regularization (right) using measurements after 5 s acquisition time of the data. Thereby, we can observe that the major structures of the object are well reconstructed by both approaches also for low count rates. However, as expected, the structures in the Bregman-FB-EM-TV result

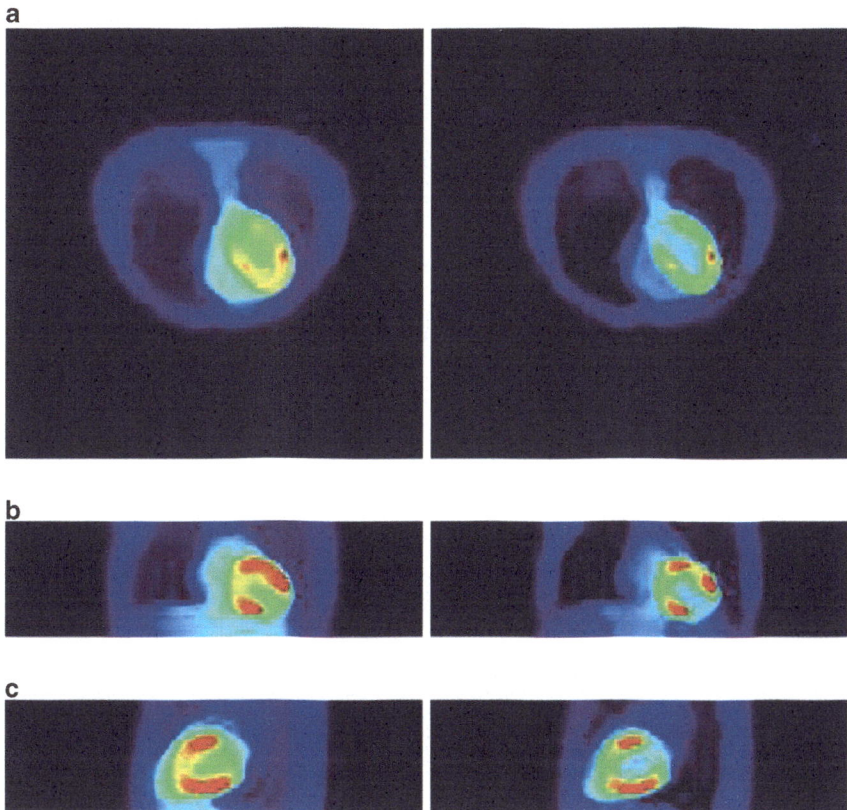

Fig. 10 Cardiac ^{18}F-FDG 3D PET measurements: tracer intensity results obtained with the (Bregman-)FB-EM-TV algorithm for measurements after 5 s data acquisition. *Left:* Reconstruction with the FB-EM-TV algorithm (28), 20 iterations. *Right:* Reconstruction with the Bregman-FB-EM-TV algorithm proposed in Sect. 3.5 at sixth refinement step. (**a**) Transversal view: FB-EM-TV reconstruction (*left*) and Bregman-FB-EM-TV result (*right*). (**b**) Coronal view: FB-EM-TV reconstruction (*left*) and Bregman-FB-EM-TV result (*right*). (**c**) Sagittal view: FB-EM-TV reconstruction (*left*) and Bregman-FB-EM-TV result (*right*)

can be identified better than in the standard FB-EM-TV reconstruction. In particular, this aspect can be observed well in Fig. 11, where we present scaled versions of both reconstructions in order to allow a quantitative comparison. In Fig. 11, the reconstructions from Fig. 10 are scaled to the maximum intensity of the EM result in Fig. 9 (left) obtained with measurements after 20 min data acquisition. There, we can observe that the result with the Bregman-FB-EM-TV algorithm has more realistic quantitative values than the reconstruction with the standard FB-EM-TV algorithm.

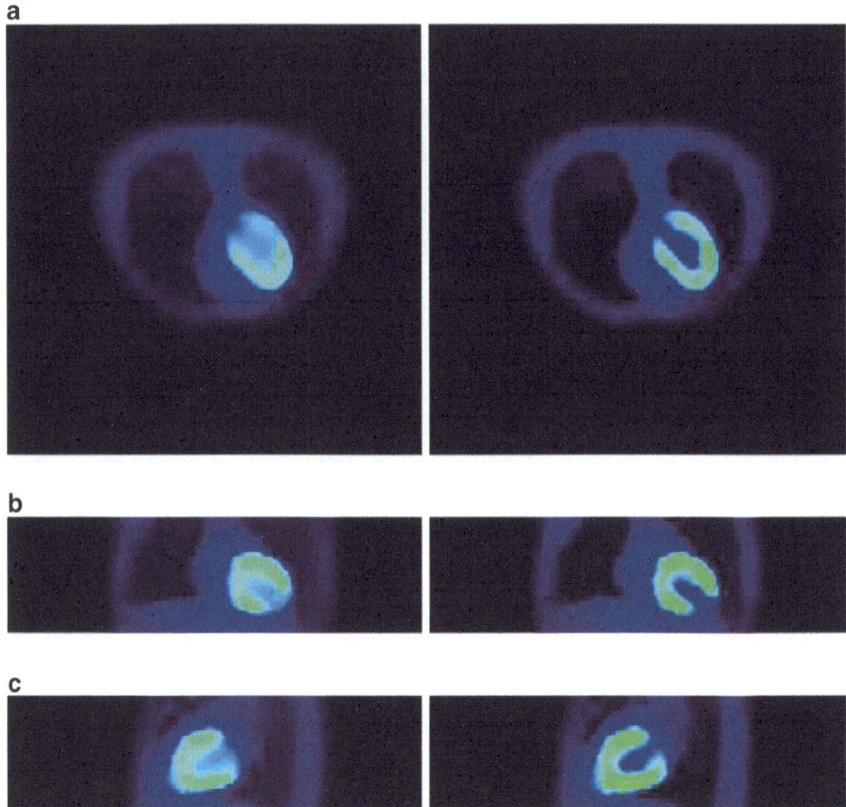

Fig. 11 Cardiac ^{18}F-FDG 3D PET measurements: quantitative comparison between Bregman- and FB-EM-TV reconstructions for measurements after 5 s data acquisition. *Left and right:* Results from Fig. 10 are scaled to the maximum intensity of ground truth in Fig. 9 (left). (**a**) Transversal view: FB-EM-TV reconstruction (*left*) and Bregman-FB-EM-TV result (*right*). (**b**) Coronal view: FB-EM-TV reconstruction (*left*) and Bregman-FB-EM-TV result (*right*). (**c**) Sagittal view: FB-EM-TV reconstruction (*left*) and Bregman-FB-EM-TV result (*right*)

Acknowledgements This work has been supported by the German Ministry of Education and Research (BMBF) through the project *INVERS: Deconvolution problems with sparsity constraints in nanoscopy and mass spectrometry*. This research was performed when the second author was with the Mathematical Imaging Group at University of Münster. C. Brune acknowledges further support by the Deutsche Telekom Foundation. The work has been further supported by the German Science Foundation DFG through the SFB 656 *Molecular Cardiovascular Imaging* and the project *Regularization with singular energies*. The authors thank Florian Büther and Klaus Schäfers (both European Institute for Molecular Imaging, University of Münster) for providing real data in PET and useful discussions. The authors thank Katrin Willig and Andreas Schönle (both Max Planck Institute for Biophysical Chemistry, Göttingen) for providing real data in optical nanoscopy.

References

1. R. Acar, C.R. Vogel, Analysis of bounded variation penalty methods for ill-posed problems. Inverse Probl. **10**(6), 1217–1229 (1994)
2. H.M. Adorf, R.N. Hook, L.B. Lucy, F.D. Murtagh, Accelerating the Richardson-Lucy restoration algorithm, in *Proceedings of the 4th ESO/ST-ECF Data Analysis Workshop*, ed. by P.J. Grosboel. (European Southern Observatory, Garching, 1992), pp. 99–103
3. L. Ambrosio, N. Fusco, D. Pallara, *Functions of Bounded Variation and Free Discontinuity Problems*. Oxford Mathematical Monographs (Oxford University Press, Oxford, 2000)
4. S. Anthoine, J.F. Aujol, Y. Boursier, C. Mélot, On the efficiency of proximal methods in CBCT and PET, in *2011 18th IEEE International Conference on Image Processing (ICIP)* (2011). doi: 10.1109/ICIP.2011.6115691
5. S. Anthoine, J.F. Aujol, Y. Boursier, C. Mélot, Some proximal methods for CBCT and PET, in *Proceedings of SPIE (Wavelets and Sparsity XIV)*, vol. 8138 (2011)
6. G. Aubert, P. Kornprobst, *Mathematical Problems in Image Processing: Partial Differential Equations and the Calculus of Variations*. Applied Mathematical Sciences, vol. 147 (Springer, New York, 2002)
7. J.F. Aujol, Some first-order algorithms for total variation based image restoration. J. Math. Imag. Vis. **34**(3), 307–327 (2009)
8. M. Bachmayr, M. Burger, Iterative total variation schemes for nonlinear inverse problems. Inverse Probl. **25**(10), 105004 (2009)
9. D. Baddeley, C. Carl, C. Cremer, 4Pi microscopy deconvolution with a variable point-spread function. Appl. Opt. **45**(27), 7056–7064 (2006)
10. D.L. Bailey, D.W. Townsend, P.E. Valk, M.N. Maisey (eds.), *Positron Emission Tomography: Basic Sciences* (Springer, New York, 2005)
11. J.M. Bardsley, An efficient computational method for total variation-penalized Poisson likelihood estimation. Inverse Probl. Imag. **2**(2), 167–185 (2008)
12. J.M. Bardsley, A theoretical framework for the regularization of Poisson likelihood estimation problems. Inverse Probl. Imag. **4**(1), 11–17 (2010)
13. J.M. Bardsley, J. Goldes, Regularization parameter selection methods for ill-posed Poisson maximum likelihood estimation. Inverse Probl. **25**(9), 095005 (2009)
14. J.M. Bardsley, J. Goldes, Regularization parameter selection and an efficient algorithm for total variation-regularized positron emission tomography. Numer. Algorithms **57**(2), 255–271 (2011)
15. J.M. Bardsley, N. Laobeul, Tikhonov regularized Poisson likelihood estimation: theoretical justification and a computational method. Inverse Probl. Sci. Eng. **16**(2), 199–215 (2008)
16. J.M. Bardsley, N. Laobeul, An analysis of regularization by diffusion for ill-posed Poisson likelihood estimations. Inverse Probl. Sci. Eng. **17**(4), 537–550 (2009)
17. J.M. Bardsley, A. Luttman, Total variation-penalized Poisson likelihood estimation for ill-posed problems. Adv. Comput. Math. **31**, 35–59 (2009)
18. M. Benning, T. Kösters, F. Wübbeling, K. Schäfers, M. Burger, A nonlinear variational method for improved quantification of myocardial blood flow using dynamic $H_2^{15}O$ PET, in *IEEE Nuclear Science Symposium Conference Record, 2008, NSS '08* (2008). doi: 10.1109/NSSMIC.2008.4774274
19. B. Berkels, M. Burger, M. Droske, O. Nemitz, M. Rumpf, Cartoon extraction based on anisotropic image classification, in *Vision, Modeling, and Visualization 2006: Proceedings*, ed. by L. Kobbelt, T. Kuhlen, T. Aach, R. Westerman (IOS Press, Aachen, 2006)
20. M. Bertero, H. Lanteri, L. Zanni, Iterative image reconstruction: a point of view, in *Mathematical Methods in Biomedical Imaging and Intensity-Modulated Radiation Therapy (IMRT)*, ed. by Y. Censor, M. Jiang, A. Louis. Publications of the Scuola Normale, CRM series, vol. 7 (2008), pp. 37–63
21. M. Bertero, P. Boccacci, G. Desiderà, G. Vicidomini, Image deblurring with Poisson data: from cells to galaxies. Inverse Probl. **25**(12), 123006 (2009)

22. M. Bertero, P. Boccacci, G. Talenti, R. Zanella, L. Zanni, A discrepancy principle for Poisson data. Inverse Probl. **26**(10), 105004 (2010)
23. S. Bonettini, V. Ruggiero, An alternating extragradient method for total variation-based image restoration from Poisson data. Inverse Probl. **27**(9), 095001 (2011)
24. S. Bonettini, R. Zanella, L. Zanni, A scaled gradient projection method for constrained image deblurring. Inverse Probl. **25**(1), 015002 (2009)
25. S. Boyd, N. Parikh, E. Chu, B. Peleato, J. Eckstein, Distributed optimization and statistical learning via the alternating direction method of multipliers. Found. Trends Mach. Learn. **3**(1), 1–122 (2010)
26. K. Bredies, A forward-backward splitting algorithm for the minimization of non-smooth convex functionals in Banach space. Inverse Probl. **25**(1), 015005 (2009)
27. L.M. Bregman, The relaxation method of finding the common point of convex sets and its application to the solution of problems in convex programming. USSR Comput. Math. Math. Phys. **7**, 200–217 (1967)
28. C. Brune, A. Sawatzky, M. Burger, Bregman-EM-TV methods with application to optical nanoscopy, in *Proceedings of the 2nd International Conference on Scale Space and Variational Methods in Computer Vision*. Lecture Notes in Computer Science, vol. 5567 (Springer, New York, 2009), pp. 235–246
29. C. Brune, A. Sawatzky, M. Burger, Primal and dual Bregman methods with application to optical nanoscopy. Int. J. Comput. Vis. **92**(2), 211–229 (2011)
30. M. Burger, S. Osher, Convergence rates of convex variational regularization. Inverse Probl. **20**, 1411–1421 (2004)
31. M. Burger, G. Gilboa, S. Osher, J. Xu, Nonlinear inverse scale space methods. Comm. Math. Sci. **4**(1), 179–212 (2006)
32. M. Burger, K. Frick, S. Osher, O. Scherzer, Inverse total variation flow. Multiscale Model. Simul. **6**(2), 366–395 (2007)
33. A. Chambolle, An algorithm for total variation minimization and applications. J. Math. Imag. Vis. **20**, 89–97 (2004)
34. A. Chambolle, Total variation minimization and a class of binary MRF models, in *Energy Minimization Methods in Computer Vision and Pattern Recognition*. Lecture Notes in Computer Science, vol. 3757 (Springer, New York, 2005), pp. 136–152
35. A. Chambolle, T. Pock, A first-order primal-dual algorithm for convex problems with applications to imaging. J. Math. Imag. Vis. **40**(1), 120–145 (2011)
36. A. Chambolle, V. Caselles, D. Cremers, M. Novaga, T. Pock, An introduction to total variation for image analysis, in *Theoretical Foundations and Numerical Methods for Sparse Recovery*. Radon Series on Computational and Applied Mathematics, vol. 9 (De Gruyter, Berlin, 2010), pp. 263–340
37. C. Chaux, J.C. Pesquet, N. Pustelnik, Nested iterative algorithms for convex constrained image recovery problems. SIAM J. Imag. Sci. **2**(2), 730–762 (2009)
38. D.Q. Chen, L.Z. Cheng, Deconvolving Poissonian images by a novel hybrid variational model. J. Vis. Comm. Image Represent. **22**(7), 643–652 (2011)
39. P. Combettes, V. Wajs, Signal recovery by proximal forward-backward splitting. Multiscale Model. Simul. **4**, 1168–1200 (2005)
40. P.L. Combettes, J.C. Pesquet, A proximal decomposition method for solving convex variational inverse problems. Inverse Probl. **24**(6), 065014 (2008)
41. I. Csiszar, Why least squares and maximum entropy? An axiomatic approach to inference for linear inverse problems. Ann. Stat. **19**(4), 2032–2066 (1991)
42. A.P. Dempster, N.M. Laird, D.B. Rubin, Maximum likelihood from incomplete data via the EM algorithm. J. R. Stat. Soc. Ser. B **39**(1), 1–38 (1977)
43. N. Dey, L. Blanc-Féraud, C. Zimmer, P. Roux, Z. Kam, J.C. Olivio-Marin, J. Zerubia, 3D microscopy deconvolution using Richardson-Lucy algorithm with total variation regularization. Technical Report 5272, Institut National de Recherche en Informatique et en Automatique (2004)

44. J. Douglas, H.H. Rachford, On the numerical solution of heat conduction problems in two and three space variables. Trans. Am. Math. Soc. **82**(2), 421–439 (1956)
45. J. Eckstein, D.P. Bertsekas, On the Douglas-Rachford splitting method and the proximal point algorithm for maximal monotone operators. Math. Program. **55**(1–3), 293–318 (1992)
46. P.P.B. Eggermont, Maximum entropy regularization for Fredholm integral equations of the first kind. SIAM J. Math. Anal. **24**(6), 1557–1576 (1993)
47. I. Ekeland, R. Temam, *Convex Analysis and Variational Problems*. Studies in Mathematics and Its Applications, vol. 1 (North-Holland, Amsterdam, 1976)
48. H.C. Elman, G.H. Golub, Inexact and preconditioned Uzawa algorithms for saddle point problems. SIAM J. Numer. Anal. **31**(6), 1645–1661 (1994)
49. H.W. Engl, M. Hanke, A. Neubauer, *Regularization of Inverse Problems*. Mathematics and Its Applications (Kluwer, Dordrecht, 2000)
50. S. Esedoglu, S.J. Osher, Decomposition of images by the anisotropic Rudin-Osher-Fatemi model. Comm. Pure Appl. Math. **57**(12), 1609–1626 (2004)
51. J.E. Esser, Primal dual algorithms for convex models and applications to image restoration, registration and nonlocal inpainting. Ph.D. thesis, University of California, Los Angeles, 2010
52. L.C. Evans, R.F. Gariepy, *Measure Theory and Fine Properties of Functions*. Studies in Advanced Mathematics (CRC Press, West Palm Beach, 1992)
53. M.A.T. Figueiredo, J. Bioucas-Dias, Deconvolution of Poissonian images using variable splitting and augmented Lagrangian optimization, in *IEEE Workshop on Statistical Signal Processing*, Cardiff (2009)
54. M.A.T. Figueiredo, J.M. Bioucas-Dias, Restoration of Poissonian images using alternating direction optimization. IEEE Trans. Image Process. **19**(12), 3133–3145 (2010)
55. M. Fortin, R. Glowinski, *Augmented Lagrangian Methods: Applications to the Numerical Solution of Boundary-Value Problems*. Studies in Mathematics and its Applications, vol. 15 (Elsevier, Amsterdam, 1983)
56. D. Gabay, Applications of the method of multipliers to variational inequalities, in *Augmented Lagrangian Methods: Applications to the Numerical Solution of Boundary-Value Problems*. Studies in Mathematics and its Applications, vol. 15 (Elsevier, Amsterdam, 1983),pp. 299–331
57. D. Gabay, B. Mercier, A dual algorithm for the solution of nonlinear variational problems via finite element approximation. Comput. Math. Appl. **2**(1), 17–40 (1976)
58. S. Geman, D. Geman, Stochastic relaxation, Gibbs distributions and the Bayesian restoration of images. J. Appl. Stat. **20**(5), 25–62 (1993)
59. S. Geman, D.E. McClure, Bayesian image analysis: an application to single photon emission tomography, in *Proceedings of Statistical Computation Section* (American Statistical Association, Alexandria, 1985), pp. 12–18
60. E. Giusti, *Minimal Surfaces and Functions of Bounded Variation*. Monographs in Mathematics, vol. 80 (Birkhäuser, Basel, 1984)
61. R. Glowinski, P. Le Tallec, *Augmented Lagrangian and Operator-Splitting Methods in Nonlinear Mechanics*. Studies in Applied Mathematics, vol. 9 (SIAM, Philadelphia, 1989)
62. T. Goldstein, S. Osher, The split Bregman method for L^1-regularized problems. SIAM J. Imag. Sci. **2**(2), 323–343 (2009)
63. T. Goldstein, B. O'Donoghue, S. Setzer, Fast alternating direction optimization methods. CAM Report 12–35, UCLA, 2012
64. C.W. Groetsch, *Inverse Problems in the Mathematical Sciences* (Vieweg, Braunschweig, 1993)
65. P.C. Hansen, J.G. Nagy, D.P. O'Leary, *Deblurring Images: Matrices, Spectra, and Filtering*. Fundamentals of Algorithms (SIAM, Philadelphia, 2006)
66. B.S. He, H. Yang, S.L. Wang, Alternating direction method with self-adaptive penalty parameters for monotone variational inequalities. J. Optim. Theor. Appl. **106**(2), 337–356 (2000)
67. S.W. Hell, Toward fluorescence nanoscopy. Nat. Biotechnol. **21**(11), 1347–1355 (2003)
68. S. Hell, A. Schönle, Nanoscale resolution in far-field fluorescence microscopy, in *Science of Microscopy*, ed. by P.W. Hawkes, J.C.H. Spence (Springer, New York, 2006)

69. S. Hell, E.H.K. Stelzer, Fundamental improvement of resolution with a 4Pi-confocal fluorescence microscope using two-photon excitation. Opt. Comm. **93**(5–6), 277–282 (1992)
70. S. Hell, E.H.K. Stelzer, Properties of a 4Pi confocal fluorescence microscope. J. Opt. Soc. Am. A **9**(12), 2159–2166 (1992)
71. S. Hell, J. Wichmann, Breaking the diffraction resolution limit by stimulated emission: stimulated-emission-depletion fluorescence microscopy. Opt. Lett. **19**(11), 780–782 (1994)
72. F.M. Henderson, A.J. Lewis, *Principles and Applications of Imaging Radar: Manual of Remote Sensing*, vol. 2 (Wiley, London, 1998)
73. A.O. Hero, J.A. Fessler, Convergence in norm for alternating expectation-maximization (EM) type algorithms. Stat. Sin. **5**, 41–54 (1995)
74. J.B. Hiriart-Urruty, C. Lemaréchal, *Convex Analysis and Minimization Algorithms I*. Grundlehren der mathematischen Wissenschaften (Fundamental Principles of Mathematical Sciences), vol. 305 (Springer, New York, 1993)
75. T.J. Holmes, Y.H. Liu, Acceleration of maximum-likelihood image restoration for fluorescence microscopy and other noncoherent imagery. J. Opt. Soc. Am. A **8**(6), 893–907 (1991)
76. K. Ito, K. Kunisch, *Lagrange Multiplier Approach to Variational Problems and Applications*. Advances in Design and Control, vol. 15 (SIAM, Philadelphia, 2008)
77. A.N. Iusem, Convergence analysis for a multiplicatively relaxed EM algorithm. Math. Meth. Appl. Sci. **14**(8), 573–593 (1991)
78. E. Jonsson, S.C. Huang, T. Chan, Total variation regularization in positron emission tomography. CAM Report 98–48, UCLA, 1998
79. C.T. Kelley, *Iterative Methods for Optimization*. Frontiers in Applied Mathematics (SIAM, Philadelphia, 1999)
80. T.A. Klar, S. Jakobs, M. Dyba, A. Egner, S.W. Hell, Fluorescence microscopy with diffraction resolution barrier broken by stimulated emission. Proc. Natl. Acad. Sci. USA **97**(15),8206–8210 (2000)
81. T. Kösters, K. Schäfers, F. Wübbeling, EMRECON: An expectation maximization based image reconstruction framework for emission tomography data, in *2011 IEEE Nuclear Science Symposium and Medical Imaging Conference (NSS/MIC)* (2011), pp. 4365–4368. doi: 10.1109/NSSMIC.2011.6153840
82. K. Lange, R. Carson, EM reconstruction algorithms for emission and transmission tomography. J. Comput. Assist. Tomogr. **8**(2), 306–316 (1984)
83. H. Lantéri, C. Theys, Restoration of astrophysical images - the case of Poisson data with additive Gaussian noise. EURASIP J. Appl. Signal Process. **15**, 2500–2513 (2005)
84. H. Lantéri, M. Roche, O. Cuevas, C. Aime, A general method to devise maximum-likelihood signal restoration multiplicative algorithms with non-negativity constraints. Signal Process. **81**(5), 945–974 (2001)
85. H. Lantéri, M. Roche, C. Aime, Penalized maximum likelihood image restoration with positivity constraints: multiplicative algorithms. Inverse Probl. **18**(5), 1397 (2002)
86. T. Le, R. Chartrand, T.J. Asaki, A variational approach to reconstructing images corrupted by Poisson noise. J. Math. Imag. Vis. **27**(3), 257–263 (2007)
87. H. Liao, F. Li, M.K. Ng, Selection of regularization parameter in total variation image restoration. J. Opt. Soc. Am. A **26**(11), 2311–2320 (2009)
88. P.L. Lions, B. Mercier, Splitting algorithms for the sum of two nonlinear operators. SIAM J. Numer. Anal. **16**(6), 964–979 (1979)
89. X. Liu, C. Comtat, C. Michel, P. Kinahan, M. Defrise, D. Townsend, Comparison of 3-D reconstruction with 3D-OSEM and with FORE + OSEM for PET. IEEE Trans. Med. Imag. **20**(8), 804–814 (2001)
90. L.B. Lucy, An iterative technique for the rectification of observed distributions. Astron. J. **79**, 745–754 (1974)
91. A. Luttman, A theoretical analysis of L^1 regularized Poisson likelihood estimation. Inverse Prob. Sci. Eng. **18**(2), 251–264 (2010)
92. M.J. Martínez, Y. Bercier, M. Schwaiger, S.I. Ziegler, PET/CT Biograph TM Sensation 16 - Performance improvement using faster electronics. Nuklearmedizin **45**(3), 126–133 (2006)

93. R.E. Megginson, *An Introduction to Banach Space Theory*. Graduate Texts in Mathematics, vol. 183 (Springer, New York, 1998)
94. Y. Meyer, *Oscillating Patterns in Image Processing and Nonlinear Evolution Equations: The Fifteenth Dean Jacqueline B. Lewis Memorial Lectures*, University Lecture Series, vol. 22 (American Mathematical Society, Boston, 2001)
95. H.N. Mülthei, Iterative continuous maximum-likelihood reconstruction method. Math. Meth. Appl. Sci. **15**(4), 275–286 (1992)
96. H.N. Mülthei, B. Schorr, On an iterative method for a class of integral equations of the first kind. Math. Meth. Appl. Sci. **9**(1), 137–168 (1987)
97. H.N. Mülthei, B. Schorr, On properties of the iterative maximum likelihood reconstruction method. Math. Meth. Appl. Sci. **11**(3), 331–342 (1989)
98. A. Myronenko, Free DCTN and IDCTN Matlab code (2011). https://sites.google.com/site/myronenko/software
99. F. Natterer, F. Wübbeling, *Mathematical Methods in Image Reconstruction* (SIAM, Philadelphia, 2001)
100. S. Osher, M. Burger, D. Goldfarb, J. Xu, W. Yin, An iterative regularization method for total variation-based image restoration. Multiscale Model. Simul. **4**(2), 460–489 (2005)
101. V.Y. Panin, G.L. Zeng, G.T. Gullberg, Total variation regulated EM algorithm [SPECT reconstruction]. IEEE Trans. Nucl. Sci. **46**(6), 2202–2210 (1999)
102. G.B. Passty, Ergodic convergence to a zero of the sum of monotone operators in Hilbert spaces. J. Math. Anal. Appl. **72**, 383–390 (1979)
103. D. Potts, G. Steidl, Optimal trigonometric preconditioners for nonsymmetric Toeplitz systems. Linear Algebra Appl. **281**(1–3), 265–292 (1998)
104. E. Resmerita, R.S. Anderssen, Joint additive Kullback-Leibler residual minimization and regularization for linear inverse problems. Math. Meth. Appl. Sci. **30**, 1527–1544 (2007)
105. E. Resmerita, H.W. Engl, A.N. Iusem, The expectation-maximization algorithm for ill-posed integral equations: a convergence analysis. Inverse Probl. **23**(6), 2575–2588 (2007)
106. W.H. Richardson, Bayesian-based iterative method of image restoration. J. Opt. Soc. Am. **62**(1), 55–59 (1972)
107. L.I. Rudin, S. Osher, E. Fatemi, Nonlinear total variation based noise removal algorithms. Phys. D **60**, 259–268 (1992)
108. A. Sawatzky, (Nonlocal) total variation in medical imaging. Ph.D. thesis, University of Münster, 2011. CAM Report 11–47, UCLA
109. A. Sawatzky, C., Brune, F. Wübbeling, T. Kösters, K. Schäfers, M. Burger, Accurate EM-TV algorithm in PET with low SNR, in *IEEE Nuclear Science Symposium Conference Record, 2008, NSS '08*. doi: 10.1109/NSSMIC.2008.4774392
110. K.P. Schäfers, T.J. Spinks, P.G. Camici, P.M. Bloomfield, C.G. Rhodes, M.P. Law, C.S.R. Baker, O. Rimoldi, Absolute quantification of myocardial blood flow with $H_2\,^{15}O$ and 3-dimensional PET: an experimental validation. J. Nucl. Med. **43**(8), 1031–1040 (2002)
111. S. Setzer, Split Bregman algorithm, Douglas-Rachford splitting and frame shrinkage, in *Proceedings of the 2nd International Conference on Scale Space and Variational Methods in Computer Vision*. Lecture Notes in Computer Science, vol. 5567 (Springer, New York, 2009), pp. 464–476
112. S. Setzer, Splitting methods in image processing. Ph.D. thesis, University of Mannheim, 2009. http://ub-madoc.bib.uni-mannheim.de/2924/
113. S. Setzer, G. Steidl, T. Teuber, Deblurring Poissonian images by split Bregman techniques. J. Vis. Comm. Image Represent. **21**(3), 193–199 (2010)
114. L.A. Shepp, Y. Vardi, Maximum likelihood reconstruction for emission tomography. IEEE Trans. Med. Imag. **1**(2), 113–122 (1982)
115. J.J. Sieber, K.I. Willig, C. Kutzner, C. Gerding-Reimers, B. Harke, G. Donnert, B. Rammner, C. Eggeling, S.W. Hell, H. Grubmüller, T. Lang, Anatomy and dynamics of a supramolecular membrane protein cluster. Science **317**, 1072–1076 (2007)

116. T.J. Spinks, T. Jones, P.M. Bloomfield, D.L. Bailey, D. Hogg, W.F. Jones, K. Vaigneur, J. Reed, J. Young, D. Newport, C. Moyers, M.E. Casey, R. Nutt, Physical characteristics of the ECAT EXACT3D positron tomograph. Phys. Med. Biol. **45**(9), 2601–2618 (2000)
117. G. Steidl, T. Teuber, Anisotropic smoothing using double orientations, in *Proceedings of the 2nd International Conference on Scale Space and Variational Methods in Computer Vision*. Lecture Notes in Computer Science, vol. 5567 (Springer, New York, 2009), pp. 477–489
118. D.M. Strong, J.F. Aujol, T.F. Chan, Scale recognition, regularization parameter selection, and Meyer's G norm in total variation regularization. Multiscale Model. Simul. **5**(1), 273–303 (2006)
119. P. Tseng, Applications of a splitting algorithm to decomposition in convex programming and variational inequalities. SIAM J. Contr. Optim. **29**(1), 119–138 (1991)
120. Y. Vardi, L.A. Shepp, L. Kaufman, A statistical model for positron emission tomography. J. Am. Stat. Assoc. **80**, 8–20 (1985)
121. L.A. Vese, S.J. Osher, Modeling textures with total variation minimization and oscillating patterns in image processing. J. Sci. Comput. **19**, 553–572 (2003)
122. C.R. Vogel, *Computational Methods for Inverse Problems*. Frontiers in Applied Mathematics (SIAM, Philadelphia, 2002)
123. C.R. Vogel, M.E. Oman, Fast, robust total variation-based reconstruction of noisy, blurred images. IEEE Trans. Image Process. **7**(6), 813–824 (1998)
124. Y. Wang, J. Yang, W. Yin, Y. Zhang, A new alternating minimization algorithm for total variation image reconstruction. SIAM J. Imag. Sci. **1**(3), 248–272 (2008)
125. M.N. Wernick, J.N. Aarsvold (eds.), *Emission Tomography: The Fundamentals of PET and SPECT* (Elsevier, Amsterdam, 2004)
126. K.I. Willig, B. Harke, R. Medda, S.W. Hell, STED microscopy with continuous wave beams. Nat. Meth. **4**(11), 915–918 (2007)
127. M. Yan, L.A. Vese, Expectation maximization and total variation based model for computed tomography reconstruction from undersampled data, in *Proceedings of SPIE 7961, Medical Imaging 2011: Physics of Medical Imaging, 79612X*, 16 March, 2011. doi:10.1117/12.878238 [From Conference Volume 7961 Medical Imaging 2011: Physics of Medical Imaging Norbert J. Pelc, Ehsan Samei, Robert M. Nishikawa, Lake Buena Vista, Florida, 12 February 2011]
128. R. Zanella, P. Boccacci, L. Zanni, M. Bertero, Efficient gradient projection methods for edge-preserving removal of Poisson noise. Inverse Probl. **25**(4), 045010 (2009)

Variational Methods in Image Matching and Motion Extraction

Martin Rumpf

Abstract In this chapter we are concerned with variational methods in image analysis. Special attention is paid on free discontinuity approaches of Mumford Shah type and their application in segmentation, matching and motion analysis. We study combined approaches, where one simultaneously relaxes a functional with respect to multiple unknowns. Examples are the simultaneous extraction of edges in two different images for joint image segmentation and image registration or the joint estimation of motion, moving object, and object intensity map. In these approaches the identification of one of the unknowns improves the capability to extract the other as well. Hence, combined methods turn out to be very powerful approaches. Indeed, fundamental tasks in image processing are highly interdependent: Registration of image morphology significantly benefits from previous denoising and structure segmentation. On the other hand, combined information of different image modalities makes shape segmentation significantly more robust. Furthermore, robustness in motion extraction of shapes can be significantly enhanced via a coupling with the detection of edge surfaces in space time and a corresponding feature sensitive space time smoothing.

Furthermore, one of the key tools throughout most of the methods to be presented is nonlinear elasticity based on hyperelastic and polyconvex energy functionals. Based on first principles from continuum mechanics this allows a flexible description of shape correspondences and in many cases enables to establish existence results and one-to-one mapping properties.

Numerical experiments underline the robustness of the presented methods and show applications on medical images and biological experimental data.

This chapter is based on a couple of recent articles (Bar et al., A variational framework for simultaneous motion estimation and restoration of motion-blurred video, 2007; Litke et al., An image processing approach to surface matching,

M. Rumpf (✉)
University of Bonn, Bonn, Germany
e-mail: martin.rumpf@uni-bonn.de

2005; Droske et al., Comput. Vis. Sci. Online First, 2008; Droske and Rumpf, SIAM Appl Math 64(2):668–687, 2004; Droske and Rumpf, IEEE Trans Pattern Anal Mach Intell 29(12):2181–2194, 2007; Rumpf and Wirth, SIAM J Imag Sci, 2008) published by the author together with Leah Bar, Benjamin Berkels, Marc Droske, Nathan Litke, Wolfgang Ring, Guillermo Sapiro, Peter Schröder, and Benedikt Wirth.

1 Some Prerequisites

In this section, we will introduce basic notion used throughout the chapter and consider the general methodology for finite element type discretization of variational free discontinuity problems.

1.1 Image Morphology

We are going to review different notions of image morphology and develop here a new one that is in particular appropriate for the use in variational approaches where images or shapes are linked to each other by nonlinear deformations.

In mathematical terms, two images $u, v : \Omega \to \mathbb{R}$ with $\Omega \subset \mathbb{R}^d$ for $d = 2, 3$ are called morphologically equivalent, if they only differ by a change of contrast, i.e., if $u(x) = (\beta \circ v)(x)$ for all $x \in \Omega$ and for some function $\beta : \mathbb{R} \to \mathbb{R}$ [1, 64]. Here, one usually restricts to contrast changes $\beta : \mathbb{R} \to \mathbb{R}$, which are strictly monotone and continuous functions. Obviously, such a contrast modulation does not change the order and the shape of level sets. Due to the enforced monotonicity, the same holds for the super level sets $l_c^+[u] = \{x : u(x) \geq c\}$. Thus, a usual description of the morphology $\mathcal{M}[u]$ of an image u is given by the upper topographic map, defined as the set of all these sets

$$\mathcal{M}[u] := \{l_c^+[u] : c \in \mathbb{R}\}.$$

Unfortunately, this set based definition is not feasible for the variational approach we intend to discuss. Furthermore, the restriction to monotone contrast changes conflicts with medical applications and medical morphology. In what follows, we derive an alternative notion based on a regular and a singular morphology. It can directly be used in variational approaches and allows us to get rid of the monotonicity assumption if appropriate. Let us suppose the image function $u : \Omega \to \mathbb{R}$ on an image domain $\Omega \subset \mathbb{R}^d$ to be in SBV [4]. Hence, we consider functions $u \in L^1(\Omega)$ of which the derivative $\mathcal{D}u$ is a vector-valued Radon measure with vanishing Cantor part. In fact, at edges we allow for jumps and thus infinitely steep gradients concentrated on a sufficiently regular, lower dimensional set, but not for jumps on sets of fractal dimensions. We consider the usual splitting

$\mathcal{D}u = \mathcal{D}^{ac}u + \mathcal{D}^s u$ [4], where $\mathcal{D}^{ac}u$ is the regular part, which is the usual image gradient apart from edges and absolutely continuous with respect to the Lebesgue measure $\mathcal{L}$ on $\Omega \subset \mathbb{R}^d$, and a singular part $\mathcal{D}^s u$, which represents the jump and is defined on the edge set $\mathcal{S}$, which consists of the edges of an image. We denote by n^s the vector valued measure representing the normal field on $\mathcal{S}$, such that the representation $\mathcal{D}^s u = (u^+ - u^-)n^s$ holds for the singular part of the derivative. Here u^+ and u^- are the upper and the lower (approximate lim sup and lim inf [4]) values at the edge $\mathcal{S}$, respectively. This normal field is defined $\mathcal{H}^{d-1}$ a.e. on $\mathcal{S}$. Obviously, n^s is a morphological invariant as long as we consider continuous strictly monotone contrast modulating functions β.

Now, we focus on the regular part of the derivative. First, we adopt the classical gradient notion $\nabla^{ac} u$ for the $\mathcal{L}$ density of $\mathcal{D}^{ac}u$, i.e., $\mathcal{D}^{ac}u = \nabla^{ac} u \, \mathcal{L}$ [4]. As long as it is defined, the normalized gradient $\frac{\nabla^{ac} u(x)}{|\nabla^{ac} u(x)|}$ is the outer normal on the upper topographic set $l^+_{u(x)}[u]$ and thus again a morphological quantity. It is undefined on the flat image region $F[u] := \{x \in \Omega : \nabla^{ac} u(x) = 0\}$. We introduce n^{ac} as the normalized regular part of the gradient

$$n^{ac} = \chi_{\Omega \setminus F[u]} \frac{\nabla^{ac} u}{|\nabla^{ac} u|}$$

with support $\Omega \setminus F$ and denote it the Gauss map of the image u. With the regular normal n^{ac} and the singular normal measure n^s at hand, we are now able to redefine the morphology $\mathcal{M}[u]$ of an image u as a vector valued Radon measure on Ω with

$$\mathcal{M}[u] = n^{ac} \mathcal{L} + n^s, \qquad (1)$$

which is up to the flat set $F[u]$ of unit length. We call $n^{ac}\mathcal{L}$ the regular morphology and n^s the singular morphology. It turns out that this new notion is equivalent to the above definition on sufficiently regular image functions. It completely describes the topographical shape information of the image u. If we skip the orientation of the vectors n^{ac} and n^s in (1) and replace them by the corresponding line subspaces we are able to treat general, not only monotone, contrast changes. This is actually reflected by our algorithm.

1.2 The Mumford–Shah Model in Image Analysis

At first, let us consider a fundamental image segmentation model, which will later pick up and combine with other techniques and integrate in joint variational approaches. In their pioneering paper, Mumford and Shah [56] proposed the minimization of the following energy functional:

$$\mathcal{E}_{\text{MS}}[u, \mathcal{S}_u] = \int_\Omega (u - u^0)^2 \, d\mathcal{L} + \mu \int_{\Omega \setminus \mathcal{S}_u} |\nabla u|^2 \, d\mathcal{L} + \eta \mathcal{H}^{d-1}(\mathcal{S}_u), \qquad (2)$$

where u^0 is the initial image defined on an image domain $\Omega \subset \mathbb{R}^d$ and μ, η are positive weights. Here, one asks for a piecewise smooth representation u of u^0 and a singularity set $\mathcal{S}_u$ consisting of the image edges, such that u approximates u^0 in a least-squares sense. The intensity function u ought to be smooth apart from the free discontinuity $\mathcal{S}_u$ and in addition $\mathcal{S}_u$ should be small with respect to the $d-1$-dimensional Hausdorff-measure [4]. Here, we will make use of this approach to regularize images in a suitable way prior to the matching and simultaneously to split image morphology into a regular part related to contour sets of the piecewise smooth portions of an image and a singular part consisting of the edge set. Even though the Mumford-Shah approach itself is not morphological (e.g. intensity scaling may lead to an identification of previously overlooked edges) prominent edges are expected to be uniformly identified basically independent of the image contrast. Mathematically, this Mumford-Shah problem has been treated in the space of functions of bounded variations BV, more precisely in the specific subset SBV [4]. A related, alternative decomposition has been proposed by Rudin et al. [61]. They suggested to minimize $|u|_{BV} + \lambda |u - u^0|^2_{L^2}$ in BV.

The explicit treatment of the edge set $\mathcal{S}$ is neither theoretically nor with respect to a numerical approximation very handsome. Hence let us review briefly the approximation of the Mumford-Shah functional (2), proposed by Ambrosio and Tortorelli [3]. They describe the edge set $\mathcal{S}_u$ by a phase field function v, which is supposed to be small on $\mathcal{S}_u$ and close to 1 elsewhere. Phase field models are widespread in physics, where they represent a material order parameter and thus describe interfaces in solids or fluids. In fact, one asks for minimizers of the energy functional

$$\mathcal{E}^\epsilon_{AT}[u, v] = \int_\Omega (u - u^0)^2 + \mu(v^2 + k_\epsilon)|\nabla u|^2 + \eta \epsilon |\nabla v|^2 + \frac{\eta}{4\epsilon}(1 - v)^2 \, d\mathcal{L}, \quad (3)$$

where ϵ is a scaling parameter and $k_\epsilon = o(\epsilon)$ a small positive regularizing parameter. For larger ϵ one obtains coarse, blurred representations of the edge sets and corresponding smoother images u. For decreasing ϵ the representation of the edges is successively refined and more and more image details are included. We will make use of this inherent multi scale in cascadic minimization algorithms.

1.3 Finite Element Approximation in Imaging

To prepare the later discussion of discrete algorithms for a variety of different variational models, we will present in this section the discrete representation of images in finite element spaces and two different types of finite element discretizations for the classical Mumford and Shah segmentation model already reviewed above. Furthermore, we will introduce some useful multi scale treatment of image data.

We consider images as piecewise multilinear (bilinear in the case of 2D applications) finite element functions on a regular image domain. We denote the corresponding finite element space by $\mathcal{V}^h$, where h represents the underlying grid size. For some applications it is more convenient to further subdivide the cells into simplices via a splitting of rectangles into two triangles and hexahedrons into six tetrahedrons. This straightforward subdivision set does not modify the underlying degrees of freedom. In fact, it only influences the numerical quadrature for the finite element assembly. Each pixel or voxel value corresponds to a node of the regular mesh and a set of nodal values uniquely identifies a finite element function in $\mathcal{V}^h$. We apply these finite element spaces not only for the representation of discrete images but also for the discretization of all other unknowns appearing in the models, such as phase fields v or the components of vector valued quantities such as elastic deformations ϕ. Throughout this chapter continuous quantities are always denoted by lower case letters, whereas upper case letters are used for discrete functions. The corresponding nodal finite element vectors are marked with a bar on top, e.g. U is considered to be the discrete counterpart of an image function u and $\bar{U}$ denotes the corresponding nodal vector. Linear systems of equations, which arise from quadratic functionals are usually solved based on a preconditioned conjugate gradient method [41]. In the assembly of these linear systems we apply on each grid cell a higher order Gaussian quadrature rule if the local integrant is not polynomial.

For the ease of implementation we suppose dyadic resolutions of the images with $2^L + 1$ pixels or voxels in each direction. Thus, we are able to build a hierarchy on grids with $2^l + 1$ nodes in each direction for $l = L, \cdots, 0$. We restrict every finite element function via a straightforward nodal value evaluation to any of these coarse grid spaces. The construction of the multigrid hierarchy allows to solve coarse scale problems in our multi scale approach on coarse grids. Indeed, scale k is resolved on the corresponding $l(k)$-th grid level (e.g. with $l(k) = k$). First of all, we apply this throughout all variational models to be discussed here in cascadic type energy minimization approaches. For this purpose, we start minimizing the underlying nonlinear functionals on a coarse scale and then successively refine the grid, prolongate the solution onto the next finer grid and take this as initial data for the further relaxation of the corresponding discrete functional. Furthermore, the grid hierarchy is used in regularized gradient descent schemes. The regularization is based on a smoothing metric in the definition of the gradient. Effectively, we consider an H^1-type metric where the smoothing is performed via multigrid V cycles corresponding to a standard finite element implementation of the differential operator $\mathbb{1} - \frac{\sigma^2}{2}\Delta$. For details we refer to [21, 29].

At various places, we have to evaluate discrete functions U at pushed forward or pulled back positions under a discrete deformation Φ. In both cases, we replace the exact evaluation of these functions by a simple and effective interpolation. Indeed, we replace $U \circ \Phi$ by $\mathcal{I}(U \circ \Phi)$, where $\mathcal{I}$ is the classical Lagrangian interpolation on the grid nodes. Thus, each grid node is mapped under the deformation Φ onto the image domain, U is evaluated at these positions, and these values define our new finite element function. Analogously, $U \circ \Phi^{-1}$ is replaced by $\mathcal{I}(U \circ (\mathcal{I} \circ \Phi)^{-1})$. Here, we proceed as follows. We map each grid cell under the deformation onto the image domain. Next we identify all grid nodes, which are located on this deformed cell. These grid nodes are then mapped back under the inverse local deformation. Now, interpolation is applied to retrieve requested values of the finite element function U. Inversion of multilinear deformation would lead to nonlinear equations. To avoid this shortcoming, we cut each cell as mentioned above virtually into simplices. On these simplices, affine functions approximate in a straightforward way the multilinear functions. Thus, we replace the regular cells in the retrieval algorithm by the simplices and end up with piecewise affine inverse mappings.

In the next two paragraphs we will describe two significantly different approaches to discretize the Mumford Shah functional. The first is based on a direct implementation of the approach and uses a level set description of the edge set, whereas the other implements the Ambrosio Tortorelli approximation in a straightforward way.

Composite Finite Elements for Level Set Described Edge Contours

In what follows, we will investigate an energy descent method directly for the Mumford Shah functional. The Mumford Shah functional can be regarded as a shape optimization problem. For given edge contour $\mathcal{S}$, the image intensity $u = u[\mathcal{S}]$ solves a quadratic variational problem. Thus, we rewrite the Mumford Shah functional as a shape functional $\hat{\mathcal{E}}[\mathcal{S}] := \mathcal{E}_{\mathrm{MS}}[u[\mathcal{S}], \mathcal{S}]$ depending solely on the edge contour $\mathcal{S}$. Hence, the shape gradient of the functional with respect to variations in $\mathcal{S}$ incorporates quantities based on $u[\mathcal{S}]$. Here, we make use of the nowadays already classical shape sensitivity calculus. For details we refer to the books of Sokołowski and Zolésio [66] or Delfour and Zolésio [27]. Furthermore, the Appendix of [42] gives a nice overview. Smooth variations of the edge contour $\mathcal{S}$ can be expressed via normal variations $x + t\vartheta\, n[\mathcal{S}]$ of points x on $\mathcal{S}$ in normal direction for small $t \in \mathbb{R}$ and a function θ on $\mathcal{S}$. Here, $n[\mathcal{S}]$ denotes the normal on $\mathcal{S}$, which coincides with the density of n^s introduced in Sect. 1.1. Let us denote by $S(t)$ the resulting family of edge contours. The shape derivative of $\hat{\mathcal{E}}[\mathcal{S}]$ is then defined as

$$\partial_{\mathcal{S}}\hat{\mathcal{E}}[\mathcal{S}](\vartheta) = \frac{d}{dt}\, \mathcal{E}[\mathcal{S}(t)]|_{t=0}\,.$$

For a given smooth function $g : \Omega \to \mathbb{R}$, the derivative of the functional $\int_{\mathcal{S}} g\, d\mathcal{H}^{d-1}$ is given by $\int_{\mathcal{S}} (\partial_{n[\mathcal{S}]} g + g\, h)\, \vartheta\, d\mathcal{H}^{d-1}$, where h is the mean curvature in $\mathcal{S}$. Given a function $f : \Omega \to \mathbb{R}$, which might jump on $\mathcal{S}$, the shape derivative of the functional $\int_{\Omega} f\, d\mathcal{L}$ is given by $\int_{\mathcal{S}} [\![f]\!]\, \vartheta\, d\mathcal{H}^{d-1}$, where $[\![f]\!](x) = \lim_{\epsilon \to 0} (f(x + \epsilon n[\mathcal{S}]) - f(x - \epsilon n[\mathcal{S}]))$ denotes the jump of f on $\mathcal{S}$. With these tools available, we obtain for the shape derivative of the Mumford Shah shape functional

$$\partial_{\mathcal{S}}\hat{\mathcal{E}}[\mathcal{S}](\vartheta) = \int_{\mathcal{S}} \left(\eta\, h + [\![|u[\mathcal{S}] - u^0|^2 + \mu |\nabla u[\mathcal{S}]|^2]\!] \right) \vartheta\, d\mathcal{H}^{d-1}\,. \qquad (4)$$

Furthermore, the minimizer $u = u[S]$ of the Mumford Shah functional for fixed S solves the corresponding Euler Lagrange equations, which is given by the linear elliptic PDE $\mu \Delta u + (u - u^0) = 0$ on $\Omega \setminus S$ and the boundary conditions $\partial_{n[S]} u^+ = 0$ and $\partial_{n[S]} u^- = 0$ where $u^\pm$ indicate the possibly discontinuous function on both sides of S separately.

Now let us briefly describe how to discretize this shape derivative in a proper way. We suppose that the edge contour can be described as the zero set $[\xi = 0]$ of a level set function ξ. Hence, S is in particular assumed to be relatively closed in Ω. Let us denote by $\Xi \in \mathcal{V}^h$ a finite element approximation of this interface. Here, we restrict to simplicial finite elements on the above already introduced subdivision of rectangular grids into triangulations. The now discrete shape $S = [\Xi = 0]$ separates the domain into two polygonal regions $\Omega^+[\Xi] = [\Xi > 0]$ and $\Omega^+[\Xi] = [\Xi < 0]$. For fixed Ξ we introduce two new finite element space $\mathcal{V}_+^h[\Xi]$ and $\mathcal{V}_-^h[\Xi]$ adapted to the two domains $\Omega^+[\Xi]$ and $\Omega^-[\Xi]$. In fact, for every finite element basis function Θ^i whose support intersects $\Omega^\pm[\Xi]$, we define the composite finite element basis function $\Theta_i^\pm = \chi_{\Omega^\pm[\Xi]} \Theta_i$ [39, 40]. The span of all these basis functions form the composite finite element spaces $\mathcal{V}_+^h[\Xi]$ and $\mathcal{V}_-^h[\Xi]$, respectively. Now, we consider discrete finite element counterparts U^+ and U^- of $u[S]$ on $\Omega^+[\Xi]$ and on $\Omega^-[\Xi]$, separately. Indeed, define $U^\pm \in \mathcal{V}_+^h[\Xi]$ as the solution of

$$0 = \int_{\Omega^\pm[\Xi]} \mu \nabla U^\pm \cdot \nabla \Theta + (U^\pm + \mathcal{I}_h^\pm u^0) \Theta \, d\mathcal{L}$$

for all functions $\Theta \in \mathcal{V}_\pm^h[\Xi]$, where $\mathcal{I}_h^\pm$ denotes a piecewise constant interpolation based on simplicial mid point evaluation. In matrix vector notation, we obtain a system of linear equations for each nodal vectors $\bar{U}^\pm$ of the finite element functions $U^\pm = U^\pm[\Xi]$:

$$\left(\mu \mathbf{L}^\pm[\Xi] + \mathbf{M}^\pm[\Xi] \right) \bar{U}^\pm = \mathbf{M}^\pm[\Xi] \overline{\mathcal{I}_h^\pm u^0}$$

The coefficients in the matrix and of the right hand side depend on the shape $S[\Xi]$ described by the current level set function Ξ, i.e. $\mathbf{L}^\pm[\Xi] = \left(\int_{\Omega^\pm[\Xi]} \nabla \Theta_j \cdot \nabla \Theta_i \, d\mathcal{L} \right)_{ij}$ and $\mathbf{M}[\Xi] = \left(\int_{\Omega^\pm[\Xi]} \Theta_j \Theta_i \, d\mathcal{L} \right)_{ij}$ are the composite finite element stiffness and mass matrix, respectively. Here, the indices i and j are running over all active indices in $\mathcal{V}_\pm^h$.

The shape variation $\partial_S \hat{\mathcal{E}}[S]$ has a singular density concentrated on the edge set. Thus, we consider a shape gradient $\text{grad}_S \mathcal{E}[S]$ based on a regularizing metric. Let us recall that a gradient always depends on an underlying metric and is defined as the representation of the variation in this metric. In our case we use the H^1 scalar product $(v, w)_\sigma = \int_\Omega \frac{\sigma^2}{2} \nabla u \cdot \nabla v + u v \, d\mathcal{L}$ for some constant $\sigma > 0$ and define

$$(\text{grad}_S \hat{\mathcal{E}}[S], \theta)_\sigma = \partial_S \hat{\mathcal{E}}[S](\theta)$$

for all smooth test functions θ on Ω. The discrete counter part $\text{grad}_{S[\Xi]} \mathcal{E}[S[\Xi]]$ is the finite element function in $\mathcal{V}^h$ such that

$$(\text{grad}_S \mathcal{E}[S[\Xi]], \Theta)_\sigma = \partial_S \mathcal{E}[S[\Xi]](\Theta)$$
$$= \int_{S[\Xi]} (\eta\, H + [\![|U[S[\Xi]] - u^0|^2 + \mu|\nabla U[S[\Xi]]|^2]\!])\, \Theta\, \mathrm{d}\mathcal{H}^{d-1}$$
(5)

for all test functions $\Theta \in \mathcal{V}^h$ and a suitable discrete mean curvature H on $S[\Xi]$. Each evaluation of $\text{grad}_S \mathcal{E}[S[\Xi]]$ requires the solution of a linear system of equations. The resulting regularized discrete gradient can now be used in a standard level set approach. Indeed, the level set equation $\partial_t \xi + |\nabla \xi| v = 0$ identifies normal variations v of level sets $[\xi = c]$ with variations $s = \partial_t \xi$ of the level set function. Hence, $-|\nabla \xi|^{-1} \text{grad}_{S[\xi]} \mathcal{E}[S[\xi]]$ is the corresponding descent direction of the functional $\xi \to \mathcal{E}[S[\xi]]$ in the spatially still continuous case. The discretization based on standard finite elements is straightforward and thus omitted here.

Finally, we have all ingredients at hand to derive an algorithm for the minimization of the discrete Mumford Shah functional. In each step of the algorithm we first compute for a given discrete level set function Ξ the discrete image intensities $U^+[\Xi]$ and $U^-[\Xi]$. Then, we can evaluate the right hand side in (5) and thus compute the discrete shape gradient $\text{grad}_S \mathcal{E}[S[\Xi]] \in \mathcal{V}^h$. From this, we deduce the discrete descent direction of the level set shape functional. Finally, a step sized controlled 1D descent strategy is applied. For details on this approach and a comparison to alternative approaches in the context of image segmentation and registration we refer to [28, 31].

Finite Elements for the Phase Field Model

Now, let us present an algorithmic alternative to the previously described discrete sharp interface model. We pick up the phase field approach (3) by Ambrosio and Tortorelli [3] and rephrase it for discrete image intensities and phase fields in the standard finite element space $\mathcal{V}^h$. Hence, we ask for discrete minimizers of the energy

$$\mathcal{E}_\epsilon[U, V] = \int_\Omega \{(U - \mathcal{I}_h u^0)^2 + \mu(V^2 + k_\epsilon)|\nabla U|^2\} + \eta \left(\epsilon|\nabla V|^2 + \frac{(1-V)^2}{4\epsilon} \right) \mathrm{d}\mathcal{L}$$

over all $U, V \in \mathcal{V}^h$. Here, $\mathcal{I}_h$ denotes the piecewise linear Lagrangian interpolation. The corresponding Euler Lagrange equations are

$$0 = \int_\Omega (U - \mathcal{I}_h u^0)\Theta + \mu(V^2 + k_\epsilon)\nabla U \cdot \nabla \Theta\, \mathrm{d}\mathcal{L},$$
(6)

$$0 = \int_\Omega \mu V \Xi |\nabla U|^2 + \eta \int_\Omega \left(\epsilon \nabla V \cdot \nabla \Xi + \frac{(V-1)}{4\epsilon} \Xi \right) \mathrm{d}\mathcal{L}$$
(7)

for all Θ, $\Xi \in \mathcal{V}^h$. The underlying mesh is supposed to consist of simplices. Hence, the different terms in the integrant are at most quadratic and can be integrated exactly using appropriate quadrature rules. Equation (6) is linear in U and (7) is linear in V. In matrix vector notation, we obtain two systems of linear equations for the nodal vectors $\bar{U}$ and $\bar{V}$ which represent the finite element functions U and V, respectively:

$$\left(\mu \mathbf{L}[V^2 + k_\epsilon] + \mathbf{M}[1]\right) \bar{U} = \mathbf{M}[1]\overline{\mathcal{I}_h u^0}$$

$$\left(\eta \epsilon \mathbf{L}[1] + \mathbf{M}[\mu |\nabla U|^2 + \frac{\eta}{4\epsilon}]\right) \bar{V} = \frac{\eta}{4\epsilon}\mathbf{M}[1]\bar{1}$$

The coefficients in the matrix and of the right hand side depend on the corresponding other discrete variable. Here $\mathbf{L}[\omega] = \left(\int_\Omega \omega \nabla \Theta_j \cdot \nabla \Theta_i \, d\mathcal{L}\right)_{ij}$ and $\mathbf{M}[\omega] = \left(\int_\Omega \omega \Theta_j \Theta_i \, d\mathcal{L}\right)_{ij}$ are the weighted stiffness and mass matrix, respectively. The weight is denoted by ω and Θ_i are the hat basis functions indexed by the nodal index i, $\overline{\mathcal{I}_h u^0}$ is the nodal vector corresponding to $\mathcal{I}_h u^0$, and $\bar{1}$ is the vector with components all equal to 1. In [10] Bourdin has shown the Γ-convergence of the discretized functionals to the corresponding continuous one. The actual algorithm now consists of an alternating minimization, which is given in pseudo code notion by the following code fragment:

DiscreteAT {
    $k = 0$; initialize $\bar{U}^0 = \overline{\mathcal{I}_h u^0}$ and $\bar{V}^0 \equiv \bar{1}$;
    do {
        Solve $\left(\mu \mathbf{L}[(V^k)^2 + k_\epsilon] + \mathbf{M}[1]\right) \bar{U}^{k+1} = \mathbf{M}[1]\overline{\mathcal{I}_h u^0}$ for $\bar{U}^{k+1}$;
        Solve $\left(\eta \epsilon \mathbf{L}[1] + \mu \mathbf{M}[|\nabla U^{k+1}|^2 + \frac{\eta}{\epsilon}]\right) \bar{V}^{k+1} = \frac{\eta}{4\epsilon}\mathbf{M}[1]\bar{1}$ for $\bar{U}^{k+1}$;
        k=k+1;
    } while($|\bar{U}^k - \bar{U}^{k-1}| + |\bar{V}^k - \bar{V}^{k-1}| \geq$ Threshold)
}

Later, we will pick up this alternating minimization strategy in combination with a gradient descent scheme for additional quantities, primarily deformations, in joint segmentation and matching approaches.

1.4 Nonlinear Elastic Deformations

The correlation or matching of images or shapes requires a mathematical description in terms of a deformation that registers image or shape structures. These matching problems are known usually to be ill-posed problems. Hence, in a variational setting a suitable regularization energy has to be considered. Here, we propose to use hyperelastic energy functionals $\mathcal{E}_{elast}[\phi]$ on deformations ϕ. In what follows we will motivate this type of regularization energy both from a physical and from a mathematical perspective.

A Physical Argument in Favor of the Hyperelastic Model

The regularization energy on deformations is based on classical concepts from continuum mechanics and in particular from the theory of elasticity. We give an exposition of the concept in three dimensions, the two dimensional case follows by analogy. For details we refer to the comprehensive introductions in the books by Ciarlet or Marsden and Hughes [19, 50]. Let Ω be an isotropic elastic body. We suppose $\psi = \mathbb{1}$ to represent the stress free deformation. Let us consider the deformation of length, area, and volume under a deformation ϕ. Here we denote by $\mathcal{H}^k$ the k-dimensional Hausdorff measure, i.e. $\mathcal{H}^d = \mathcal{L}$. We observe that the length of a curve $\gamma : [0, 1] \to \Omega$ undergoing the deformation ϕ is transformed via

$$\mathcal{H}^1(\phi(\gamma)) = \int_0^1 \left| \frac{\mathcal{D}\phi \circ \gamma}{dt} \right| d\mathcal{H}^1 = \int_0^1 \sqrt{\mathcal{D}\phi^T \mathcal{D}\phi \dot\gamma \cdot \dot\gamma} \, d\mathcal{H}^1.$$

Hence, $\mathcal{D}\phi^T \mathcal{D}\phi$ controls the change of length under the deformation, and in case of isotropic elasticity we confine with the norm $|\mathcal{D}\phi|_2$ as the term controlling the change of length, where $|A|_2 := \mathrm{tr}(A^T A) = \sum_{i,j} A_{ij} A_{ij}$ for $A \in \mathbb{R}^{d,d}$. Secondly, the local volume transformation under a deformation ϕ is obviously controlled by $\det \mathcal{D}\phi$, i.e.

$$\mathcal{H}^d(\phi(V)) = \int_V |\det \mathcal{D}\phi(x)| \, d\mathcal{L}$$

for a volume $V \subset \Omega$. If $\det \mathcal{D}\phi < 0$, self penetration may be observed. Finally, let us consider the transformation of area elements. Let $\mathcal{S}$ be a hypersurface patch with normal n and $T_x \mathcal{S}$ the tangent space of $\mathcal{S}$ at a point x on $\mathcal{S}$. Suppose n^ϕ is the deformed normal on $T_x \phi[\mathcal{S}]$ at position $\phi(x)$. Hence, from $n^\phi \cdot \mathcal{D}\phi \, v = 0$ for all $v \in T_x \phi[\mathcal{S}]$ we derive

$$n^\phi = \frac{\mathrm{cof}\, \mathcal{D}\phi \, n}{|\mathrm{cof}\, \mathcal{D}\phi \, n|},$$

where $\mathrm{cof}\, A = \det A \, A^{-T}$ for $A \in \mathbb{R}^{d,d}$. Thus the deformed area element is given by $\frac{\det \mathcal{D}\phi}{n^\phi \cdot \mathcal{D}\phi \, n}$ and we obtain

$$\mathcal{H}^{d-1}(\phi[\mathcal{S}]) = \int_{\varphi[\mathcal{S}]} d\mathcal{H}^{d-1} = \int_{\mathcal{S}} \frac{|\det \mathcal{D}\phi|}{|n^\phi \cdot \mathcal{D}\phi \, n|} \, d\mathcal{H}^{d-1}$$

$$= \int_{\mathcal{S}} \frac{|\det \mathcal{D}\phi| \, |\mathcal{D}\phi^{-T} n|}{|\mathcal{D}\phi^{-T} n \cdot \mathcal{D}\phi \, n|} \, d\mathcal{H}^{d-1} = \int_{\mathcal{S}} \left| \det \mathcal{D}\phi \, \mathcal{D}\phi^{-T} n \right| d\mathcal{H}^{d-1}$$

$$= \int_{\mathcal{S}} |\mathrm{cof}\, \mathcal{D}\phi \, n| \, d\mathcal{H}^{d-1}.$$

In analogy to the case of length control, $|\text{cof}\,\mathcal{D}\phi|_2 = \text{tr}\,(\text{cof}\,\mathcal{D}\phi^T \text{cof}\,\mathcal{D}\phi)$ is the proper measure for the change of area in an isotropic elastic body.

Based on these considerations we can define a simple physically reasonable isotropic elastic energy for $d = 3$, which separately cares about length, area and volume deformation and especially penalizes volume shrinkage:

$$\mathcal{E}_{elast}[\phi] := \int_\Omega W(|\mathcal{D}\phi|^2, |\text{cof}\,\mathcal{D}\phi|^2, \det \mathcal{D}\phi)\,d\mathcal{L} \tag{8}$$

where $W : \mathbb{R} \times \mathbb{R} \times \mathbb{R} \to \mathbb{R}$ is supposed to be convex. In particular, we will consider

$$W(I_1, I_2, I_3) =:= \alpha_1 |\mathcal{D}\phi|^{\frac{p}{2}} + \alpha_2 |\text{cof}\,\mathcal{D}\phi|^{\frac{q}{2}} + \alpha_3 \Gamma(\det \mathcal{D}\phi)\,d\mathcal{L}, \tag{9}$$

with $\Gamma(D) = D^r + \frac{s}{r} D^{-s}$, $r, s > 0$ and $\alpha_1, \alpha_2, \alpha_3 > 0$. In nonlinear elasticity such materials laws have been proposed by Ogden [57], and for $p = q = 2$ we obtain the Mooney-Rivlin model [19]. More general, we can consider a so called polyconvex energy functional [25]. Hence, W and thereby $\mathcal{E}_{elast}[\phi]$ penalize volume shrinkage, i.e. $W(I_1, I_2, I_3) \xrightarrow{I_3 \to 0} \infty$ as reflected by the property of Γ above. The arguments of W are in fact the principle invariants of the matrix Jacobian of the deformation ϕ. It is worth to underline that under these assumptions the function $\hat{W} : \mathbb{R}^{d,d} \to \mathbb{R}$ with

$$\hat{W}(A) := W(|A|^2, |\text{cof}\,A|^2, \det A)$$

is not convex [19]. Physically,

$$\sigma[\phi] = (\sigma_{ij}[\phi])_{i,j=1,\cdots,d} := \left(\mathcal{D}_{A_{ij}} \hat{W}(\mathcal{D}\phi)\right)_{i,j=1,\cdots,d}$$

plays the role of the elastic Piola–Kirchhoff stress tensor. The built-in penalization of volume shrinkage, i.e. $\bar{W}(I_1, I_2, I_3) \xrightarrow{I_3 \to 0} \infty$, enables us to control local injectivity (cf. [5] and the next paragraph). Furthermore, a deformation that is locally isometric, i.e. $\nabla \phi_i^T(x) \nabla \phi_i(x) = \mathbb{1}$, is a local minimizer of the energy.

The physical advantages of the nonlinear elastic model are the following:

- It allows to incorporate large deformations with strong material and geometric nonlinearities, which cannot be treated by a linear approach.
- The dependency of the energy density $\hat{W}$ follows from first principle and measures the physical effects of length, area, and volume distortion, which reflect the local distance from an isometry.
- Finally, it balances in an intrinsic way expansion and collapse of the elastic objects and hence frees us to impose artificial conditions on the expected image shape.

A Mathematical Argument in Favor of the Hyperelastic Model

The matching of 2D and 3D images—also known as image registration—is one of the fundamental tasks in image processing. Here, as we will see below, even for one of simplest possible matching energies standard regularization techniques fail, and the problem turns out not to be well posed. With the help of a hyperelastic and polyconvex regularization we will be able to establish existence results.

One aims to correlate two images—a reference image u_R and a template image u_T—via an energy relaxation over a set of in general non rigid spatial deformations. Let us denote the reference image by $u_R : \Omega \to \mathbb{R}$ and the template image by $u_T : \Omega \to \mathbb{R}$. Both images are supposed to be defined on a bounded domain $\Omega \in \mathbb{R}^d$ for $d = 1, 2$, or 3 with Lipschitz boundary. We ask for a deformation $\phi : \Omega \to \Omega$ such that $u_T \circ \phi$ is optimally correlated to u_R. In case of matching the intensities we search for ϕ such that $u_T \circ \phi \approx u_R$ reflected by the basic matching energy

$$\mathcal{E}_m[\phi] := \int_\Omega |u_T \circ \phi - u_R|^2 \, \mathrm{d}\mathcal{L}.$$

As boundary condition we require $\phi = \mathbb{1}$ on $\partial\Omega$. Here, $\mathbb{1}$ indicates the identity mapping on Ω and simultaneously the identity matrix. This corresponding minimization problem is known to be ill-posed [12, 67]. Thus, we ask for a suitable regularization. Different regularization approaches have been considered in the literature [16, 18, 26]. Usually, the results presented so far rely on the assumption that the images are smooth. To our knowledge, there are no analytic results for images with sharp edges as they frequently appear in the applications. We are going to address this problem here. As a simple regularization energy the Dirichlet energy

$$\mathcal{E}_{reg}[\phi] := \frac{1}{2} \int_\Omega |\mathcal{D}\phi|^2 \, \mathrm{d}\mathcal{L}$$

of the deformation ϕ for $\alpha > 0$ is at hand. We are aiming to apply the direct method from the calculus of variations [25] to prove existence of a minimizing deformation for the energy $\mathcal{E} := \mathcal{E}_m + \mathcal{E}_{reg}$. Caratheodory's condition is known to be the essential assumption to ensure existence of minimizers for such functionals. This especially requires the continuity of the energy integrand with respect to the argument ϕ and therefore in our case $u_T \in C^0(\overline{\Omega})$. Real images frequently contain edges where the image intensity jumps, and hence such images are not continuous and the Caratheodory condition is not fulfilled. Indeed, a more appropriate function space for images would be a space that allows images to have edge discontinuities (cf. the Mumford and Shaw approach to image processing [53, 56]). Here, we consider the space of bounded functions I being continuous up to a singular set S_I. We suppose that for the Lebesque measure $\mathcal{L}(\cdot)$

Variational Methods in Image Matching and Motion Extraction 155

$$\mathcal{L}(B_\epsilon(S_I)) \xrightarrow{\epsilon \to 0} 0$$

holds. Let us introduce a corresponding function space

$$\mathcal{I}^0(\Omega) := \left\{ u : \Omega \to \mathbb{R} \,\middle|\, u \in L^\infty, \exists S_u \subset \Omega \text{ s. t. } u \in C^0(\Omega \setminus S_u), \mathcal{L}(B_\epsilon(S_u)) \xrightarrow{\epsilon \to 0} 0 \right\}.$$

To exemplify the difficulties of the above stated basic matching problem we will study the following illustrative 1D problem in detail:

Example 1.2.1. Let us consider images

$$u_T^\delta = \text{sign}_\delta(x) := \begin{cases} -1 \,;\, x < -\delta \\ \frac{x}{\delta} \,;\, x \in [-\delta, \delta] \\ 1 \,;\, x > \delta \end{cases}$$

for some fixed $\delta \geq 0$ and $u_R \equiv 0$ on $\Omega = (-1, 1)$ for $d = 1$. Obviously, the infimum zero for the energy $\mathcal{E}_m$ can only be attained if $\phi = 0$ a. e., and therefore $\mathcal{E}_m$ has no minimizer ϕ which is C^0 and fulfills the boundary conditions. Indeed, a minimizing sequence $\{\phi_k\}_{k=0,1,\dots} \subset C^0$ can be selected such that $\mathcal{E}_m[\phi_k] \to 0$. But the convergence $\phi_k \to 0$ is not uniform. If we now consider the regularized images u_T^δ with $\delta > 0$, we observe that the Euler Lagrange equation for a solution $\phi_{\alpha,\delta}$ of the minimization problem

$$\mathcal{E}_m[\phi] + \alpha \mathcal{E}_{reg}[\phi] \to \min$$

leads to $\phi_{\alpha,\delta}'' = 0$ for $|\phi_{\alpha,\delta}| > \delta$ and $\phi_{\alpha,\delta}'' = \frac{2}{\alpha \delta^2} \phi_{\alpha,\delta}$ for $|\phi_{\alpha,\delta}| < \delta$. Hence, $\phi_{\alpha,\delta}(x) = \gamma_{\alpha,\delta} \sinh(\frac{\sqrt{2}x}{\sqrt{\alpha\delta}})$ on $[-y_{\alpha,\delta}, y_{\alpha,\delta}]$ for some constant $\gamma_{\alpha,\delta}$ and $\phi_{\alpha,\delta}(x) = \text{sign}(x)(\delta + \frac{1-\delta}{1-y_{\alpha,\delta}}(|x| - y_{\alpha,\delta}))$ elsewhere. Here, $y_{\alpha,\delta}$ is the pre-image of δ, i.e. $\phi_{\alpha,\delta}(y_{\alpha,\delta}) = \delta$, and finally we have in addition $\phi_{\alpha,\delta}'(y_{\alpha,\delta} + 0) = \phi_{\alpha,\delta}'(y_{\alpha,\delta} - 0)$. One easily verifies that these conditions uniquely define $y_{\alpha,\delta}$ as the root of

$$q(y, \alpha, \delta) = \frac{1-\delta}{1-y} - \frac{\sqrt{2}}{\sqrt{\alpha}} \text{cotanh}\left(\frac{\sqrt{2}y}{\sqrt{\alpha\delta}}\right)$$

on $(0, 1)$. Indeed, because $\lim_{y \to 0+0} q(y, \alpha, \delta) = -\infty$ and $\lim_{y \to 1-0} q(y, \alpha, \delta) = \infty$, there is a root for any $\alpha > 0$ and any $\delta \in (0, 1)$, and from the uniqueness of the solution of the Euler Lagrange equation we deduce that this root is unique. To explore the solution behavior for different values of α and δ let us derive upper and lower bounds for $y_{\alpha,\delta}$. We estimate

$$\bar{\mathcal{E}} := \mathcal{E}[\mathbb{1}] \geq \mathcal{E}[\phi_{\alpha,\delta}] \geq \alpha \mathcal{E}_{reg}[\phi_{\alpha,\delta}]$$

$$= \frac{\alpha}{2} \int_{-1}^{1} |\phi_{\alpha,\delta}'|^2 = -\frac{\alpha}{2} \int_{-1}^{1} \phi_{\alpha,\delta}'' \phi_{\alpha,\delta} + \frac{\alpha}{2} [\phi_{\alpha,\delta}' \phi_{\alpha,\delta}]_{-1}^{1} \geq \alpha \frac{1-\delta}{1-y_{\alpha,\delta}} - 2$$

and claim $y_{\alpha,\delta} \leq 1 - \alpha \frac{1-\delta}{\mathcal{E}+2}$. Furthermore, the total energy of the function

$$\psi = \begin{cases} 0 & ; |x| \leq 1 - \sqrt{\frac{\alpha}{2}} \\ \text{sign}(x)(1 - \sqrt{\frac{2}{\alpha}}(1 - |x|)) & ; \text{else} \end{cases}$$

turns out to be bounded from above by $2\sqrt{2\alpha}$, and the matching energy $\mathcal{E}_m[\phi_{\alpha,\delta}]$ is bounded from below by $2(1 - y_{\alpha,\delta})$. Thus $y_{\alpha,\delta} \geq 1 - \sqrt{2\alpha}$. As a consequence we observe that on $[-1 + \sqrt{2\alpha}, 1 - \sqrt{2\alpha}]$, the family of minimizers $\phi_{\alpha,\delta} \to 0$ uniformly for $\delta \to 0$. This clearly outlines a disadvantage of the regularization via a Dirichlet integral: If we approximate u_T^0 by the continuous images u_T^δ, the Lebesque measure of the pre-image of a neighborhood of the template singularity set $S_{u_T^0}$ cannot be controlled uniformly for the corresponding minimizing deformation.

Example 1.2.1 underlines that in case of a Dirichlet regularization energy it seems to be impossible to control the measure of the pre-image of arbitrarily small edge neighborhoods of the image u_T. Hence, it cannot be ruled out that small regions containing the singularities of the image u_T are referred by large regions with respect to their pre-image under the mapping ϕ. Beyond this, one cannot rule out self-intersection on the image domain. We ask for a new regularization energy, which in particular allows to control volume shrinkage and simultaneously ensures continuity and injectivity for the minimizing deformation. The nonlinear polyconvex elastic energies of the above type fulfill these requirements. The existence proof for minimizers of nonlinear elastic energies via the calculus of variations and direct methods dates back to the work of Ball [6]. Especially the incorporated control of the volume transformation in this theory turns out to be the key to prove existence of minimizing, continuous, and injective deformations for the image matching problem discussed here. Let us denote by L^p for $p \in [1, \infty]$ the usual Lebesque spaces of functions on Ω into $\mathbb{R}$, $\mathbb{R}^d$ respectively, by $|\cdot|_p$ the corresponding norm, and by $H^{1,p}$ the Hilbert space of functions in L^p with weak first derivatives also in L^p. For the ease of presentation, we do not exploit the full generality of the corresponding existence theory. Here the reader is for instance referred to [6, 7, 20, 35, 68]. We confine to a basic model and state the following theorem:

Theorem 1.2.2 (Existence of Minimizing Deformations). *Suppose $d = 3$,*

$$u_T, u_R \in \mathcal{I}^0(\Omega),$$

and consider a matching energy $\mathcal{E}_m[\phi] = \int_\Omega |u_T \circ \phi - u_R|^2$, a regularization energy

$$\mathcal{E}_{elast}[\phi] = \int_\Omega W(|D\phi|^2, |\text{cof } D\phi|^2, \det D\phi) \, d\mathcal{L},$$

and the total energy $\mathcal{E}[\phi] = \mathcal{E}_m[\phi] + \mathcal{E}_{elast}[\phi]$ for deformations ϕ in the set of admissible deformations

$$\mathcal{A} := \{\phi : \Omega \to \Omega \mid \phi \in H^{1,p}(\Omega), \operatorname{cof} \mathcal{D}\phi \in L^q(\Omega),$$
$$\det \mathcal{D}\phi \in L^r(\Omega), \det \mathcal{D}\phi > 0 \text{ a.e. in } \Omega, \phi = \mathbb{1} \text{ on } \partial\Omega\}$$

where $p, q > 3$ and $r > 1$. Furthermore, suppose $W : \mathbb{R} \times \mathbb{R} \times \mathbb{R} \to \mathbb{R}$ is convex, and there exist constants $\beta, s \in \mathbb{R}$, $\beta > 0$, and $s > \frac{2q}{q-3}$ such that

$$W(I_1, I_2, I_3) \geq \beta(I_1^{\frac{p}{2}} + I_2^{\frac{q}{2}} + I_3^r + I_3^{-s}) \quad \forall I_1, I_2 \in \mathbb{R}, I_3 \in \mathbb{R}^+. \tag{10}$$

Then $\mathcal{E}[\phi]$ attains its minimum over all deformation $\phi \in \mathcal{A}$, the minimizing deformation ϕ is a homeomorphism and in particular $\det \mathcal{D}\phi > 0$ a.e. in Ω.

Proof. The proof of this result is based on the observation that the volume of the pre-image $\phi^{-1}(B_\epsilon(S_T))$ of an ϵ-neighborhood of the singularity set S_T of the image u_T can be controlled. At first, let us recall some well known, fundamental weak convergence results:

Let $(\phi^k)_k$ be a sequence of deformations in $H^{1,p}$ with $\operatorname{cof} \mathcal{D}\phi^k \in L^q$ and $\det \mathcal{D}\phi^k \in L^r$, such that the sequence convergence weakly in the sense $\phi^k \rightharpoonup \phi$ in $H^{1,p}$, $\operatorname{cof} \mathcal{D}\phi^k \rightharpoonup C$ in L^q and $\det \mathcal{D}\phi^k \rightharpoonup D$ in L^q, then $C = \operatorname{cof} \mathcal{D}\phi$ and $D = \det \mathcal{D}\phi$ (weak continuity).

For the proof of these fundamental results we refer to Ball [6] or the book of Ciarlet [19, Sects. 7.5 and 7.6]. The proof of the theorem proceeds in four steps:

Step 1. We observe $\mathbb{1} \in \mathcal{A}$, thus $\underline{\mathcal{E}} := \inf_{\phi \in \mathcal{A}} \mathcal{E}[\phi] < \infty$. Next we observe that $\mathcal{E}[\cdot]$ is well defined for every $\phi \in \mathcal{A}$. Let us consider a minimizing sequence $(\phi^k)_{k=0,1,\cdots}$ in $\mathcal{A}$ with $\mathcal{E}[\phi^k] \to \inf_{\phi \in \mathcal{A}} \mathcal{E}[\phi]$. We denote by $\overline{\mathcal{E}}$ an upper bound of the energy $\mathcal{E}$ on this sequence. Due to the growth condition on W we get that

$$\{(\mathcal{D}\phi^k, \operatorname{cof} \mathcal{D}\phi^k, \det \mathcal{D}\phi^k)\}_{k=0,1,\cdots}$$

is uniformly bounded in $L^p \times L^q \times L^r$. By Poincaré's inequality applied to $(\phi^k - \mathbb{1})$ we obtain that $\{\phi^k\}_{k=0,1,\cdots}$ is uniformly bounded in $H^{1,p}(\Omega)$. Because of the reflexivity of $L^p \times L^q \times L^r$ for $p, q, r > 1$ we can extract a weakly convergent subsequence, again denoted by an index k, such that

$$(\mathcal{D}\phi^k, \operatorname{cof} \mathcal{D}\phi^k, \det \mathcal{D}\phi^k) \rightharpoonup (\mathcal{D}\phi, C, D)$$

in $L^p \times L^q \times L^r$ with $C : \Omega \to \mathbb{R}^{3 \times 3}$, $D : \Omega \to \mathbb{R}$. Applying the above results on weak convergence we achieve $C = \operatorname{cof} \mathcal{D}\phi$ and $D = \det \mathcal{D}\phi$. In addition, by Rellich's embedding theorem we know that $\phi^k \to \phi$ strongly in $L^p(\Omega)$ and by Sobolev's embedding theorem we claim $\phi \in C^0(\bar{\Omega})$.

Step 2. Next, we control the set where the volume shrinks by a factor of more than ϵ^{-1} for the limit deformation. Let us define

$$S_\epsilon = \{x \in \Omega : \det \mathcal{D}\phi \leq \epsilon\}$$

for $\epsilon \leq \delta_0$. Without loss of generality we assume that the sequence of energy values $\mathcal{E}[\phi^k]$ is monotone decreasing and for given $\epsilon > 0$ we denote by $k(\epsilon)$ the smallest index such that

$$\mathcal{E}[\phi^k] \leq \mathcal{E}[\phi^{k(\epsilon)}] \leq \underline{\mathcal{E}} + \epsilon \qquad \forall k \geq k(\epsilon).$$

From Step 1 we know that $\Psi^k := (\mathcal{D}\phi^k, \operatorname{cof} \mathcal{D}\phi^k, \det \mathcal{D}\phi^k)$ converges weakly to $\Psi := (\mathcal{D}\phi, \operatorname{cof} \mathcal{D}\phi, \det \mathcal{D}\phi)$ in $L^2 \times L^2 \times L^r$. Hence, applying Mazur's Lemma we obtain a sequence of convex combinations of Ψ^k, which converges strongly to Ψ. I.e. there exists a family of weights $((\lambda_i^k)_{k(\epsilon) \leq i \leq k})_{k \geq k(\epsilon)}$ with $\lambda_i^k \geq 0$, $\sum_{k(\epsilon)}^k \lambda_i^k = 1$, such that

$$\lambda_i^k \Psi^i \to \Psi \text{ and } \lambda_i^k \phi^i \to \phi.$$

Here and in what follows we make use of the summation convention. Now, using especially the properties of Γ, the convexity of W and Fatou's lemma we estimate

$$\beta \epsilon^{-s} \mathcal{L}(S_\epsilon) \leq \beta \int_{S_\epsilon} (\det \mathcal{D}\phi)^{-s} \, d\mathcal{L} \leq \int_{S_\epsilon} W(\Psi) \, d\mathcal{L}$$

$$= \int_{S_\epsilon} \liminf_{k \to \infty} W(\lambda_i^k \Psi^i) \, d\mathcal{L} \leq \int_{S_\epsilon} \liminf_{k \to \infty} \lambda_i^k W(\Psi^i) \, d\mathcal{L}$$

$$\leq \liminf_{k \to \infty} \lambda_i^k \int_{S_\epsilon} W(\Psi^i) \, d\mathcal{L} \leq \liminf_{k \to \infty} \lambda_i^k \int_\Omega W(\Psi^i) + (u_T \circ \phi^i - u_R)^2 \, d\mathcal{L}$$

$$\leq \overline{\mathcal{E}}$$

and claim $\mathcal{L}(S_\epsilon) \leq \frac{\overline{\mathcal{E}} \epsilon^s}{\beta}$. As a consequence S_0 is a null set and we know that $\det \mathcal{D}\phi > 0$ a. e. on Ω. Thus, together with the results from Step 1 we deduce that the limit deformation ϕ is in the set of admissible deformations $\mathcal{A}$. Following Ball [7] we furthermore obtain that ϕ is injective. Hence, ϕ is a homeomorphism.

Step 3. Now, we deal with the singularity sets of the images u_T, u_R. By our assumption on the image space $\mathcal{I}^0(\Omega)$ we know that for given $\delta > 0$ there exist $\epsilon_T, \epsilon_R > 0$ such that $\mathcal{L}(B_{\epsilon_T}(S_T))$, $\mathcal{L}(B_{\epsilon_R}(S_R)) \leq \delta$. From this and the injectivity (cf. Theorem 1 (ii) in [7]) we deduce the estimate

$$\mathcal{L}\left(\phi^{-1}(B_{\epsilon_T}(S_T)) \setminus S_\epsilon\right) \leq \frac{1}{\epsilon} \int_{\phi^{-1}(B_{\epsilon_T}(S_T))} \det \mathcal{D}\phi \, d\mathcal{L} = \frac{1}{\epsilon} \int_{B_{\epsilon_T}(S_T)} d\mathcal{L} \leq \frac{\delta}{\epsilon}.$$

Hence, we can control the pre-image of $B_\epsilon(S_T)$ with respect to ϕ restricted to $\Omega \setminus S_\epsilon$, i.e.

$$\mathcal{L}(\phi^{-1}(B_{\epsilon_T}(S_T) \setminus S_\epsilon)) \leq \frac{\delta}{\epsilon}$$

Step 4. Due to Egorov's theorem and the strong convergence of ϕ^k in $L^p(\Omega)$ there is a set K_ϵ with $\mathcal{L}(K_\epsilon) < \epsilon$ such that ϕ^k converges uniformly on $\Omega \setminus K_\epsilon$. Let us now define the set

$$R_{\delta,\epsilon} := \phi^{-1}(B_{\epsilon_T}(S_T)) \cup B_{\epsilon_R}(S_R) \cup S_\epsilon \cup K_\epsilon,$$

whose measure can be estimated in terms of ϵ and δ, i.e.

$$\mathcal{L}(R_{\delta,\epsilon}) \leq \frac{\delta}{\epsilon} + \delta + \frac{\bar{\mathcal{E}} \epsilon^s}{\beta} + \epsilon.$$

On $\Omega \setminus R_{\delta,\epsilon}$ the sequence $(u_T \circ \phi^k - u_R)_{k=0,1,\ldots}$ converges uniformly to $u_T \circ \phi - u_R$. From the above control of the pre-image of $B_{\epsilon_T}(S_T)$ we deduce the measurability of $u_T \circ \phi - u_R$ (cf. 1.4). Obviously, the integral $\int_\Omega |u_T \circ \phi - u_R|^2$ is bounded by $C_\infty \mathcal{L}(\Omega)$ with

$$C_\infty := 2(|u_T|_\infty^2 + |u_R|_\infty^2).$$

We choose $k(\epsilon)$ large enough to ensure that

$$||u_T \circ \phi^k(x) - u_R(x)|^2 - |u_T \circ \phi(x) - u_R(x)|^2| \leq \epsilon$$

for $x \in \Omega \setminus R_{\delta,\epsilon}$ and $\forall k \geq k(\epsilon)$. Now we are able to estimate $\mathcal{E}[\phi]$ using especially the convexity of W and Fatou's lemma:

$$\mathcal{E}[\phi] = \int_\Omega W(\Psi) + |u_T \circ \phi - u_R|^2 \, d\mathcal{L}$$

$$\leq \int_\Omega W(\lim_{k \to \infty} \lambda_i^k \Psi^i) \, d\mathcal{L} + \int_{\Omega \setminus R_{\delta,\epsilon}} |u_T \circ \phi - u_R|^2 \, d\mathcal{L} + C_\infty \mathcal{L}(R_{\delta,\epsilon})$$

$$\leq \int_\Omega \liminf_{k \to \infty} \lambda_i^k W(\Psi^i) \, d\mathcal{L} + \liminf_{k \to \infty} \lambda_i^k \int_{\Omega \setminus R_{\delta,\epsilon}} |u_T \circ \phi^i - u_R|^2 \, d\mathcal{L}$$

$$+ \epsilon \mathcal{L}(\Omega) + C_\infty \mathcal{L}(R_{\delta,\epsilon})$$

$$\leq \liminf_{k \to \infty} \lambda_i^k \int_\Omega W(\Psi^i) + |u_T \circ \phi^i - u_R|^2 \, d\mathcal{L} + \epsilon \mathcal{L}(\Omega) + 2 C_\infty \mathcal{L}(R_{\delta,\epsilon})$$

$$= \liminf_{k \to \infty} \lambda_l^k \mathcal{E}[\phi^i] + \epsilon \mathcal{L}(\Omega) + 2 C_\infty \mathcal{L}(R_{\delta,\epsilon})$$
$$\leq \underline{\mathcal{E}} + \epsilon + \epsilon \mathcal{L}(\Omega) + 2 C_\infty \mathcal{L}(R_{\delta,\epsilon}).$$

Finally, for given $\bar{\epsilon}$ we choose ϵ and then δ and the dependent ϵ_T, ϵ_R small enough to ensure that

$$\epsilon + \epsilon \mathcal{L}(\Omega) + 2 C_\infty \left(\epsilon + \frac{\bar{\mathcal{E}} \epsilon^s}{\beta} \right) \leq \frac{\bar{\epsilon}}{2}, \quad 2 C_\infty \left(\frac{\delta}{\epsilon} + \delta \right) \leq \frac{\bar{\epsilon}}{2}.$$

and get $\mathcal{E}[\phi] \leq \underline{\mathcal{E}} + \bar{\epsilon}$. This holds true for an arbitrary choice of $\bar{\epsilon}$. Thus we conclude with the desired result

$$\mathcal{E}[\phi] \leq \underline{\mathcal{E}} = \inf_{\varphi \in \mathcal{A}} \mathcal{E}[\varphi].$$

□

Let us remark that for the case of two dimensional image matching problems an analogous results holds true. Hence, we skip the dependency of the regularization energy on $\operatorname{cof} \mathcal{D} \phi$ and occasionally weaken the growth conditions. Theorem 1.2.2 can be regarded as a generalization of Theorem 3 in [7] with respect to forces with discontinuities. Indeed, some further simple generalization of the external force term—here encoding the derivative of the matching energy—is straightforward.

Space Discretization of Nonlinear Elasticity

In what follows we will briefly comment on the variation and the spatial discretization of the energy $\mathcal{E}_{elast}[\cdot]$. For the variation we obtain

$$\langle \delta_\phi \mathcal{E}_{elast}[\phi], \psi \rangle = \int_\Omega W_{,A}(\mathcal{D}\phi) : \mathcal{D}\psi \, \mathrm{d}\mathcal{L}$$

for a vector valued displacement type test functions ψ. Concerning the polyconvex energy integrand given in (8) and (9), we obtain

$$W_{,A}(A) : B = 2 \partial_{I_1} \bar{W}(|A|^2, |\operatorname{cof} A|^2, \det A) \, A : B$$
$$+ 2 \partial_{I_2} \bar{W}(|A|^2, |\operatorname{cof} A|^2, \det A) \operatorname{cof} A : \partial_A \operatorname{cof}(A)(B)$$
$$+ \partial_{I_3} \bar{W}(|A|^2, |\operatorname{cof} A|^2, \det A) \, \partial_A \det(A)(B),$$

where

$$\partial_A \det(A)(B) = \det(A) \operatorname{tr}(A^{-1} B),$$

$$\partial_A \text{cof}(A)(B) = \det(A)\text{tr}(A^{-1}B)A^{-T} - \det AA^{-T}B^T A^{-T},$$

$$\partial_{I_1}\bar{W}(I_1, I_2, I_3) = \frac{p}{2}\alpha_1 (I_1 - d)^{\frac{p-2}{2}},$$

$$\partial_{I_2}\bar{W}(I_1, I_2, I_3) = \frac{q}{2}\alpha_2 (I_2 - d)^{\frac{q-2}{2}},$$

$$\partial_{I_3}\bar{W}(I_1, I_2, I_3) = s\alpha_3 \left(I_3^{r-1} - I_3^{-s-1}\right).$$

If we now consider a discrete deformation ϕ in the finite element space $\mathcal{V}^d$, we can evaluate the nodal vector of the L^2 gradient of the energy

$$\mathcal{E}_{elast}[\Phi] = \int_\Omega \hat{W}(D\Phi) \, d\mathcal{L}$$

by

$$(\text{grad}_{L^2}\mathcal{E}_{elast}[\bar{\Phi}])_{ij} = (M[\mathbb{1}](\int_\Omega \hat{W}_{,A}(D\Phi) : e_i \Theta_j \, d\mathcal{L})_j)_i,$$

where $M[\mathbb{1}]$ is the usual mass matrix, e_i the ith canonical unit vector, and Θ_j one of the basis functions of $\mathcal{V}$.

2 Matching Geometries and Image Morphologies

Now we will start the discussion of morphological image matching, where instead of a direct comparison of image intensities we compare local image morphologies. At first we focus on a matching of the regular morphology, followed by a discussion of the treatment of the edge sets. Finally, full joint models will be investigated.

2.1 Matching Ensembles of Level Sets

In this section we will construct a suitable matching energy, which measures the defect of the morphology of the reference image u_R and the deformed template image u_T. Thus, with respect to the above identification of morphologies and normal fields we ask for a deformation ϕ such that

$$n_T \circ \phi \parallel n_R^\phi, \tag{11}$$

where n_R^ϕ is the transformed normal n_R of the reference image u_R on $T_{\phi(x)}\phi([u_R(x)])$ and n_T the normal of the template image u_T; both are evaluated at position x. We have already seen that the deformed normal is given by

$$n_R^\phi = \frac{\operatorname{cof} \mathcal{D}\phi\, n_R}{|\operatorname{cof} \mathcal{D}\phi\, n_R|},$$

In a variational setting, optimality can be expressed in terms of energy minimization. Hence, we consider a matching energy

$$\mathcal{E}_{GM}[\phi] := \int_\Omega g(n_T \circ \phi, n_R, \operatorname{cof} \mathcal{D}\phi)\, d\mathcal{L}$$

for some function $g : S^{d-1} \times S^{d-1} \times \mathbb{R}^{d,d} \to \mathbb{R}^+;\ (u, v, A) \mapsto g(u, v, A)$. Here S^{d-1} denotes the unit sphere in $\mathbb{R}^d$. As boundary condition we require $\phi = \mathbb{1}$ on $\partial\Omega$. To be not too restrictive with respect to the space of images we have to take into account the problem of degenerate Gauss maps. Hence, let us recall the set $F[u] := \{x \in \Omega : \nabla u = 0\}$ for $u = u_T$ or $u = u_R$, where no image normal can be defined. At first, we resolve this problem of undefined normals at least formally by introducing a zero-homogeneous extension $g_0 : \mathbb{R}^d \times \mathbb{R}^d \times \mathbb{R}^{d,d} \to \mathbb{R}^+$ of g in the first and second argument:

$$g_0(v, w, A) = \begin{cases} 0 & ;\ v = 0 \text{ or } w = 0 \\ g(\frac{v}{|v|}, \frac{w}{|w|}, A) & ;\ \text{else} \end{cases}. \tag{12}$$

Based on g_0 we can redefine the matching energy $\mathcal{E}_{GM}$ and obtain

$$\mathcal{E}_{GM}[\phi] := \int_\Omega g_0(\nabla u_T \circ \phi, \nabla u_R, \operatorname{cof} \mathcal{D}\phi)\, d\mathcal{L}. \tag{13}$$

In the later analysis we have to take special care of the singularity of g_0 for vanishing first or second argument. Indeed, we will assume that the measure of $F[u_T]$ and $F[u_R]$ is in a suitable sense sufficiently small. Furthermore, in the existence theory we will explicitly control the impact of these sets on the energy. As a first choice for the energy density g let us consider

$$g(v, w, A) := \left(v - \frac{Aw}{|Aw|}\right)^2 \tag{14}$$

for $v, w \in S^{d-1}$, which corresponds to the energy

$$\int_\Omega |n_T \circ \phi - n_R^\phi|^2.$$

We observe that the energy $\mathcal{E}_{GM}$ vanishes if $u_T \circ \phi = \gamma \circ u_R$ for a monotone grey value transformation $\gamma : \mathbb{R} \to \mathbb{R}$. If we want $\mathcal{E}_{GM}$ to vanish also for non-monotone transformations γ we are lead to the symmetry assumption:

Variational Methods in Image Matching and Motion Extraction

$$g(v, w, A) = g(-v, w, A) = g(v, -w, A). \tag{15}$$

Example 1.3.1. A useful class of matching functionals $\mathcal{E}_{GM}$ is obtained by choosing functions g that depend on the scalar product $v \cdot u$ or alternatively on $(\mathbb{1} - v \otimes v)u$ (where $\mathbb{1} - v \otimes v = (\delta_{ij} - v_i v_j)_{ij}$ denotes the projection of u onto the plane normal to v) for $u = \frac{Aw}{|Aw|}$ and $v, w \in S^{d-1}$, i.e.

$$g(v, w, A) = \hat{g}\left((\mathbb{1} - v \otimes v)\frac{Aw}{|Aw|}\right). \tag{16}$$

Let us remark that $\hat{g}((\mathbb{1} - v \otimes v)u)$ is convex in u, if $\hat{g}$ is convex. With respect to arbitrary grey value transformations mapping morphologically identical images onto each other, we might consider $\hat{g}(s) = |s|^\gamma$ for some $\gamma \geq 1$.

Let us now discuss under which conditions there exists a minimizing deformation of the total energy $\mathcal{E}[\cdot]$ if we add a hyperelastic regularization energy $\mathcal{E}_{elast}[\phi]$. Let us emphasize that the problem stated here significantly differs from most other regularized image registration problems, e.g., all intensity based approaches, where the matching energy is defined solely in terms of the deformation ϕ, and the regularization energy is of higher order and considers the Jacobian $\mathcal{D}\phi$ of the deformation. In our case already the matching energy incorporates the cofactor of the Jacobian. Thus, with respect to the direct method in the calculus of variations, we cannot use standard compactness arguments due to Rellich's Embedding Theorem to deal with the matching energy on a minimizing sequence [25]. Instead, we will need suitable convexity assumptions on the function g.

Remark 1.3.2 (Lack of Lower Semicontinuity for Certain Functionals $\mathcal{E}_{GM}$). Recalling Example 1.3.1 we might wish to choose a matching energy with an integrand $g_0(v, w, A) := \hat{g}((\mathbb{1} - \frac{v}{|v|} \otimes \frac{v}{|v|})\frac{Aw}{|Aw|})$ for $\hat{g} \in C^0(\mathbb{R}^d, \mathbb{R}_0^+)$. It is well known that the essential condition to ensure weak sequential lower semicontinuity of functionals depending on the Jacobian of a deformation is quasiconvexity [54, 55]. With our special choice of the class of energies (13) this requires the convexity of g in the argument A (cf. [25, Sect. 5.1]). Indeed, we easily verify that a function

$$f : \mathbb{R}^{d,d} \to \mathbb{R}; \; A \mapsto f(A),$$

which is zero-homogeneous on $\mathbb{R}^{d,d}$ and convex has to be constant and thus an existence result for our image matching problem via the direct calculus of variations can only be expected for trivial matching energies, i.e., for $\hat{g} \equiv const$. To see this, suppose $A, B \in \mathbb{R}^{d,d}$ with $f(A) - f(B) = \delta > 0$. Using the definition $A_{\alpha,r} := r A + \alpha (A - B)$ for $r > 0, \alpha > 0$ and setting $s = \frac{\alpha}{r}$, we obtain

$$f(A_{\alpha,r}) = f(r A + s r (A - B))$$
$$\geq f(r A) + s (f(r A) - f(r B))$$

$$= f(A) + \frac{\alpha}{r}(f(A) - f(B)) = f(A) + \frac{\alpha\delta}{r} \to \infty$$

for $r \to 0$. Finally, we deduce $f(A - B) = \infty$ which contradicts our assumptions on f. Thus, the definition of the matching energy via the above integrand $\hat{g}((\mathbb{1} - v \otimes v)\frac{Aw}{|Aw|})$ and especially our first choice of a matching energy in (14) is not appropriate with respect to a positive answer to the question of existence of minimizers via direct methods.

Next, we take into account the singularities of the normal fields. We require that

$$\mathcal{L}(B_\epsilon(\partial F[u])) \xrightarrow{\epsilon \to 0} 0,$$

for u equal to either u_R or u_T. The corresponding set of image intensity functions is then given by

$$\mathcal{I}(\Omega) := \left\{ u : \Omega \to \mathbb{R} : u \in C^1(\bar{\Omega}), \ \mathcal{L}(B_\epsilon(\partial F[u])) \xrightarrow{\epsilon \to 0} 0 \right\}.$$

The existence proof for minimizers of nonlinear elastic energies via the calculus of variations and direct methods dates back to the work of Ball [6]. Especially the incorporated control of the volume transformation in this theory turns out to be the key to prove existence of minimizing, continuous and injective deformations for the image matching problem discussed here. We consider the following energy (cf. (13) and (8)):

$$\mathcal{E}[\phi] := \mathcal{E}_{GM}[\phi] + \mathcal{E}_{elast}[\phi]. \tag{17}$$

Theorem 1.3.3 (Existence of Minimizing Deformations). *Suppose $d = 3$, $u_T, u_R \in \mathcal{I}(\Omega)$ and consider the total energy defined in (17) for deformations ϕ in the set of admissible deformations*

$$\mathcal{A} := \{ \phi : \Omega \to \Omega \mid \phi \in H^{1,p}(\Omega), \operatorname{cof} \mathcal{D}\phi \in L^q(\Omega),$$
$$\det \mathcal{D}\phi \in L^r(\Omega), \det \mathcal{D}\phi > 0 \text{ a.e. in } \Omega, \phi = \mathbb{1} \text{ on } \partial\Omega \},$$

where $p, q > 3$ and $r > 1$. Suppose $W : \mathbb{R} \times \mathbb{R} \times \mathbb{R}^+ \to \mathbb{R}$ is convex, and there exist constants $\beta, s \in \mathbb{R}$, $\beta > 0$, and $s > \frac{2q}{q-3}$ such that $W(I_1, I_2, I_3) \geq \beta(I_1^{\frac{2}{p}} + I_1^{\frac{2}{q}} + I_3^r + I_3^{-s})$.

Furthermore, assume that $g_0(v, w, A) = g(\frac{v}{|v|}, \frac{w}{|w|}, A)$, for some function $g : S^2 \times S^2 \times \mathbb{R}^{3,3} \to \mathbb{R}_0^+$, which is continuous in $\frac{v}{|v|}, \frac{w}{|w|}$, convex in A and for a constant $m < q$ and a constant $C_g > 0$ the estimate

$$g(v, w, A) - g(u, w, A) \leq C_g |v - u| (1 + |A|^m)$$

holds for a constant $m < q$, for all $u, v, w \in S^2$ and $A \in \mathbb{R}^{3,3}$. Then $\mathcal{E}[\cdot]$ attains its minimum over all deformations $\phi \in \mathcal{A}$, and the minimizing deformation ϕ is a homeomorphism and in particular $\det \mathcal{D}\phi > 0$ a.e. in Ω.

Proof. The proof of this result proceeds first along the lines of the proof of Theorem 1.2.2. Here, the critical issue is not the set of discontinuities of the images but the boundaries of the sets of degenerate normals $\partial F[u_T] \cap \partial F[u_R]$. We observe that the total energy is polyconvex. Furthermore, as in Theorem 1.2.2 the volume of the neighborhood sets of the singularity sets can be controlled. Hence, we can basically confine to a set where the integrand fulfills Carathéodory's conditions. The proof of the theorem proceeds in three steps:

Step 1. Due to the assumption on the image, the set $\mathcal{I}(\Omega) \mathcal{E}_{GM}[\phi]$ is well defined for $\phi \in \mathcal{A}$. In particular $g_0(\nabla u_T \circ \phi, \nabla u_R, \operatorname{cof} \mathcal{D}\phi)$ is measurable. Obviously, $\mathbb{1} \in \mathcal{A}$ and $\mathcal{E}[\mathbb{1}] < \infty$, thus $\underline{\mathcal{E}} := \inf_{\phi \in \mathcal{A}} \mathcal{E}[\phi] < \infty$, and due to the growth conditions and the assumption on g we furthermore get $\underline{\mathcal{E}} \geq 0$. Let us consider a minimizing sequence $(\phi^k)_{k=0,1,\ldots} \subset \mathcal{A}$ with $\mathcal{E}[\phi^k] \to \inf_{\phi \in \mathcal{A}} \mathcal{E}[\phi]$. We denote by $\overline{\mathcal{E}}$ an upper bound of the energy $\mathcal{E}$ on this sequence. At first, let us recall as in the proof of Theorem 1.2.2 that for a minimizing sequence, $(\phi^k)_k$ we have $\phi^k \rightharpoonup \phi$ in $H^{1,p}$, $\operatorname{cof} \mathcal{D}\phi^k \rightharpoonup \operatorname{cof} \mathcal{D}\phi$ in L^q and $\det \mathcal{D}\phi^k \rightharpoonup \det \mathcal{D}\phi$ in L^r. In addition, $\phi^k \to \phi$ strongly in $L^p(\Omega)$ and by Sobolev's embedding theorem we obtain $\phi \in C^0(\bar{\Omega})$.

Step 2. Again, we control the set $S_\epsilon = \{x \in \Omega : \det \mathcal{D}\phi \leq \epsilon\}$, where the volume shrinks by a factor of more than ϵ for the limit deformation. Let us assume without loss of generality that the sequence of energy values $\mathcal{E}[\phi^k]$ is monotone decreasing and that for given $\epsilon > 0$ we let $k(\epsilon)$ be the smallest index such that $\mathcal{E}[\phi^k] \leq \mathcal{E}[\phi^{k(\epsilon)}] \leq \underline{\mathcal{E}} + \epsilon$ for all $k \geq k(\epsilon)$. Then as in the proof of Theorem 1.2.2 we deduce $\mathcal{L}(S_\epsilon) \leq \frac{\bar{\mathcal{E}}\epsilon^s}{\beta}$. As one consequence, S_0 is a null set and we know that $\det \mathcal{D}\phi > 0$ a. e. on Ω and hence the limit deformations ϕ is in the set of admissible deformation $\mathcal{A}$. In addition, ϕ is injective and a homeomorphism. Further on, as in the proof of Theorem 1.2.2 we obtain that $\mathcal{L}\left(\phi^{-1}(B_{\epsilon_T}(\partial F[u_T]) \setminus S_\epsilon\right) \leq \frac{\delta}{\epsilon}$. Due to the continuous differentiability of both images u_T and u_R we can assume that

$$|\nabla u_T(x)| \geq \gamma(\epsilon) \text{ on } \Omega \setminus B_\epsilon(\partial F[u_T]), \tag{18}$$

where $\gamma : \mathbb{R}_0^+ \to \mathbb{R}$ is a strictly monotone function with $\gamma(0) = 0$.

Step 3. Due to Egorov's theorem and the strong convergence of ϕ^k in $L^p(\Omega)$ there is a set K_ϵ with $\mathcal{L}(K_\epsilon) < \epsilon$ such that a subsequence, again denoted by ϕ^k, converges uniformly on $\Omega \setminus K_\epsilon$. Let us now define the set

$$R_{\delta,\epsilon} := \phi^{-1}(B_{\epsilon_T}(\partial F[u_T])) \cup S_\epsilon \cup K_\epsilon,$$

whose measure can be estimated in terms of ϵ and δ, i.e.

$$\mathcal{L}(R_{\delta,\epsilon}) \leq \frac{\delta}{\epsilon} + \frac{\bar{\mathcal{E}}\epsilon^s}{\beta} + \epsilon.$$

On $\Omega \setminus R_{\delta,\epsilon}$ the sequence $(\nabla u_T \circ \phi^k)_{k=0,1,\cdots}$ converges uniformly to $\nabla u_T \circ \phi$. Next, from the assumption on g and the 0-homogeneous extension property of g_0 we deduce that

$$|g_0(u,w,A) - g_0(v,w,A)| \leq C_\gamma |u-v| (1+|A|^m) \qquad (19)$$

for $u,v,w \in \mathbb{R}^3$, $A \in \mathbb{R}^{3,3}$ and $|u|,|v|,|w| \geq \gamma$. Here, C_γ is a constant depending on C_g and on γ. To use this estimate for $u = \phi^k$ and $v = \phi$ below, we assume that $k(\epsilon)$ is large enough, such that $\phi^k(x) \in \Omega \setminus B_{\frac{\epsilon_T}{2}}(S_T)$ for $x \in \Omega \setminus R_{\delta,\epsilon}$ and

$$C_{\gamma(\frac{\epsilon_T}{2})} \left|\nabla u_T \circ \phi^k - \nabla u_T \circ \phi\right|_{\infty,\Omega \setminus K_\epsilon} \leq \epsilon$$

for all $k \geq k(\epsilon)$. Now we are able to estimate $\mathcal{E}[\phi]$ using especially the convexity of W and $g(v,w,\cdot)$, the estimate (19) and Fatou's lemma

$$\mathcal{E}[\phi] = \int_\Omega W(\Psi) + g_0(\nabla u_T \circ \phi, \nabla u_R, \operatorname{cof} \mathcal{D}\phi) \, d\mathcal{L}$$

$$\leq \int_\Omega \liminf_{k\to\infty} \lambda_i^k W(\Psi^i) \, d\mathcal{L} + 2C_g \int_{R_{\delta,\epsilon}} 1 + |\operatorname{cof} \mathcal{D}\phi|^m \, d\mathcal{L}$$

$$+ \int_{\Omega \setminus R_{\delta,\epsilon}} \liminf_{k\to\infty} \lambda_i^k g_0(\nabla u_T \circ \phi, \nabla u_R, \operatorname{cof} \mathcal{D}\phi^i) \, d\mathcal{L}$$

$$\leq \liminf_{k\to\infty} \lambda_i^k \int_\Omega W(\Psi^i) \, d\mathcal{L} + b(\mathcal{L}(R_{\delta,\epsilon}))$$

$$+ \liminf_{k\to\infty} \lambda_i^k \int_{\Omega \setminus R_{\delta,\epsilon}} g_0(\nabla u_T \circ \phi, \nabla u_R, \operatorname{cof} \mathcal{D}\phi^i) - g_0(\nabla u_T \circ \phi^i, \nabla u_R, \operatorname{cof} \mathcal{D}\phi^i)$$

$$+ g_0(\nabla u_T \circ \phi^i, \nabla u_R, \operatorname{cof} \mathcal{D}\phi^i) \, d\mathcal{L},$$

where $b(s) := 2C_g (s + (\frac{\bar{\mathcal{E}}}{\beta})^{\frac{m}{q}} s^{1-\frac{m}{q}})$. Here we have in particular used the a priori estimate $|\operatorname{cof} \mathcal{D}\phi|_{q,\Omega} \leq (\frac{\bar{\mathcal{E}}}{\beta})^{\frac{1}{q}}$. We estimate further and obtain

$$\mathcal{E}[\phi] \leq \liminf_{k\to\infty} \lambda_i^k \int_\Omega W(\Psi^i) + g_0(\nabla u_T \circ \phi^i, \nabla u_R, \operatorname{cof} \mathcal{D}\phi^i) \, d\mathcal{L} + 2b(\mathcal{L}(R_{\delta,\epsilon}))$$

$$+ C_{\gamma(\frac{\epsilon_T}{2})} \sup_{k\to\infty} \int_{\Omega \setminus R_{\delta,\epsilon}} \left|\nabla u_T \circ \phi - \nabla u_T \circ \phi^k\right| \left(1 + \left|\operatorname{cof} \mathcal{D}\phi^k\right|^m\right) d\mathcal{L}$$

$$\leq \liminf_{k\to\infty} \lambda_i^k \mathcal{E}[\phi^i] + 2b(\mathcal{L}(R_{\delta,\epsilon})) + \epsilon b(\mathcal{L}(\Omega))$$

$$\leq \underline{\mathcal{E}} + \epsilon + 2b(\mathcal{L}(R_{\delta,\epsilon})) + \epsilon b(\mathcal{L}(\Omega)).$$

Finally, for given $\bar{\epsilon}$ we choose ϵ, δ, the dependent ϵ_T small enough and $k(\bar{\epsilon})$ large enough to ensure that

$$\epsilon + 2\,b(\mathcal{L}(R_{\delta,\epsilon})) + \epsilon\,b(\mathcal{L}(\Omega)) \leq \bar{\epsilon}$$

and get $\mathcal{E}[\phi] \leq \underline{\mathcal{E}} + \bar{\epsilon}$. This holds true for an arbitrary choice of $\bar{\epsilon}$. Thus we conclude

$$\mathcal{E}[\phi] \leq \underline{\mathcal{E}} = \inf_{\phi \in \mathcal{A}} \mathcal{E}[\phi],$$

which is the desired result. □

From the proof we have seen that the assumptions on the reference image could be weakened considerably compared to the template image. With respect to the applications we do not detail this difference here. A suitable matching energy density is given by

$$g(v, w, A) = |(\mathbb{1} - v \otimes v)Aw|^\gamma, \qquad (20)$$

for $1 \leq \gamma < q$. Hence, we obtain an admissible matching energy

$$\mathcal{E}_{GM}[\phi] = \int_\Omega |(\mathbb{1} - (n_T \circ \phi) \otimes (n_T \circ \phi))\mathrm{cof}\,\mathcal{D}\phi\,n_R|^\gamma.$$

Applying Theorem 1.3.3, the existence of a minimizing deformation can be established. Recalling Remark 1.3.2, we recognize that scaling the original energy density by an additional factor $|\mathrm{cof}\,\mathcal{D}\phi n_R|^\gamma$ turns the minimization task into a feasible problem. This corresponds to a modification of the area element on the level sets $[u_R = c]$. Indeed, $|\mathrm{cof}\,\mathcal{D}\phi n_R|$ is the change of the area element at a position x on $[u_R = c]$ under the deformation.

As an example for the performance of the resulting algorithm the registration of real MR and CT images of a human spine has been considered (cf. Fig. 1).

2.2 Matching Edge Sets

Now, we will consider the matching of the singular morphologies of two different images and couple this matching with edge segmentation and image denoising. In the last decade, different approaches to couple segmentation with registration have been proposed. Young and Levy [71] used segmentation results for one image to guide the search for edges in consecutive images to resolve boundaries even though they are not well defined in all images. Yezzi et al. [44] have simultaneously segmented contours in different images based on an affine matching deformation. Feron and Mohammad-Djafari [33] proposed a Bayesian approach for the joint

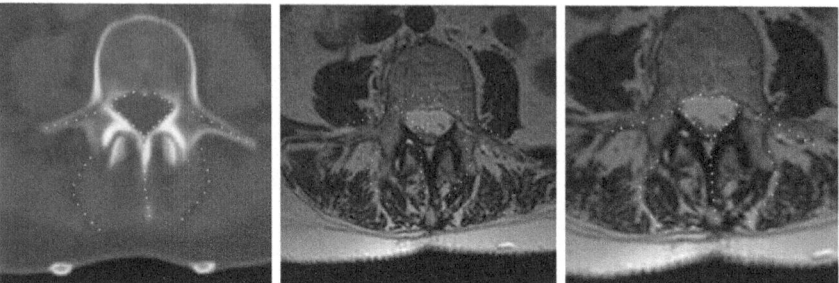

Fig. 1 Sectional morphological registration on a pair of MR and CT images of a human spine. *Dotted lines* mark certain features visible in the reference image. They are repeatedly drawn at the same position in the other images. The reference CT image (*left*), the template MR (*middle*) and the registered template (*right*) are rendered. In the *middle image* the misfit of structures of the CT image marked with corresponding *dotted lines* is clearly visible. For further details on this example we refer to [29]

segmentation and fusion of images via a coupling of suitable hidden Markov Models for multi modal images.

If we would minimize the Mumford-Shah functional (2) for u_T^0 and u_R^0 separately, we would obtain smooth representations u_T and u_R together with singularity sets $\mathcal{S}_T$ and $\mathcal{S}_R$. Now, we sum up these two functionals and replace either the reference edge set $\mathcal{S}_R$ by the pull back $\phi^{-1}(\mathcal{S}_T)$ of the template edge set or the template edge set $\mathcal{S}_T$ by the push forward $\phi(\mathcal{S}_R)$ of the reference edge set. Thus, a deformation ϕ with $\mathcal{S}_R \subset \phi^{-1}(\mathcal{S}_T)$ or $\mathcal{S}_T \subset \phi(\mathcal{S}_R)$, respectively, is a suitable candidate for the minimization of the resulting combined energy. In the first case, where we use $\phi(\mathcal{S}_R)$ instead of $\mathcal{S}_T$, the $\mathcal{H}^{d-1}$ measure [4] of $\phi^{-1}(\mathcal{S}_T)$ can be controlled by the $\mathcal{H}^{d-1}$ measure of $\mathcal{S}_T$ and the deformation ϕ, i.e., $\mathcal{H}^{d-1}(\phi(\mathcal{S}_T)) = \int_{\mathcal{S}_T} \det D\phi^{-1} \left| D\phi D\phi^T n_T^s \cdot n_T^s \right| d\mathcal{H}^{d-1}$ [25]. A similar result holds in the second case. Indeed, the control of the deformation on such lower dimensional sets is analytically and numerically difficult. Hence, we omit the corresponding energy term in our joint segmentation and registration model. Finally, in the first case of the *pull back model* the energy for the coupled Mumford-Shah segmentation in the reference and the template image is given by

$$\mathcal{E}_{\mathrm{MS}}[u_R, u_T, \mathcal{S}_T, \phi] = \int_\Omega (u_R - u_R^0)^2 \, d\mathcal{L} + \int_\Omega (u_T - u_T^0)^2 \, d\mathcal{L} \qquad (21)$$

$$+ \mu \int_{\Omega \setminus \phi^{-1}(\mathcal{S}_T)} |\nabla u_R|^2 \, d\mathcal{L} + \mu \int_{\Omega \setminus \mathcal{S}_T} |\nabla u_T|^2 \, d\mathcal{L} + \eta \mathcal{H}^{d-1}(\mathcal{S}_T)$$

with $\mu, \eta > 0$; whereas in the second case for the *push forward model* we consider the energy

$$\mathcal{E}_{MS}[u_R, u_T, \mathcal{S}_T, \phi] = \int_\Omega (u_R - u_R^0)^2 \, d\mathcal{L} + \int_\Omega (u_T - u_T^0)^2 \, d\mathcal{L} \qquad (22)$$

$$+ \mu \int_{\Omega \setminus \mathcal{S}_R} |\nabla u_R|^2 \, d\mathcal{L} + \mu \int_{\Omega \setminus \phi(\mathcal{S}_R)} |\nabla u_T|^2 \, d\mathcal{L} + \eta \mathcal{H}^{d-1}(\mathcal{S}_R).$$

Let us remark that these energies do not care about the orientation of the normals on the singularity sets $\mathcal{S}_R$ and $\phi(\mathcal{S}_R)$. Thus, they are invariant not only under monotone contrast changes.

A Level Set Method

In this section we will review a level set model for the coupled free discontinuity problem (22). Thereby, we restrict ourselves to edge sets which are the union of finitely many Jordan-curves. In this case, the feature set can be viewed as the boundary of detected segments, which are mapped to similar segment boundaries in the second image. For a large class of images, this is a very suitable and convenient approach, since images can often be decomposed into a finite set of independent objects. However this is not always the case. Crack tips might occur not only due to weak edge information but due to the fact that the image contains interrupted discontinuity sets (cf. the phase field approximation below).

In a shape optimization framework, we start with an initial shape describing the edge set and evolve it based on a suitable energy descent. The edge set may be elegantly described and propagated by the level set approach of Osher and Sethian [58,59]. In [42] a level set based, Newton-Type regularized optimization algorithm has been derived for the minimization of the original Mumford-Shah functional. That work is the algorithmical basis for our method. For related approaches we refer to [13]. In explicit, we consider $\mathcal{S}_R$ to be given as the zero level set of the level set function $\Xi : \Omega \to \mathbb{R}$, i.e., $\mathcal{S}_R = \{x : \Xi(x) = 0\}$.

The functional (22) depends on the variables u_R, u_T, ϕ and $\mathcal{S}_R$. In the process of minimization we develop different strategies for the different variables. Fortunately, the functional is quadratic with respect to the variables u_R and u_T. Hence, we may minimize (22) for fixed $\mathcal{S}_R$ and ϕ over image spaces of u_R and u_T. Let us now denote by $u_R[\mathcal{S}_R]$ and $u_T[\mathcal{S}_R, \phi]$ the corresponding minimizers. They are obtained solving the Euler Lagrange equations with respect to u_R and u_T:

$$\begin{aligned} -\mu \Delta u_R + u_R &= u_R^0 & & \text{in } \Omega \setminus \mathcal{S}_R \\ \partial_{n[\mathcal{S}_R]} u_R &= 0 & & \text{on } \mathcal{S}_R, \\ -\mu \Delta u_T + u_T &= u_T^0 & & \text{in } \Omega \setminus \phi[\mathcal{S}_R], \\ \partial_{n[\phi(\mathcal{S}_R)]} u_T &= 0 & & \text{on } \phi(\mathcal{S}_R). \end{aligned} \qquad (23)$$

It is obvious that the minimizer with respect to u_R depends only on $\mathcal{S}_R$, whereas the minimizer with respect to u_T depends also on ϕ via the domain of integration $\Omega \setminus \phi[\mathcal{S}_R]$. Now we can define the *reduced* functional

$$\hat{\mathcal{E}}[\mathcal{S}_R, \phi] = \mathcal{E}[u_R[\mathcal{S}_R], u_T[\mathcal{S}_R, \phi], \mathcal{S}_R, \phi] + \mathcal{E}_{elast}[\phi]. \quad (24)$$

Via an integral transform, we first decouple $\mathcal{S}_R$ and ϕ and obtain

$$\hat{\mathcal{E}}[\mathcal{S}_R, \phi] = \int_\Omega (u_R[\mathcal{S}_R] - u_R^0)^2 \, d\mathcal{L} + \mu \int_{\Omega \setminus \mathcal{S}_R} |\nabla u_R[\mathcal{S}_R]|^2 \, d\mathcal{L}$$

$$+ \int_\Omega \left((u_T[\mathcal{S}_R, \phi] - u_T^0)^2 \circ \phi\right) |\det \mathcal{D}\phi| \, d\mathcal{L}$$

$$+ \mu \int_{\Omega \setminus \mathcal{S}_R} \left(|\nabla u_T[\mathcal{S}_R, \phi]|^2 \circ \phi\right) |\det \mathcal{D}\phi| \, d\mathcal{L} + \eta \mathcal{H}^{d-1}[\mathcal{S}_R] + \mathcal{E}_{elast}[\phi].$$

Now we can apply (4), where we have to integrate along the boundaries from both sides of the contour, which leads to corresponding jump terms. We obtain

$$\partial_{\mathcal{S}_R} \hat{\mathcal{E}}[\mathcal{S}_R, \phi](\vartheta) = \int_{\mathcal{S}_R} \left(\llbracket (u_R[\mathcal{S}_R] - u_R^0)^2 \rrbracket + \mu \llbracket |\nabla u_R[\mathcal{S}_R]|^2 \rrbracket\right) \vartheta \, d\mathcal{H}^{d-1}$$

$$+ \int_{\mathcal{S}_R} \left(\llbracket (u_T[\mathcal{S}_R, \phi] - u_T^0)^2 \rrbracket + \mu \llbracket |\nabla u_T[\mathcal{S}_R, \phi]|^2 \rrbracket\right) \circ \phi \, |\det \mathcal{D}\phi| \, \vartheta \, d\mathcal{H}^{d-1}$$

$$+ \eta \int_{\mathcal{S}_R} h \vartheta \, d\mathcal{H}^{d-1}.$$

Recall that $u_R[\mathcal{S}_R]$ and $u_T[\mathcal{S}_R, \phi]$ are defined as the solutions of the corresponding elliptic boundary value problems (23). For the Fréchet derivative of $\hat{\mathcal{E}}$ with respect to the deformation ϕ in a direction ψ we obtain

$$\partial_\phi \hat{\mathcal{E}}[\phi](\psi) = \int_{\phi[\mathcal{S}_R]} \left(\llbracket |u_T[\mathcal{S}_R, \phi] - u_T^0|^2 \rrbracket + \mu \llbracket |\nabla u_T[\mathcal{S}_R, \phi]|^2 \rrbracket\right) (\psi \circ \phi^{-1} \cdot n_{\phi[\mathcal{S}_R]}) \, d\mathcal{H}^{d-1}$$

$$+ \partial_\phi \mathcal{E}_{reg}[\phi](\psi),$$

where the transformed normal is again given by $n_{\phi[\mathcal{S}_R]} = |\text{cof}\, \mathcal{D}\phi n[\mathcal{S}_R]|^{-1} \text{cof}\, \mathcal{D}\phi n[\mathcal{S}_R]$.

Hence, we have all the ingredients at hand to construct a gradient descent algorithm for this shape optimization problem. As outlined in Sect. 1.3 for the classical Mumford Shah model, the level set function ξ is assumed to be approximated by a finite element function Ξ in the space $\mathcal{V}^h$ on the domain Ω. This discrete function splits the domain into two discrete domains $\Omega^+[\mathcal{S}_R]$ and $\Omega^-[\mathcal{S}_R]$. On these two domains the discrete counterparts of (23) are solved via a composite finite element approach. Given the discrete solutions U_R and U_T we are then able to compute both a regularized shape gradient and a regularized gradient with respect to a discrete

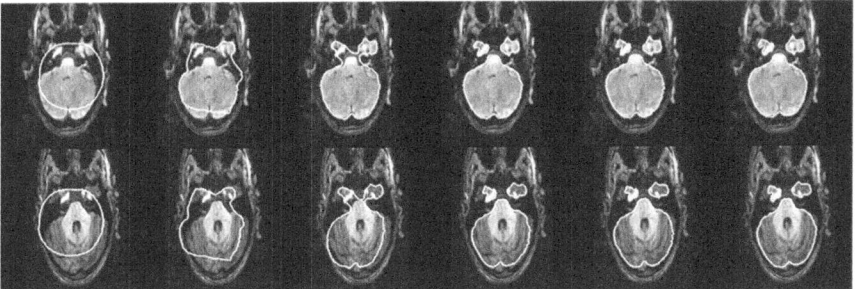

Fig. 2 For the joint segmentation approach, the evolution of level sets in a reference (*top*) and a template (*bottom*) image during the numerical energy relaxation are shown

deformation $\Phi \in (\mathcal{V}^h)^2$ based on the above formulas for the variation of the total energy with respect to $\mathcal{S}_R$ and ϕ. We demonstrate the performance of the resulting methods for 2D brain images. Figure 2 shows the relaxation of level sets segmenting brain structures jointly in two images.

A Phase Field Model

Now, we suggest for the joint edge segmentation and registration problem a coupled phase field formulation. This model picks up the pull back model (21) and represents the template edge set by a diffused interface described by a phase field variable v. This phase field, describing the edge set $\mathcal{S}_T$ of the image u_T and simultaneously $v \circ \phi$, represents a super set of the edge set $\mathcal{S}_R$ in the image u_R. The corresponding variational formulation in the spirit of the Ambrosio Tortorelli model is then given by the functional

$$\mathcal{E}_{AT}^\epsilon [u_R, u_T, v, \phi] := \int_\Omega \left\{ (u_R - u_R^0)^2 + (u_T - u_T^0)^2 \right\} d\mathcal{L}$$

$$+ \mu \int_\Omega \left\{ (v^2 \circ \phi + k_\epsilon) |\nabla u_R|^2 + (v^2 + k_\epsilon) |\nabla u_T|^2 \right\} d\mathcal{L}$$

$$+ \eta \int_\Omega \left\{ \epsilon |\nabla v|^2 + \frac{1}{4\epsilon} (v-1)^2 \right\} d\mathcal{L} \qquad (25)$$

with $k_\epsilon = o(\epsilon)$. Here, the phase field function v corresponds to the contour Γ^ϕ and the contour Γ is described by $v \circ \phi$. The first integral measures the deviation of u_R and u_T to the data in the L^2-sense. The second integral now forces the signature v^2 to be small where u_T has steep gradients and, correspondingly, $v^2 \circ \phi$ to be small where u_R has steep gradients. On the other hand, this determines ϕ to align the signature function in the reference domain to line up with the edges of u_R, and finally, for fixed signature and deformation, the smoothness of the images u_R and u_T

is controlled, i.e., steep gradients of u_T are penalized where $v \not\approx 0$ and analogously for u_R.

Again, the deformation ϕ will mainly be determined along the discontinuity sets. Indeed, as outlined above, away from the contours the phase field v will approximately be identical to 1, and hence variations of ϕ will not change the energy in these regions. Hence, we again consider a nonlinear hyperelastic regularization given by the additional energy function $\mathcal{E}_{elast}[\phi]$ (8) and finally define

$$\mathcal{E}^\epsilon[u_R, u_T, v, \phi] =:= \mathcal{E}^\epsilon_{AT}[u_R, u_T, v, \phi] + \alpha \mathcal{E}_{elast}[\phi]$$

and ask for minimizers. In what follows, let us calculate the variations of this functional with respect to the variables u_R, u_T and v in directions ζ and ϑ, respectively:

$$\partial_{u_R} \mathcal{E}^\epsilon_{AT}[u_R, u_T, v, \phi](\zeta) = 2 \int_\Omega (u_R - u_R^0) \cdot \zeta \, d\mathcal{L} + 2\mu \int_\Omega (v^2 \circ \phi + k_\epsilon) \nabla u_R \cdot \nabla \zeta \, d\mathcal{L}$$

$$\partial_{u_T} \mathcal{E}^\epsilon_{AT}[u_R, u_T, v, \phi](\zeta) = 2 \int_\Omega (u_T - u_T^0) \cdot \zeta \, d\mathcal{L} + 2\mu \int_\Omega (v^2 + k_\epsilon) \nabla u_T \cdot \nabla \zeta \, d\mathcal{L}$$

$$\partial_v \mathcal{E}^\epsilon_{AT}[u_R, u_T, v, \phi](\vartheta) = 2\mu \int_\Omega |\nabla u_T|^2 v \cdot \vartheta \, d\mathcal{L} + 2\mu \int_\Omega |\nabla u_R|^2 (v \circ \phi) \cdot (\vartheta \circ \phi)$$
$$+ 2\eta \int_\Omega \epsilon \nabla v \cdot \nabla \vartheta + \frac{1}{4\epsilon}(v-1)\vartheta \, d\mathcal{L}.$$

We rewrite this via the transformation formula:

$$\partial_v \mathcal{E}^\epsilon_{AT}[u_R, u_T, v, \phi](\vartheta) = 2\mu \int_\Omega |\nabla u_T|^2 v \cdot \vartheta \, d\mathcal{L}$$

$$+ 2\mu \int_\Omega |\nabla u_R|^2 \circ \phi^{-1} v \cdot \vartheta \det \mathcal{D}\phi^{-1} \, d\mathcal{L}$$

$$+ 2\eta \int_\Omega \epsilon \nabla v \cdot \nabla \vartheta + \frac{1}{4\epsilon}(v-1)\vartheta \, d\mathcal{L}.$$

Hence, for fixed v and ϕ, the reconstructed images u_R and u_T can be computed by solving the elliptic problems

$$u_R - \mu \mathrm{div}\left((v^2 \circ \phi + k_\epsilon)\nabla u_R\right) = u_R^0,$$
$$u_T - \mu \mathrm{div}\left((v^2 + k_\epsilon)\nabla u_T\right) = u_T^0$$

in Ω with boundary conditions $\partial_n u_R = \partial_n u_T = 0$ on $\partial\Omega$. Since $v \geq 0$, the corresponding bilinear forms are coercive. Furthermore, we are able to find for each u_T, u_R and ϕ the optimal phase field v as the solution of the Euler-Lagrange equation with respect to the variation in the variable v, i.e.,

$$\mu|\nabla u_T|^2 v + \mu|\nabla u_R|^2 \circ \phi^{-1} v \det \mathcal{D}\phi^{-1} + \frac{\eta}{4\epsilon}(v-1) - \eta\epsilon\Delta v = 0$$

in Ω and $\partial_n v = 0$ on $\partial\Omega$. Finally, the variation of the energy with respect to the deformation in a direction ψ is given by

$$\partial_\phi \mathcal{E}_{AT}^\epsilon[u_R, u_T, v, \phi](\psi) = 2\mu \int_\Omega |\nabla u_R|^2 v \circ \phi \, (\nabla v \circ \phi \cdot \psi) \, d\mathcal{L}$$

$$= 2\mu \int_\Omega |\nabla u_R|^2 \circ \phi^{-1} v \, (\nabla v \cdot \psi \circ \phi^{-1}) \det \mathcal{D}\phi^{-1} \, d\mathcal{L}.$$

Analogously to the approach chosen in the above sharp interface model, the energy functional can be reduced to depend only on ϕ, where $u_R[\phi]$, $u_T[\phi]$ and $v[\phi]$ are determined as the unique solutions of the quadratic minimization problem for fixed deformation ϕ:

$$\hat{\mathcal{E}}^\epsilon[\phi] = \mathcal{E}^\epsilon[u_R[\phi], u_T[\phi], v[\phi], \phi]. \tag{26}$$

Based on these variations of the energy we can now construct an alternating minimization algorithm for this diffusive shape optimization problem. Hence, the image intensity functions u_R, u_T, the phase field function v and the deformation ϕ are approximated by corresponding finite element functions U_R, U_T, V, and Φ in the finite element space $\mathcal{V}^h$ and $(\mathcal{V}^h)^d$ on the domain Ω. Given the discrete solutions U_R and U_T we are then able to compute both a regularized shape gradient and a regularized gradient with respect to the deformation ϕ based on the above formulas for the variation of the total energy with respect to $\mathcal{S}$ and ϕ. Figure 3 depicts an application where brain structures of MR scans of two different patients with varying image contrast are to be matched. The underlying 3D images consist of 257^3 voxels and thus $3 \cdot 257^3$ unknowns in the nodal vector of deformation. Our results demonstrate that without any pre-registration the algorithm is able to generate a fairly good match. Nevertheless, due to the structural differences in the two brains the capabilities of the algorithm are locally limited basically by the built in regularity control of the elastic deformation. The algorithm applied to raw data without any preprocessing is still capable to generate a reasonable overall matching for instance of the cortex outline or of the skull. But it significantly suffers from the local deviation of medical morphology, which requires prior knowledge on non local anatomy from the underlying mathematical morphology with its purely local definition.

2.3 Matching Singular and Regular Morphology

Now, we have all the ingredients at hand to formulate the variational problem for a matching of the singular and regular image morphology combined with a simultaneous edge segmentation and denoising in the template and the reference

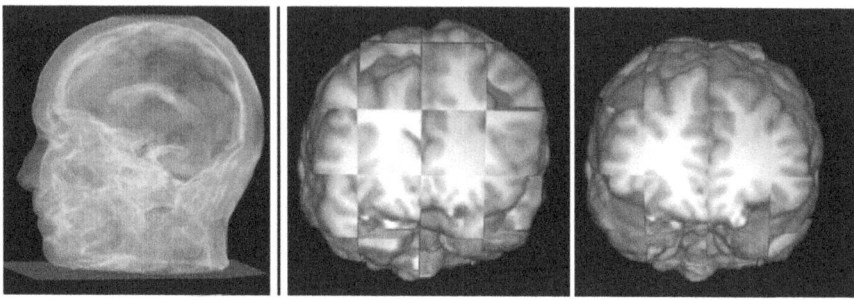

Fig. 3 On the *left*, the 3D phase field corresponding to an MR image is shown. Furthermore, the matching of two MR brain images of different patients is depicted. We use a volume renderer based on ray casting (VTK) for a 3D checkerboard with alternating boxes of the reference and the pull back of the template image to show the initial mismatch of MR brain images of two different patients (*middle*) and the results of our matching algorithm (*right*)

image. We collect the matching energy (22) for the singular morphology, the matching energy (13) for the regular morphology, the elastic regularization energy (8) and define the global energy

$$\mathcal{E}[u_R, u_T, \mathcal{S}_T, \phi] := \mathcal{E}_{MS}[u_R, u_T, \mathcal{S}_T, \phi] + \mathcal{E}_{GM}[u_R, u_T, \phi] + \mathcal{E}_{reg}[\phi]. \quad (27)$$

Even for very simple image pairs u_R^0 and u_T^0 we expect the resulting energy landscape to be very complicated. To address this issue, we will not restrict to a single fine scale problem but as above consider an embedding into a scale of problems to be solved from coarse to fine. This scale will be induced by the phase field approximation of the energy $\mathcal{E}_{MS}$. The width of the edge regions indicated by small values of v is expected to be proportional to ϵ. For decreasing ϵ we will obtain successively sharper regularized images u_T and u_R. This implicitly introduces a scale in the energy $\mathcal{E}_{GM}$ as well. In explicit the gradients ∇u_T and ∇u_R corresponding to u_T and u_R are expected to be smoother for larger ϵ. Thus, we no longer have to distinguish regular and singular gradients. To focus only on the regular morphology in this energy contribution—in particular not measuring edges—we mask out a gradient comparison in the vicinity of edges. Therefore, the integrand is multiplied by $v^2 \circ \phi$ and we obtain

$$\mathcal{E}_{GM}^\epsilon[u_T, u_R, v, \phi] = \int_\Omega v^2 \circ \phi \; g_0(\nabla u_T \circ \phi, \nabla u_R, \operatorname{cof} \mathcal{D}\phi) \, d\mathcal{L}. \quad (28)$$

Finally, gathering this energy and the energy contributions from (25) and (8) we define a scale of global approximate energies

$$\mathcal{E}^\epsilon[u_R, u_T, v, \phi] := \mathcal{E}_{AT}^\epsilon[u_R, u_T, v, \phi] + \mathcal{E}_{GM}^\epsilon[u_R, u_T, v, \phi] + \mathcal{E}_{reg}[\phi]$$

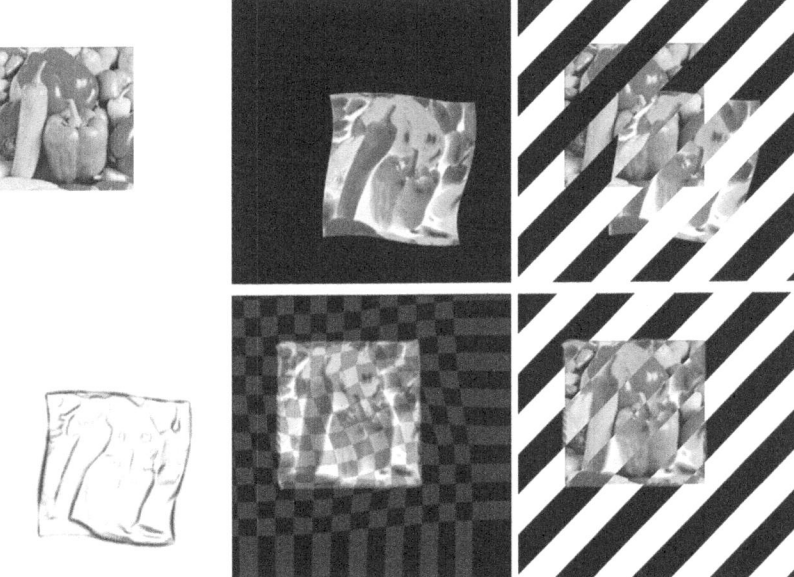

Fig. 4 A contrast invariant matching is shows. In the *first row* we have inverted, moved and distorted the peppers image (*left*) to obtain a template image (*middle*). On the *right* the initial misfit is shown. The registration result for this input data are depicted in the *second row*. The final phase field function v is depicted on the *left*. The image in the *middle* shows a plot of the deformation due to a relaxation of the combined energy, i.e., the registration of discontinuity sets and level sets. On the *right* alternating slices of the reference and the pulled back template image allow a visual validation of the matching result

depending on the scale parameter ϵ. We refer to Fig. 4 for results achieved via a relaxation of this energy. Apart from $\mathcal{E}_{GM}$ the energy depends quadratically on u_T, u_R and v. Thus, the corresponding necessary conditions to be fulfilled by a minimizer, i.e., the Euler Lagrange equations with respect to these variables, turn into linear problems. In the relaxation scheme for the deformation, which actually describes the image matching, we treat u_T, u_R and v in a quasi stationary way. In fact, the iterative relaxation proceeds as follows: For given images and deformation, we optimize w.r.t. the phase field v. In a next step, we then optimize for the regularized images u_T and u_R for given ϕ and already optimized v. Finally, we consider one gradient descent step for the global energy w.r.t. the deformation. Here, we pick up a regularized gradient descent as described above. This procedure is repeated until convergence. In the current implementation we neglect the impact of the ongoing segmentation process on the variation of the energy concerned with the regular morphology and consider the following simplification in the method:

$$(u_T, u_R) = \arg\min_{u_T, u_R} \mathcal{E}_{AT}^{\epsilon}[u_T, u_R, v, \phi].$$

Fig. 5 A facial texture matching problem. Initial reference texture map u_R (*left*), initial template u_T (*middle*) and the initial misfit plot on the (*right*)

Even though we no longer actually minimize the global energy, the proposed restricted energy relaxation already leads to satisfying edge segmentation and matching results.

We have applied our method also to 3D medical image registration problems and present here some results, where we concentrate on a matching only of the singular morphology. In particular in 3D a cascadic multiscale approach turned out to be indispensable to ensure an efficient numerical implementation.

Finally, we demonstrate the applicability of the method by registering two different facial texture maps. Figure 5 depicts the input data, whereas Fig. 6 pinpoints the differences of the different matching approaches. As in the previous example, Fig. 7 illustrates the energetic improvement due to the interplay of the deformation and the phase field function, reducing the length of the overall interface by alignment of edges.

2.4 A Parametric Approach to Surface Matching

Establishing a correspondence between two surfaces is a basic ingredient in many geometry processing applications. Motivated by the ability to scan geometry with high fidelity, a number of approaches have been developed in the graphics literature to bring such scans into correspondence. Early work used parameterizations of the meshes over a common parameter domain to establish a direct correspondence between them [48]. Typically these methods are driven by user-supplied feature correspondences which are then used to drive a mutual parametrization. The main difficulty is the proper alignment of selected features during the parameterizations process [47, 60, 65] and the algorithmic issues associated with the management of irregular meshes and their effective overlay. Here, we address this problem on the background of deformations of elastic shells. Recently, Gu and Vemuri [38] considered matches of topological spheres through conformal maps with applications to brain matching. Their energy measures the defect of the conformal factor and—similar to our approach—the defect of the mean curvature. However they

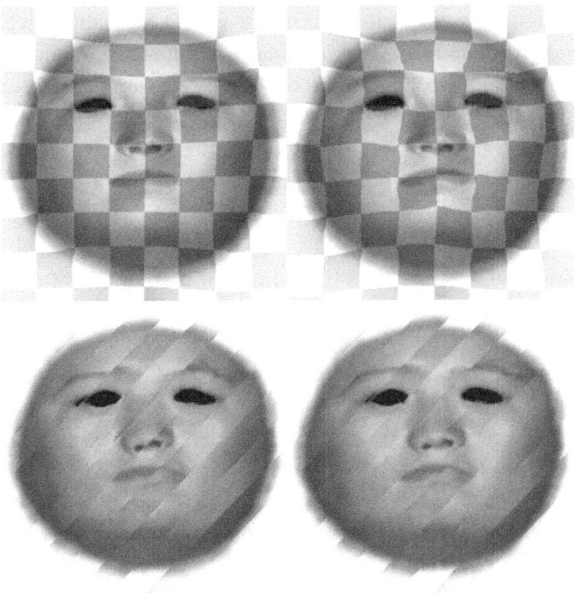

Fig. 6 Results for the facial texture matching problem given in Fig. 5. The *top row* shows the deformed template that has been overlaid by a uniform checkerboard pattern. On the *left* the regular morphology has not been taken into account, hence, mainly the face outline and strong edges are matched properly. Considering the entire energy functional significantly improves the result (*right*). The deformation is characterized by a much higher variability. The *bottom row* displays with alternating *stripes* the corresponding reference and pulled back template images to enable a validation of the matching results

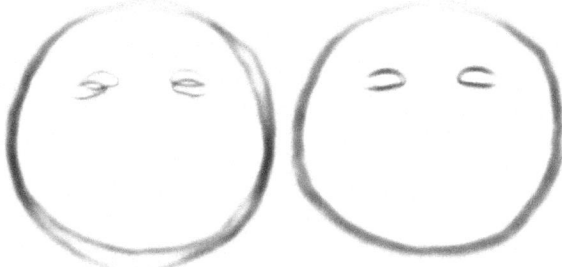

Fig. 7 The phase field v corresponding to the matching results in Fig. 5. *Left:* Initial phase field. *Right:* Phase field after alignment

do not measure the correspondence of feature sets or tangential distortion, and thus do not involve a regularization energy for the ill-posed energy minimization. Furthermore, they seek a one-to-one correspondence, whereas we must address the difficult problem of partial correspondences between surfaces with boundaries.

Instead of matching two surfaces directly in 3D, we apply well-established matching methods from *image processing* in the parameter domains of the surfaces. A matching energy is introduced that can depend on curvature, feature demarcations or surface textures, and a regularization energy controls length and area changes in the induced non-rigid deformation between the two surfaces. The metric on both surfaces is properly incorporated into the formulation of the energy. This approach reduces all computations to the 2D setting while accounting for the original geometries. Consequently a fast multiresolution numerical algorithm for regular image grids can be used to solve the global optimization problem. The final algorithm is robust, generically much simpler than direct matching methods, and very fast for highly resolved triangle meshes.

Our goal is to correlate two surface patches, $\mathcal{S}_A$ and $\mathcal{S}_B$, through a non-rigid spatial deformation

$$\phi_\mathcal{S} : \mathcal{S}_A \to \mathbb{R}^3,$$

such that corresponding regions of $\mathcal{S}_A$ are mapped onto regions of $\mathcal{S}_B$. In doing so, we want to avoid the general difficulty of formulating these maps directly in $\mathbb{R}^3$ and the particularly tedious algorithmic issues in the application, where the two surface patches are given as distinct, arbitrary triangulations. Instead, we match parameter domains covered with geometric and user-defined feature characteristics. The main benefit of this approach is that it simplifies the problem of finding correspondences for surfaces embedded in $\mathbb{R}^3$ to a matching problem for images in two dimensions. To ensure that the actual geometry of the surface patches is treated properly here, the energy on the deformations from one parameter space to the other will measure:

- *Elastic distortion energy:* smoothness of the deformation in terms of tangential distortion,
- *Bending energy:* bending of normals through the proper correspondence of curvature, and
- *Feature energy:* the proper correspondence of important surface and texture features.

Furthermore, it will consistently take into account the proper metrics on the parameter domains, which ensures that we are actually treating a deformation from one surface onto the other even though all computations are performed in 2D.

Physical Motivation

Consider the first surface to be a thin shell, which we press into a mould of the second surface (cf. Fig. 8). One can distinguish between stresses induced by stretching and compression and stressed induced by bending that occurs in the surface as it is being pressed. Thus $\phi_\mathcal{S}$ can be regarded as the deformation of such a thin shell. We assume this deformation to be elastic. The regularization energy in (29) will measure the induced in-plane stresses, and the concrete energy density

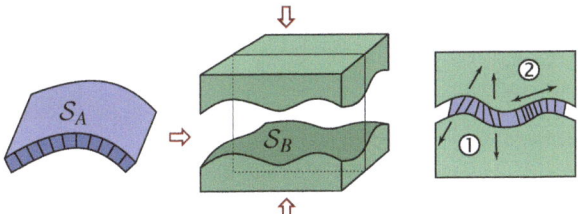

Fig. 8 A physical interpretation of ϕ_M as pressing a thin shell $\mathcal{S}_A$ into a mould of the surface $\mathcal{S}_B$ being matched. The bending (1) and stretching (2) of the thin shell is measured in our matching energy and minimized by the optimal match ϕ_S

in (30) allows control over length and area-distortion in this surface-to-surface deformation. Since we are aiming for a proper correspondence of shape, we will incorporate the bending of normals in our energy with (31). Finally, the matching of feature sets in (32) will provide user-specified landmarks to guide the surface deformation. In what follows, we will develop the variational approach step-by-step.

To begin with, let us set up proper parameterizations. A parametrization is a mapping from the plane onto a given surface, or in the case of its inverse, from the surface onto the plane. Consider a smooth surface $\mathcal{S} \subset \mathbb{R}^3$, and suppose $x : \omega \to \mathcal{S}$ is a parameterizations of $\mathcal{S}$ on a parameter domain ω. For a parameterizations to be properly defined, its inverse x^{-1} cannot allow the surface to fold onto itself in the plane. In this case x is a bijection and a metric g is properly defined on ω,

$$g = \mathcal{D}x^T \mathcal{D}x,$$

where $\mathcal{D}x \in \mathbb{R}^{3,2}$ is the Jacobian of the parameterization x. The metric g acts on tangent vectors v, w on the parameter domain ω with $(g\,v) \cdot w = \mathcal{D}x\,v \cdot \mathcal{D}x\,w$, which is simply the inner product of tangent vectors $\mathcal{D}x\,v, \mathcal{D}x\,w$ on the surface. Thus, it follows that the metric describes how length, area and angles are distorted under the parametrization function.

Let us now focus on the distortion from the surface $\mathcal{S}$ onto the parameter domain ω under the inverse parametrization x^{-1}. This distortion is measured by the inverse metric $g^{-1} \in \mathbb{R}^{2,2}$. Just as $\sqrt{\operatorname{tr}(A^T A)}$ measures the average *change of length* under a linear mapping A, $\sqrt{\operatorname{tr} g^{-1}}$ measures the average change of length of tangent vectors under the mapping from the surface onto the parameter plane. Additionally, $\sqrt{\det g^{-1}}$ measures the corresponding *change of area*. We will use these quantities in the following sections to account for the distortion of length and area on the surface as we formulate our matching energy in the parameter domain.

Measuring Distortion via a Deformation

The above discussion now applies to the parameter maps x_A and x_B of the surfaces $\mathcal{S}_A$ and $\mathcal{S}_B$. We suppose that these parameterizations are defined in an initial step and

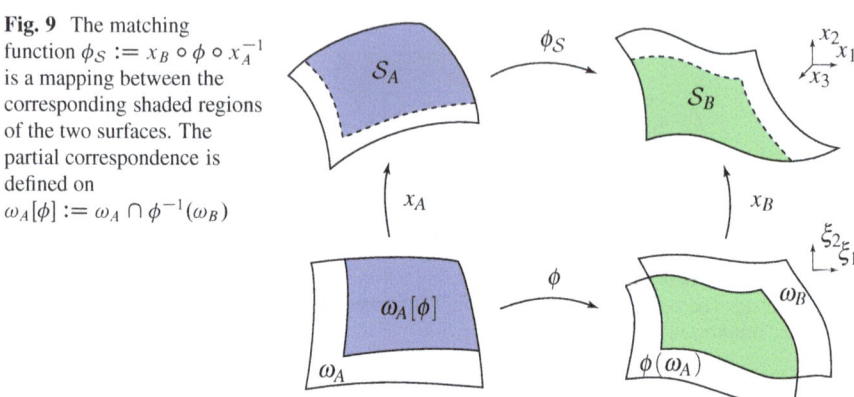

Fig. 9 The matching function $\phi_S := x_B \circ \phi \circ x_A^{-1}$ is a mapping between the corresponding shaded regions of the two surfaces. The partial correspondence is defined on $\omega_A[\phi] := \omega_A \cap \phi^{-1}(\omega_B)$

we assume that x_A and x_B as well as the corresponding parameter domains ω_A and ω_B are fixed from now on. Their metrics are denoted by g_A and g_B, respectively. We will now study the distortion, which arises from a deformation of the first parameter domain onto the second parameter domain. First, let us consider deformations $\phi : \omega_A \to \omega_B$ which are one-to-one. This deformation between parameter domains induces a deformation between the surface patches $\phi_S : S_A \to S_B$ defined by

$$\phi_S := x_B \circ \phi \circ x_A^{-1}.$$

Let us emphasize that we do not actually expect a one-to-one correspondence between surface patches. Later we will relax this assumption and in particular allow for deformations ϕ with $\phi(\omega_A) \not\subset \omega_B$. The complete mapping is illustrated in Fig. 9.

Now let us focus on the distortion from the surface S_A onto the surface S_B. In elasticity, the distortion under an elastic deformation ϕ is measured by the Cauchy-Green strain tensor $\mathcal{D}\phi^T \mathcal{D}\phi$. We wish to adapt this definition to measure distortion between tangent vectors on the two surfaces, as we did with the metric g in the previous section. Therefore, we properly incorporate the metrics g_A and g_B at the deformed position and obtain the distortion tensor $\mathcal{G}[\phi] \in \mathbb{R}^{2,2}$ given by

$$\mathcal{G}[\phi] = g_A^{-1} \mathcal{D}\phi^T (g_B \circ \phi) \mathcal{D}\phi,$$

which acts on tangent vectors on the parameter domain ω_A, where products are given in matrix notation. Mathematically, this tensor is defined implicitly via the identity $(g_A \mathcal{G}[\phi] v) \cdot w = (g_B \circ \phi) \mathcal{D}\phi v \cdot \mathcal{D}\phi w$ for tangent vectors v, w on the surface S_A and their images as tangent vectors $\mathcal{D}\phi v$, $\mathcal{D}\phi w$ on S_B, where here we have identified tangent vectors on the surfaces with vectors in the parameter domains.

As in the parametrization case, one observes that $\operatorname{tr} \mathcal{G}[\phi]$ measures the average *change of length* of tangent vectors from S_A when being mapped to tangent vectors on S_B, and $\sqrt{\det \mathcal{G}[\phi]}$ measures the *change of area* under the deformation ϕ_S. Thus tr $\mathcal{G}[\phi]$ and $\sqrt{\det \mathcal{G}[\phi]}$ are natural variables for an energy density in a variational approach measuring the *regularity* of a surface deformation,

$$\mathcal{E}_{elast}[\phi] = \int_{\omega_A} W(\operatorname{tr} \mathcal{G}[\phi], \sqrt{\det \mathcal{G}[\phi]}) \sqrt{\det g_A} \, d\xi \,. \tag{29}$$

This simple class of energy functionals was rigorously derived in [22] from a set of natural axioms for measuring the distortion of a single parametrization. In particular, the following energy density

$$W(I_1, I_2) = \alpha_l I_1 + \alpha_a \left(I_2^2 + (1 + \tfrac{\alpha_l}{\alpha_a}) I_2^{-2} \right) \tag{30}$$

accounts for length distortion with $I_1 = \operatorname{tr} \mathcal{G}[\phi]$, area expansion with $I_2 = \sqrt{\det \mathcal{G}[\phi]}$ and area compression with I_2^{-1}. The weights $\alpha_l, \alpha_a > 0$ are typically chosen by the user according to the relative importance of length and area distortion.

Measuring Bending in a Deformation

When we press a given surface $\mathcal{S}_A$ into the thin mould of the surface $\mathcal{S}_B$, a major source of stress results from the bending of normals. We assume these stresses to be elastic as well and to depend on changes in normal variations under the deformation. Variations of normals are represented in the metric by the shape operator. For the derivation of the shape operators S_A and S_B of the surface patches $\mathcal{S}_A$ and $\mathcal{S}_B$ we refer to [49], where we end up with $\operatorname{tr}(S_B \circ \phi) - \operatorname{tr}(S_A)$ as a *measure for the bending of normals*. Since the trace of the shape operator is the mean curvature, we can instead aim at comparing the mean curvature $h_B = \operatorname{tr}(S_B)$ of the surface $\mathcal{S}_B$ at the deformed position $\phi_S(x)$ and the mean curvature $h_A = \operatorname{tr}(S_A)$ of the surface $\mathcal{S}_A$. A similar observation was used by [37] to define a bending energy for discrete thin shells. This proposed simplification neglects any rotation of directions due to the deformation, *e.g.* if the deformation aligns a curve with positive curvature on the first surface to a curve with negative curvature on the second surface and vice versa, an energy depending solely on $h_B \circ \phi - h_A$ does not recognize this mismatch. Nevertheless, in practice the bending energy

$$\mathcal{E}_{bend}[\phi] = \int_{\omega_A} (h_B \circ \phi - h_A)^2 \sqrt{\det g_A} \, d\xi \tag{31}$$

turns out to be effective and sufficient. By minimizing this energy, we ensure that the deformation properly matches mean curvature on the surfaces.

Matching Features

Frequently, surfaces are characterized by similar geometric or texture features, which should be matched properly as well. Therefore we will incorporate a correspondence between one-dimensional feature sets in our variational approach to match characteristic lines drawn on the surface. In particular, we prefer feature lines to points for the flexibility afforded to the user, as well as to avoid the theoretical problems introduced by point constraints [19]. We will denote the feature sets by $\mathcal{F}_{S_A} \subset \mathcal{S}_A$ and $\mathcal{F}_{S_B} \subset \mathcal{S}_B$ on the respective surfaces. Furthermore, let $\mathcal{F}_A \subset \omega_A$ and $\mathcal{F}_B \subset \omega_B$ be the corresponding sets on the parameter domains. We are aiming for a proper match of these sets via the deformation, i.e.

$$\phi_S(\mathcal{F}_{S_A}) = \mathcal{F}_{S_B}$$

or in terms of differences, $\mathcal{F}_{S_A} \setminus \phi_S^{-1}(\mathcal{F}_{S_B}) = \emptyset$ and $\mathcal{F}_{S_B} \setminus \phi_S(\mathcal{F}_{S_A}) = \emptyset$. A rigorous way to reflect this in our variational approach is with a third energy contribution,

$$\mathcal{E}_\mathcal{F}[\phi] = \mathcal{H}^1(\mathcal{F}_{S_A} \setminus \phi_S^{-1}(\mathcal{F}_{S_B}))$$
$$+ \mathcal{H}^1(\mathcal{F}_{S_B} \setminus \phi_S(\mathcal{F}_{S_A})), \tag{32}$$

where $\mathcal{H}^1(\mathcal{A})$ is the one-dimensional Hausdorff measure of a set $\mathcal{A}$ on the corresponding surface. Roughly speaking, this gives a symmetric measurement of the size of the mismatch of the features. This type of energy does not lend itself to a robust numerical minimization. Therefore, we will instead consider a suitable approximation of (32) that involves the *distance on the surface to the feature sets* and define

$$\tilde{\mathcal{E}}_\mathcal{F}^\epsilon[\phi] = \int_{\omega_A} \left(\eta^\epsilon \circ d_A(\xi)\right)\left(\vartheta^\epsilon \circ d_B(\phi(\xi))\right) \sqrt{\det g_A}\, d\xi +$$
$$\int_{\omega_B} \left(\eta^\epsilon \circ d_B(\xi)\right)\left(\vartheta^\epsilon \circ d_A(\phi^{-1}(\xi))\right) \sqrt{\det g_B}\, d\xi, \tag{33}$$

where $d_A(\cdot) = \mathrm{dist}_A(\cdot, \mathcal{A})$ and $d_B(\cdot) = \mathrm{dist}_B(\cdot, \mathcal{A})$ are distance functions on the parameter domains ω_A and ω_B with respect to some set $\mathcal{A}$ on the corresponding surface. Note that we measure distance either in the metric g_A on ω_A or in the metric g_B on ω_B. Additionally, we define the localization functions

$$\eta^\epsilon(s) = \tfrac{1}{\epsilon} \max\left(1 - \tfrac{s}{\epsilon}, 0\right), \quad \vartheta^\epsilon(s) = \min\left(\tfrac{s^2}{\epsilon}, 1\right),$$

which act as cut-off functions. For Lipschitz continuous feature sets and bi-Lipschitz continuous deformations, we observe that $\tilde{\mathcal{E}}_\mathcal{F}^\epsilon[\phi] \to \mathcal{E}_\mathcal{F}[\phi]$ as $\epsilon \to 0$, which motivates our approximation. In view of the later discretization, we can reformulate the second term in (33) as

$$\int_{\omega_A} \left(\eta^\epsilon \circ d_B(\phi(\xi))\right)\left(\vartheta^\epsilon \circ d_A(\xi)\right) \sqrt{\det g_B(\phi(\xi))} \det \mathcal{D}\phi\, d\xi.$$

Usually, we cannot expect that $\phi_S(S_A) = S_B$, particularly near the boundary where certain subregions of S_A will have no corresponding counterpart on S_B and vice versa. Therefore, we must allow for points on S_B with no pre-image in S_A under a matching deformation ϕ_S and points on S_A, which are not correlated to points on S_B via ϕ_S (cf. Fig. 9). Thus we must adapt the variational formulation accordingly. If $\phi(\omega_A) \neq \omega_B$, then ϕ_S is now defined on $x_A(\omega_A[\phi])$ only, where

$$\omega_A[\phi] := \phi^{-1}(\phi(\omega_A) \cap \omega_B)$$

is the corresponding subset of the parameter domain ω_A. Furthermore, we define new energies (with modifications marked in red):

$$\mathcal{E}_{bend}[\phi] = \int_{\omega_A[\phi]} (h_B \circ \phi - h_A)^2 \sqrt{\det g_A}\, d\xi,$$

$$\mathcal{E}_{\mathcal{F}}[\phi] = \mathcal{H}^1(\omega_A[\phi] \cap \mathcal{F}_{S_A} \setminus \phi_S^{-1}(\mathcal{F}_{S_B}))$$
$$+ \mathcal{H}^1(\mathcal{F}_{S_B} \setminus \phi_S(\omega_A[\phi] \cap \mathcal{F}_{S_A})).$$

For an energy that controls tangential distortion, it is still helpful to control the regularity of the deformation outside the actual matching domain $\omega_A[\phi]$, where we would like to allow significantly larger deformations by using a "softer" elastic material. Hence we will suppose that g_B, which is initially only defined on ω_B, is extended to $\mathbb{R}^2$ and takes values that are relatively small to allow for greater stretching.

In the minimization algorithm, we need descent directions, which will involve derivatives of these energies with respect to the deformation ϕ. In taking these derivatives, integrals over the variable boundary $\partial \omega_A[\phi]$ will appear. Since these are tedious to treat numerically, we will rely on another approximation for the sake of simplicity. Our strategy here is to change the domain of integration $\omega_A[\phi]$ to a superset ω, which extends beyond the boundary $\partial \omega_A[\phi]$. Doing so means that special treatment of boundary integrals is no longer necessary, although we are now required to evaluate the integrands of the energies outside of ω_A, and similarly for deformed positions outside of ω_B. To achieve this, we will extend our surface quantities onto $\omega \setminus \omega_A$ and $\omega \setminus \omega_B$, respectively, by applying a harmonic extension with natural boundary conditions on $\partial \omega$ to g_A, g_B and h_A, h_B (e.g., we define h_A as the solution of Laplace's equation on $\omega \setminus \omega_A$ with vanishing flux on $\partial \omega$). Additionally, we introduce a regularized characteristic function

$$\chi_A^\epsilon(\xi) = \max(1 - \epsilon^{-1} \text{dist}(\xi, \mathcal{A}), 0)$$

to cause the energy contributions to be ignored at some distance ϵ away from $\omega_A[\phi]$. Thus, instead of dealing with a deformation dependent domain $\omega[\phi]$ in the definition of our different energy contributions, we always integrate over the *whole* image domain ω and insert the product of the two regularized characteristic functions

$$\chi^\epsilon(\xi) = \chi_{\omega_A}^\epsilon(\xi)\, \chi_{\omega_B}^\epsilon(\phi(\xi))$$

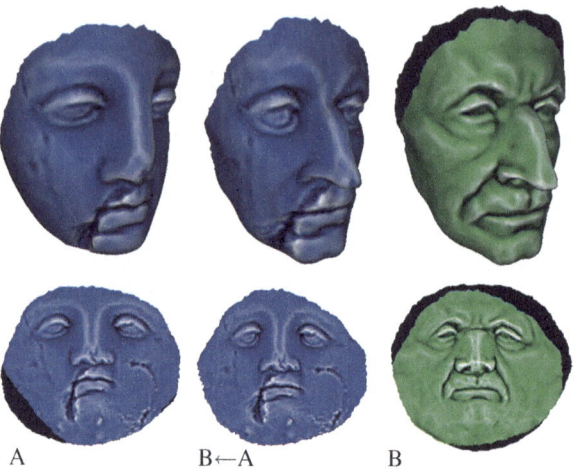

Fig. 10 For surfaces with boundaries, a partial correspondence is often desired. The correspondence is defined where their parameter domains intersect under the matching deformation (*bottom*). In this domain, quantities such as texture maps can be mapped between the surfaces (*center*). The unmatched regions are in *black*

as an additional factor in the energy integrand. We apply this modification to the energy $\mathcal{E}_{bend}$ (31) and the already regularized energy $\tilde{\mathcal{E}}_{\mathcal{F}}^{\epsilon}$ (33) and denote the resulting energies by $\mathcal{E}_{bend}^{\epsilon}$ and $\mathcal{E}_{\mathcal{F}}^{\epsilon}$, respectively. Furthermore, for the elastic energy $\mathcal{E}_{elast}[\phi]$ we assume a very soft elastic material outside $\omega_A[\phi]$ and the actual material parameters inside this partial matching domain. Hence, as an approximate model we consider

$$\mathcal{E}_{elast}^{\epsilon,\delta}[\phi] = \int_{\omega_A} (\delta + (1-\delta)\chi^{\epsilon}(\xi)) \, W(\operatorname{tr} \mathcal{G}[\phi], \sqrt{\det \mathcal{G}[\phi]}) \sqrt{\det g_A} \, d\xi$$

for some fixed, small δ.

Definition of the Matching Energy

We are now ready to collect the different cost functionals and define the global matching energy. Depending on the user's preference, we introduce weights β_{bend}, β_{elast}, $\beta_{\mathcal{F}}$ for the energies $\mathcal{E}_{bend}^{\epsilon}$, $\mathcal{E}_{elast}$ and $\mathcal{E}_{\mathcal{F}}^{\epsilon}$, respectively, and define the global energy

$$\mathcal{E}^{\epsilon}[\phi] = \beta_{bend} \, \mathcal{E}_{bend}^{\epsilon}[\phi] + \beta_{elast} \, \mathcal{E}_{elast}^{\epsilon,\delta}[\phi] + \beta_{\mathcal{F}} \, \mathcal{E}_{\mathcal{F}}^{\epsilon}[\phi] \,,$$

which measures the quality of a matching deformation ϕ on the domain ω. In the limit for $\epsilon, \delta \to 0$ we aim at an approximation of the energy $\mathcal{E}[\phi] = \beta_{bend} \, \mathcal{E}_{bend}[\phi] + \beta_{elast} \, \mathcal{E}_{elast}[\phi] + \beta_{\mathcal{F}} \, \mathcal{E}_{\mathcal{F}}[\phi]$. Figure 10 demonstrates the capability of the presented approach to describe a partial matching problem in variational form.

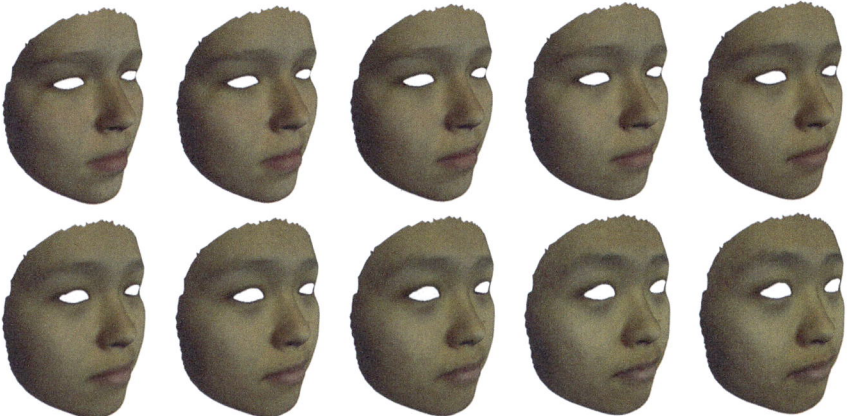

Fig. 11 A morphing between two different faces is shown

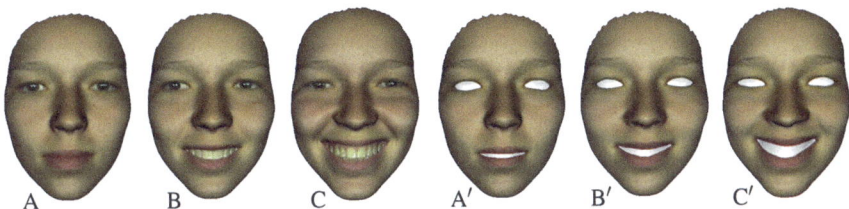

Fig. 12 Morphing through keyframe poses A, B, C is accomplished through pair-wise matches $A \to B$ and $B \to C$. Starting with A we blend its shape into B using $A \to B$, and then morph to C by applying $A \to B$ followed by $B \to C$. The skin texture from A is used throughout. Because of the close similarity in the poses, one can expect the intermediate blends A', B', C' to correspond very well with the original keyframes A, B, C, respectively

Based on the matching we can generate warps between different surface based on a linear blend between matched points on the different geometries (cf. Figs. 11 and 12). Figure 13 shows the distortion of a checkerboard pattern on the parameter domain and the corresponding distortion on the surface.

3 Joint Motion Estimation, Restoration, and Deblurring

In this section we introduce a variational method, which jointly handles motion estimation, moving object detection and motion blur deconvolution (cf. Fig. 17). The proposed framework is again a Mumford-Shah type variational formulation, which includes an explicit modeling of the motion-blur process as well as shape and image regularization terms. The input to the variational formulation are two consecutive frames, while the output are the corresponding reconstructed frames, the segmented

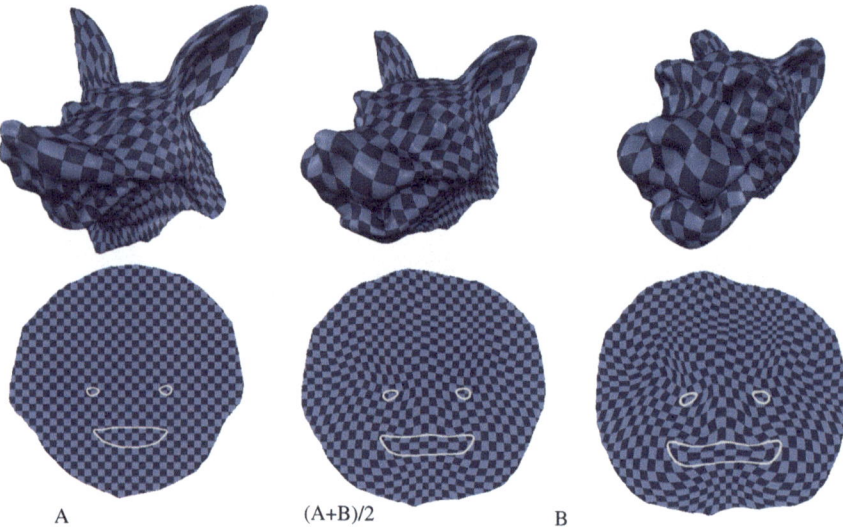

Fig. 13 Large deformations are often needed to match surfaces that have very different shapes. A checkerboard is texture mapped onto the first surface as it morphs to the second surface (*top*). The matching deformation shown in the parameter domain (*bottom*) is smooth and regular, even where the distortion is high (e.g., around the outlines of the mouth and eyes)

moving object, and the actual motion velocity. As in the problems discussed so far the joint estimation of motion, moving object region and reconstructed images outperforms techniques where each individual unknown is individually handled.

3.1 The Motion Blur Model

Images from an image sequence captured with a video camera are integrated measurements of light intensity emitted from moving objects over the aperture time interval of the camera. Let $u : [-T, T] \times \Omega; (t, x) \mapsto \mathbb{R}$ denote a continuous sequence of scene intensities over a time interval $[-T, T]$ and on a spatial image domain Ω observed via the camera lens. The video sequence recorded with the camera consists of a set of images $g_i : \Omega \to \mathbb{R}$ associated with times t_i, for $i = 1, \cdots, m$, given as the convolution

$$g_i(x) = \frac{1}{\tau} \int_{t_i - \frac{1}{2}\tau}^{t_i + \frac{1}{2}\tau} u(t + s, x) \, ds \qquad (34)$$

over the aperture time τ. For the time integral, we propose a box filter, which realistically approximates the mechanical shutters of film cameras and the electronic read out of modern CCD video recorders. In the simplest case, where the sequence f

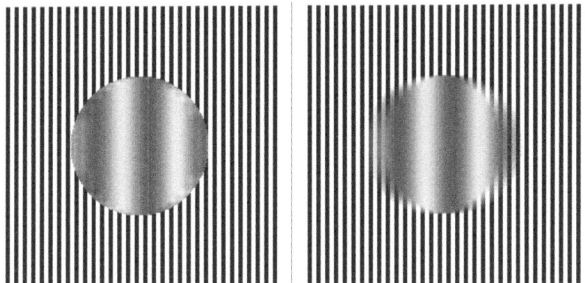

Fig. 14 We consider a moving circle with *black* and *white stripes* in front of a similarly textured background. For this test case a comparison is shown between the wrong (*left*) motion blur model which ignores the motion discontinuity at the boundary and our realistic, consistent model (*right*) given in (36)

renders an object moving at constant velocity $w \in \mathbb{R}^2$, i.e. $u(x - sw) = u(t + s, x)$, we can transform integration in time to an integration in space and obtain for the recorded images

$$g_i(x) = \frac{1}{\tau} \int_{-\frac{1}{2}\tau}^{\frac{1}{2}\tau} u(x - sw) \, ds = (u * h_w)(x), \qquad (35)$$

for a one dimensional filter kernel $h_w = \delta_0(\frac{w^\perp}{|w|} \cdot y) h(\frac{w}{|w|} \cdot y)$ with filter width $\tau |w|$ in the direction of the motion trajectory $\{y = x + sw : s \in \mathbb{R}\}$. Here $w^\perp$ denotes w rotated by 90°, δ_0 is the usual 1D Dirac distribution and h the 1D block filter with $h(s) = \frac{1}{\tau |w|}$ for $s \in [-\frac{\tau |w|}{2}, \frac{\tau |w|}{2}]$ and $h(s) = 0$, else. In case of an object moving in front of a (still) background the situation is somewhat more complicated. At a point x close to the boundary of the object, the convolution (34) decomposes into a spatial convolution of object intensities along the motion path for the subinterval of the aperture interval where the object covers the background at position x, and a retrieval of the background intensity for the remaining opening time of the lens. Figure 14 shows a comparison between the actually observed motion blur and results obtained by a (wrongly) direct application of the spatial convolution formula (35) on a moving circular object in front of a textured background (more specifics on this below). This observation is particularly important for the reliable recovery of boundaries of moving objects from recorded video frames g_i and subsequently for the proper restoration of image frames (cf. Fig. 15 for a corresponding comparison).

In what follows we consider an object moving with speed $w \in \mathbb{R}^2$ in front of a still background $u_{\text{bg}} : \Omega \to \mathbb{R}$. The object at time 0 is represented by an intensity function $u_{\text{obj}} : \mathcal{O} \to \mathbb{R}$ defined on an object domain $\mathcal{O}$. From u_{obj} and u_{bg} one assembles the actual scene intensity function

$$u(t, x) = u_{\text{obj}}(x - tw)\chi_{\text{obj}}(x - wt) u_{\text{bg}}(x)(1 - \chi_{\text{obj}}(x - wt)) \qquad (36)$$

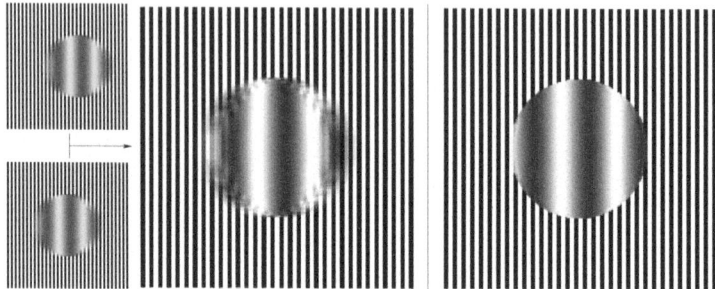

Fig. 15 Given two frames for the realistic motion blur showing the moving circle on the texture background from Fig. 14 (*left*), computational results for the deblurring are depicted based on the wrong motion blur model built into G_i (*middle*), and on our consistent model (*right*). This clearly outlines the importance of a proper handling of the motion discontinuity in the considered motion blur model

at time t, where $\chi_{\text{obj}} : \mathbb{R}^2 \to \mathbb{R}$ denotes the characteristic function of $\mathcal{O}$. Now, inserting (36) in (34) and then using (35) on $\mathcal{O}$, we deduce the correct formula for the theoretically observed motion blur at time t_i,

$$G_i[\mathcal{O}, w, u_{\text{obj}}, u_{\text{bg}}](x) = ((u_{\text{obj}}\chi_{\text{obj}}) * h_w)(x - t_i w) + u_{\text{bg}}(x)(1 - (\chi_{\text{obj}} * h_w)(x - t_i w)),$$

for given object domain $\mathcal{O}$, motion velocity w, and object and image (background) intensity functions u_{obj} and u_{bg} respectively. If we do not carefully model the observed intensities while the moving object occludes and uncovers the background, we would observe $(u_{\text{obj}}(t, \cdot) * h_w)$ on the object domain and u_{bg} elsewhere (cf. the combination of (14) and (3) in [32]). Given the more precise motion blur model proposed here, we now proceed to derive a variational formulation to simultaneously estimate all parameters in this equation based on two consecutive frames.

3.2 Joint Variational Approach

Given two frames g_1 and g_2 of a video sequence with motion blur recorded at times t_1 and t_2, respectively, we construct a variational model to extract from these frames the domain $\mathcal{O}$, the image intensity u_{obj} of a moving object and the motion velocity w. Here, we propose that the background intensity u_{bg} can a priori be extracted from the video sequence, for example, by averaging pixels with stable values over a sequence of frames. We aim at formulating a joint energy for these degrees of freedom. Modeling this energy we take into account the following observations:

Given w and intensity maps $u_{\text{obj}}, u_{\text{bg}} : \Omega \to \mathbb{R}$ (extended on the whole domain in a suitable way), we phrase the identification problem of the object boundary $\mathcal{S} = \partial\mathcal{O}$ in terms of a piecewise constant Mumford–Shah model. This appears to be well-suited, in particular because the unknown contour is significantly smeared out due

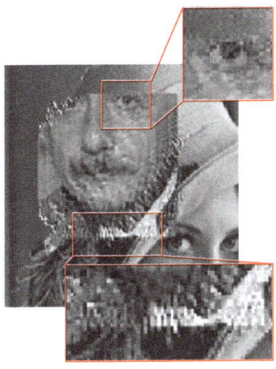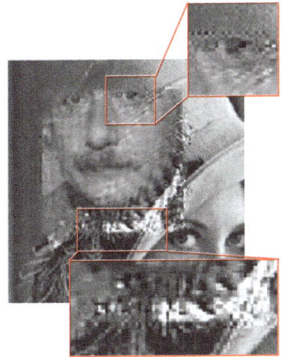

Fig. 16 A comparison of our joint method with a non-joint method and with a method not taking into account the consistent motion blur model is shown. A restored frame with two zoom up areas is depicted for a straightforward scale variant motion deblurring, where the contour is extracted a priori based on pure motion competition (*left*), for the non-consistent motion blur model on the same a priori computed contour (*middle*) and for the fully joint method with the consistent model (*right*)

to the motion blur. Hence, a comparison of the expected motion blur G_i with the observed time frames g_i in a least square sense $\int_\Omega (G_i[\mathcal{O}, w, u_{\text{obj}}, u_{\text{bg}}] - g_i)^2 \, d\mathcal{L}$ is considered as the fidelity energy, where the length of the boundary contour $\mathcal{H}^1(\mathcal{S})$ acts as the corresponding prior.

For known w and $\mathcal{O}$, we obtain an almost classical deblurring problem for u_{obj} with the modification of the blurring kernel given in (36), which is already reflected in the above fidelity term. We expect u_{obj} to be characterized by edges (cf. Fig. 17). As a suitable prior for these intensity maps we select the total variation functional $\int_\Omega |\nabla u_{\text{obj}}| \, d\mathcal{L}$ [61], which at the same time guarantees a suitable extension onto the whole space.

Finally, given $\mathcal{O}$ and the two intensities u_{obj}, u_{bg}, the extraction of the motion velocity w is primarily an optical flow problem. The transport of the object intensity u_{obj} from time t_1 to t_2 described in G_1 and G_2 provides us with information on w. In the case of limited intensity modulations on the moving object, it is the comparison of the expected transition profile $\chi_{\text{obj}} * h_w$, encoded in G_i, with the observed profile in g_i that will act as a guidance for the identification of the motion velocity.

Based on these modeling aspects we finally obtain the energy

$$\mathcal{E}[u_{\text{obj}}, w, \mathcal{S}] = \sum_{i=1,2} \int_\Omega (G_i[\mathcal{O}, w, u_{\text{obj}}, u_{\text{bg}}] - g_i)^2 \, d\mathcal{L} + \mu \int_\Omega |\nabla u_{\text{obj}}| \, d\mathcal{L} + \eta \mathcal{H}^1(\mathcal{S}),$$

and ask for a minimizing set of the degrees of freedom $\mathcal{O}$, w, and u_{obj}. Once a minimizer is known, we can retrieve the deblurred images $u(t_1, \cdot)$ and $u(t_2, \cdot)$ applying (36). In the application, the joint approach for all three unknowns—the motion velocity w, the object intensity u_{obj} and the shape of the object $\mathcal{S}$—turns out to be crucial for a proper reconstruction of blurred video frames. This interdependence is demonstrated by the results in Fig. 16 where we compare our joint approach

Fig. 17 From two real blurred frames (*left*), we automatically and simultaneously estimate the motion region, the motion vector and the image intensity of the foreground (*middle*). Based on this and the background intensity we reconstruct the two frames (*right*)

with a two step method which first tries to identify $\mathcal{O}$ and w based on a motion competition algorithm [24], followed by the actual deblurring in a second step. Note that the proposed method can be regarded as a motion competition method if we skip the convolution with the convolution kernel h_w in the variational formulation. Figure 16 also shows the importance of the consistent motion blur model for a proper reconstruction in the vicinity of motion singularities. Finally, we have applied our model to a true motion sequence recorded with a hand held video camera. The sequence shows a toy car moving in front of a puzzle (background; cf. Fig.17). We choose a textured object moving in front of a textured background to demonstrate the interplay between the deblurring steered by the fidelity functional and the reconstruction of sharp edges due to the total variation built into the prior. Finally, a real world application is shown in Fig. 17.

4 Elastic Shape Averaging

This section is concerned with a physically motivated notion of shape averages. As shapes we consider object contours or image morphologies, both encoded as edge sets in images. We again propose to employ concepts from nonlinear elasticity, which reflect first principles and to define shape averages that incorporate natural local measures of the underlying deformation and its dissimilarity from an isometry. Furthermore, we will introduce a robust approximation based on a diffusive phase field description of shapes. Averaging is a fundamental task for the quantitative analysis of ensembles of shapes and has already been extensively studied in the literature.

Very basic notions of averaging include the arithmetic mean of landmark positions [23] and the image obtained by the arithmetic mean of the matching deformations [9, 62]. For general images there are various qualitatively different notions of averaging. The intension is a fusion or blending to simultaneously represent complementary information of different but related images [43, 45]. It appears natural to study relations between shapes or more general image structures via deformations which transform one shape or image onto the another one [16,23,62]. Conceptually, in the last decade correlations of shapes have been studied on the basis of a general framework of a space of shapes and its intrinsic structure.

Variational Methods in Image Matching and Motion Extraction

The notion of a shape space was introduced by Kendall [46] already in 1984. Charpiat et al. [14, 15] discuss shape averaging and shape statistics based on the notion of the Hausdorff distance of sets. Understanding shape space as an infinite-dimensional Riemannian manifold, Miller et al. [51, 52] defined the average $\mathcal{S}$ of samples $\mathcal{S}_i$, $i = 1, \ldots, n$, as the minimizer of

$$\mathcal{E}[\mathcal{S}] = \sum_{i=1}^{n} d(\mathcal{S}_i, \mathcal{S})^2$$

for some metric $d(\cdot, \cdot)$, e.g. a geodesic distance in the space of shapes. Fuchs et al. [36] proposed a viscoelastic notion of the distance between shapes $\mathcal{S}$ given as boundaries of physical objects $\mathcal{O}$. The elasticity paradigm for shape analysis on which our approach is founded differs significantly from the metric approach in shape space. Given two shapes $\mathcal{S}_1$ and $\mathcal{S}_2$ and an elastic deformation ϕ from $\mathcal{S}_1$ to $\mathcal{S}_2$, we have to distinguish a usually stress free reference configuration $\mathcal{S}_1$ from the deformed configuration $\mathcal{S}_2 = \phi(\mathcal{S}_1)$, which is under stress and not in equilibrium. Due to the axiom of elasticity the energy at the deformed configuration $\mathcal{S}_2$ is independent of the path which connects $\mathcal{S}_1$ with $\mathcal{S}_2$. Hence, there is no notion of shortest paths if we consider a purely elastic shape model. As outlined above assumed viscous or visco plastic materials forming a shape, the underlying dissipation allows to measure length of connecting paths as long as the final configuration is again stress free. Fletcher et al. [34] propose to use a shape median instead of the geometric shape mean. The median is defined as the minimizer of the functional $\mathcal{E}[\mathcal{S}] = \sum_{i=1}^{n} d(\mathcal{S}_i, \mathcal{S})$.

4.1 A Nonlinear Elastic Spring Model

Here, we consider shapes encoded in images. In the simplest case, such an image $u : \Omega \to \mathbb{R}$ is a characteristic function $u = \chi_{\mathcal{O}}$ representing an object $\mathcal{O}$ as an open set on some domain $\Omega \subset \mathbb{R}^d$ with $d = 2, 3$, and we define the object shape $\mathcal{S} := \partial \mathcal{O}$. More generally we are interested in a shape $\mathcal{S}$ defined as the morphology of an image and represented via the image edge set.

Let us assume that n images $u_i : \Omega \to \mathbb{R}$ for $i = 1, \ldots, n$ are given with a sufficiently regular $(d-1)$-dimensional edge sets $\mathcal{S}_i$. We are interested in an average shape that reflects the geometric characteristics of the n given shapes in a physically intuitive manner. Suppose $\mathcal{S} \subset \mathbb{R}^d$ denotes a candidate for this unknown set. Now, we take into account elastic deformations $\phi_i : \Omega \to \mathbb{R}^d$ with $\phi_i(\mathcal{S}_i) = \mathcal{S}$. Assigned to each of these deformations is an elastic energy $\mathcal{W}[\mathcal{O}_i, \phi_i]$, and we ask for a shape $\mathcal{S}$ such that the total energy given as the sum over all the energies $\mathcal{W}[\mathcal{O}_i, \phi_i]$ for $i = 1, \ldots, n$ is minimal. Here, we will again consider a polyconvex elastic energy $\mathcal{W}[\mathcal{O}_i, \phi_i] = \int_{\mathcal{O}_i} \hat{W}(\mathcal{D}\phi) \, d\mathcal{L}$ given in (8) with the energy density $\hat{W}$ from (9). As described below we will have to consider in addition a further

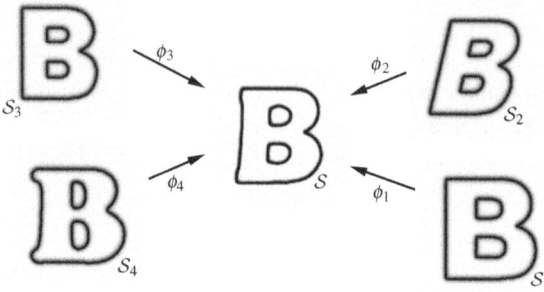

Fig. 18 A sketch of the elastic shape averaging is shown. Input shapes S_i ($i = 1, \ldots, 4$) extracted from input images u_i are mapped onto a shape S via elastic deformations ϕ_i. The shape S which minimizes primarily the elastic deformation energy plus some shape prior to be discussed later is called the shape average of the given input shapes. (Displayed are actual numerical results obtained by the algorithm.) The resolution of the underlying grid is 257×257 and the values for the involved parameters are $\gamma = 10^7$, $\mu = 1$, $(a_1, a_2, a_3) = (10^6, 0, 10^6)$

energy contribution which acts as a prior on the shape S in this variational approach. Obviously, this is a constrained variational problem. We simultaneously have to minimize over n deformations ϕ_i and the unknown shape S given n constraints $\phi_i(S_i) = S$. The model is related to the physical interpretation of the arithmetic mean of n points $x_1, \cdots, x_n$ in $\mathbb{R}^d$. Indeed, the arithmetic mean $x \in \mathbb{R}^d$ minimizes $\sum_{i=1,\ldots,n} \alpha\, d(x, x_i)^2$, where $d(x, x_i)$ is the distance between x and x_i and $\alpha > 0$ is the elasticity constant. Due to Hooke's law the stored energy $\alpha d(x, x_i)^2$ in the spring connecting x_i and x is proportional to the squared distance. Let us restrict the set of admissible deformations for each object $\mathcal{O}_i$ imposing the constraint $\phi_i(S_i) = S$ to deduce a suitable energy on shapes S being candidates for the shape average and sets of deformations $(\phi_i)_{i=1,\ldots,n}$ matching given shapes S_i with S (cf. Fig. 18):

$$\mathcal{E}[S, (\phi_i)_{i=1,\ldots,n}] = \begin{cases} \frac{1}{n} \sum_{i=1,\ldots,n} \mathcal{W}[\mathcal{O}_i, \phi_i] \; ; \; \phi_i(S_i) = S \text{ for } i = 1,\ldots,n \\ \infty \qquad\qquad\qquad\qquad ; \text{ else } . \end{cases}$$

Finally, we define the shape average S as the minimizer over a suitable set of admissible shapes $\mathcal{A}_S$, i.e.

$$S = \arg\min_{\tilde{S} \in \mathcal{A}_S} \left(\left(\arg\min_{\phi_i : \mathcal{O}_i \to \mathbb{R}^d} \mathcal{E}[\tilde{S}, (\phi_i)_{i=1,\ldots,n}] \right) + \eta \mathcal{H}^{d-1}[\tilde{S}] \right).$$

Here $\eta \mathcal{H}^{d-1}[\tilde{S}]$ with $\eta > 0$ acts as a prior on admissible shapes. In fact, we interpret the elastic energy $\mathcal{W}[\mathcal{O}_i, \phi_i]$ associated with each deformation ϕ_i which maps one of the shapes S_i onto the shape S as a nonlinear counterpart of the energy stored in a spring in the above classical interpretation of an averaged position. It measures in a physically rigorous way locally the lack of isometry as already mentioned above (cf. Fig. 20). A necessary condition for a set of minimizing deformations are the corresponding Euler Lagrange equations. As usual, inner variations of one of the deformations lead to the classical system of partial differential equations div $\hat{W}_A(\mathcal{D}\phi_i) = 0$ for every deformation ϕ_i on $\mathcal{O}_i \setminus \mathcal{S}_i$. Due to the set of constraints $(\phi_i(\mathcal{S}_i) = \mathcal{S})_{i=1,...,n}$ the conditions on $\mathcal{S}_i$ are interlinked. For simplicity let us assume that $\mathcal{S}_i = \partial \mathcal{O}_i$ for all $i = 1, \ldots, n$. Then we can deduce a balance relation between deformation stresses on the averaged shape $\mathcal{S}$. Namely, for $x \in \mathcal{S}$ we obtain

$$\sum_{i=1,...,n} \sigma[\phi_i](\phi_i^{-1}(x)) n[\mathcal{S}_i](\phi_i^{-1}(x)) \mathcal{H}^{d-1}(\mathcal{S}_i) = 0, \tag{37}$$

where $n[\mathcal{S}_i]$ is the outer normal on $\mathcal{S}_i$, $\mathcal{H}^{d-1}(\mathcal{S}_i)$ the surface element on $\mathcal{S}_i$, and $\phi_i^{-1}(x)$ the pre-image of x under the deformation ϕ_i. From elasticity theory we know that the forces $\sigma[\phi_i](\phi_i^{-1}(x)) n[\mathcal{S}_i](\phi_i^{-1}(x)) \mathcal{H}^{d-1}(\mathcal{S}_i)$ in the reference configuration equal the corresponding forces $\sigma^{\text{def}}[\phi_i](x) n[\mathcal{S}](x) \mathcal{H}^{d-1}[\mathcal{S}]$ in the deformed configuration so that (37) directly implies a balance of all normal stresses on the average shape $\mathcal{S}$,

$$0 = \sum_{i=1,...,n} \sigma^{\text{def}}[\phi_i](x) n[\mathcal{S}](x),$$

where $n[\mathcal{S}](x)$ is the outer normal in $\mathcal{S}$ and $\sigma^{\text{def}}[\phi_i] \circ \phi_i = \sigma[\phi_i] \det(\mathcal{D}\phi_i)^{-1} \mathcal{D}\phi_i^T$ the usual Cauchy stress tensor corresponding to the deformation ϕ_i in the deformed configuration (cf. Fig. 19). This gives a refined physical interpretation of the shape average as the stable shape on which all surface forces implied by the elastic deformations are balanced. Obviously, there is a straightforward generalization involving jumps of normal stresses on interior interfaces in case of components of $\mathcal{S}_i$, which are interior edges in $\mathcal{O}_i$. The hard constraint $\phi_i(\mathcal{S}_i) = \mathcal{S}$ is often inadequate in applications. Due to local shape fluctuations, for example, or noise in the shape acquisition there are frequently local details like spurious edges in the input shapes which should not be encountered in the shape average. Hence, we relax the constraint and introduce a penalty functional, which measures the symmetric difference of the input shapes $\mathcal{S}_i$ and the pull back $\phi_i^{-1}[\mathcal{S}]$ of the shape $\mathcal{S}$ and is given by

$$\mathcal{F}[\mathcal{S}_i, \phi_i, \mathcal{S}] = \mathcal{H}^{d-1}(\mathcal{S}_i \triangle \phi_i^{-1}[\mathcal{S}]),$$

where $A \triangle B = A \setminus B \cup B \setminus A$. Finally, to sum up, our shape averaging model is based on the energy

$$\mathcal{E}^\gamma[\mathcal{S}, (\phi_i)_{i=1,...,n}] = \frac{1}{n} \sum_{i=1}^{n} \left(\int_{\mathcal{O}_i} \hat{W}(\mathcal{D}\phi_i) \, d\mathcal{L} + \gamma \mathcal{F}[\mathcal{S}_i, \phi_i, \mathcal{S}] \right) + \eta \mathcal{H}^{d-1}[\mathcal{S}].$$

(38)

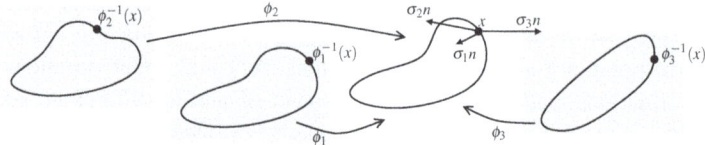

Fig. 19 Sketch of the stress balance relation on the averaged shape. $\sigma_i n$ is short for $\sigma^{\mathrm{def}}[\phi_i]n[\mathcal{S}]$

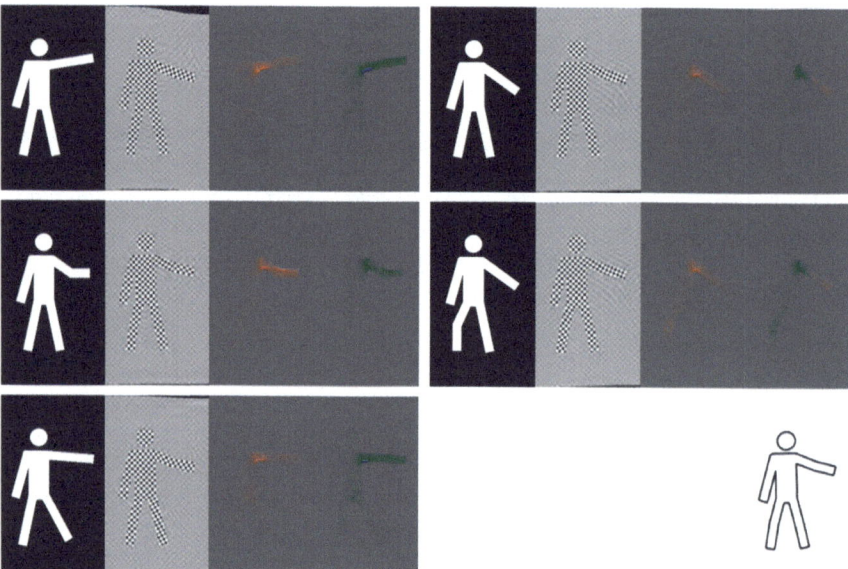

Fig. 20 Given five silhouettes of a person as input shapes a shape average (*bottom right*) is computed based on our elastic averaging approach. The original images are depicted along with their deformations ϕ_i (continued to the region outside the object and displayed acting on a checkerboard on the image domain Ω) and the distribution of local change of length $|\mathcal{D}\phi_i|_2$ and local change of area $\det(\mathcal{D}\phi_i)$ (from *left* to *right*). These local densities ranging over $[0.97, 1.03]$ and $[0.97\sqrt{2}, 1.03\sqrt{2}]$, respectively, are color-coded using the *color map* ▬▬▬. The underlying image resolution is 513×513, and the energy parameters are $\gamma = 10^7$, $\mu = 10^{-2}$. The phase field parameter in the implementation is chosen equal to the grid size

This approach is related to groupwise registration and segmentation results [69, 70]. Here, $\mathcal{F}$ acts as a fidelity term measuring the quality of the registration of the shapes $\mathcal{S}_i$ with a given shape $\mathcal{S}$ under the deformations ϕ_i. Figure 20 shows results for the averaging of some human figure sketches and underlines that the proposed method measures locally the isometry defect.

4.2 A Phase Field Approximation

As in the registration applications considered so far, we pick up the phase field approximation by Ambrosio and Tortorelli [2, 3] and encode the edge set $\mathcal{S}$ by a smooth phase field function $v : \Omega \to \mathbb{R}$. For the input shapes $\mathcal{S}_i$ we assume the corresponding phase field description v_i to be given a priori. Usually, v_i can be computed beforehand, minimizing the original Ambrosio Tortorelli type energy for given input images u_i, or explicitly constructed for a given edge set using the comparison function from the slicing argument in [3]. Given a phase field parameter ϵ and corresponding phase field representations v of $\mathcal{S}$ and v_i of $\mathcal{S}_i$, respectively, we define an approximate mismatch penalty

$$\mathcal{F}^\epsilon[v_i, \phi_i, v] = \frac{1}{\epsilon} \int_\Omega (v \circ \phi_i)^2 (1 - v_i)^2 + v_i^2 (1 - v \circ \phi_i)^2 \, \mathrm{d}\mathcal{L}.$$

Here, we suppose v to be extended by 1 outside the computational domain Ω. The first term in the integrand is close to 1 on $\mathcal{S}_i \setminus \phi_i^{-1}[\mathcal{S}]$, because $(1 - v_i) \approx 1$ on $\mathcal{S}_i$ and $v \circ \phi_i \approx 1$ away from the vicinity of $\phi_i^{-1}[\mathcal{S}]$. It tends to 0 with increasing distance from this set. Analogously, the second term acts as an approximate indicator function for $\phi_i^{-1}[\mathcal{S}] \setminus \mathcal{S}_i$. Let us emphasize that $\mathcal{F}^\epsilon[v_i, \phi_i, v]$ is expected to be a true approximation of $\mathcal{F}[\mathcal{S}_i, \phi_i, \mathcal{S}]$ only, if ϕ_i is neither distending nor compressive orthogonally to the shape, i.e. $\mathcal{D}\phi_i \, n[\phi_i^{-1}[\mathcal{S}]] \cdot n[\phi_i^{-1}[\mathcal{S}]] = 1$ on $\phi_i^{-1}[\mathcal{S}]$. Nevertheless, because we are primarily interested in the limit for $\gamma \to \infty$, $\mathcal{F}^\epsilon[v_i, \phi_i, v]$ acts as a proper penalty functional.

Next, we have to describe the phase field v, which is not given a priori, in an implicit variational form and consider the usual energy

$$\mathcal{H}^{d-1}_\epsilon[v] = \int_\Omega \epsilon |\nabla v|^2 + \frac{1}{4\epsilon}(v - 1)^2 \, \mathrm{d}\mathcal{L},$$

which additionally acts as a regularization energy measuring an approximation of the $\mathcal{H}^{d-1}$ measure of the shape $\mathcal{S}$ represented by the phase field v.

So far the elastic energy is evaluated on the object domains $\mathcal{O}_i$ only. For practical reasons of the later numerical discretization, we now let the whole computational domain behave elastically with an elasticity several orders of magnitude softer outside the object domains $\mathcal{O}_i$ on the complement set $\Omega \setminus \mathcal{O}_i$. We suppose that, based on a prior segmentation of the images u_i, a smooth approximation $\chi^\epsilon_{\mathcal{O}_i}$ of the characteristic function $\chi_{\mathcal{O}_i}$ is given and define a corresponding approximate elastic energy

$$\mathcal{W}^\epsilon[\mathcal{O}_i, \phi_i] = \int_\Omega \left((1 - \delta)\chi^\epsilon_{\mathcal{O}_i} + \delta\right) \hat{W}(\mathcal{D}\phi_i) \, \mathrm{d}\mathcal{L},$$

where in our application $\delta = 10^{-4}$. Also, in the above we implicitly assumed that deformations ϕ_i map the domain Ω onto itself; for numerical implementation

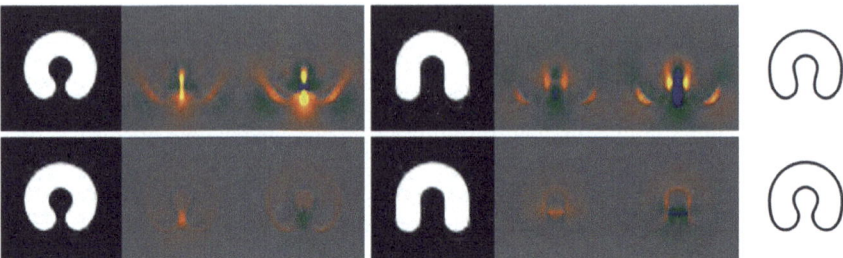

Fig. 21 Input images together with $\|\mathcal{D}\phi_i\|$ and $\det(\mathcal{D}\phi_i)$ (ranges of $[0.6, 1.4]$ and $[0.6\sqrt{2}, 1.4\sqrt{2}]$ color-coded as) and the average phase field (*rightmost*). In the *top row* only the interior of the two shapes behave elastic, whereas in the *bottom row* the whole computational domain is considered to be homogeneously elastic. Obviously, in the upper case far stronger strains are visible in the region of the gap and in the lower case it is much more expensive to pull the lobes apart in the first shape than to push them together in the second shape. Hence, the resulting average in the *second row* is characterized by stronger bending of the two lobes similar than in the *first row*. (Results are obtained for a grid resolution 1025×1025 and parameter values $\gamma = 10^7$, $\mu = 0.1$, $\epsilon = 6h$.)

we will relax this assumption and perform integrations only in regions where all integrands are defined. Finally, the resulting approximation of the total energy functional to be minimized reads

$$\mathcal{E}^{\gamma,\epsilon}[v, (\phi_i)_{i=1,\ldots,n}] = \frac{1}{n}\sum_{i=1}^{n}\left(\mathcal{W}^{\epsilon}[\mathcal{O}_i, \phi_i] + \gamma\mathcal{F}^{\epsilon}[v_i, \phi_i, v]\right) + \eta\mathcal{H}_{\epsilon}^{d-1}[v]. \quad (39)$$

Let us remark that we are particularly interested in the case where $\mathcal{F}^{\epsilon}$ acts as a penalty with $\gamma \gg 1$ and $\mathcal{H}_{\epsilon}^{d-1}$ ensures a mild regularization of the averaged shape with $\eta \ll 1$.

The structure of the penalty functional $\mathcal{F}^{\epsilon}$ tries to match the shapes of the given phase field functions v_i, and the pull back $v \circ \phi_i$ of the phase field v has to be determined. This implies a particular stiffness of the deformations ϕ_i on the diffuse interface around the shapes $\mathcal{S}_i$. Indeed, there the set of deformations $\phi_1 \ldots, \phi_n$ tries to minimize stretch of compression normal to the shape contour measured in terms of $\mathcal{D}\phi\, n[\mathcal{S}_i] \cdot n[\mathcal{S}_i] - 1$. In the limit for $\gamma \to \infty$ this does not hamper the elastic deformation, because the other (tangential) components of the deformation tensor can relax freely. Figure 21 shows the impact of the choice of the elastic domain on the average shape. Here, we once consider the whole computational domain as homogeneously elastic, and alternatively—and in many cases physically more sound—only the object domain is assumed to be elastic and considerably stiff. The region between both lobes is more severely dilated if the elastic energy is weighted with a small factor outside the shape, which becomes obvious especially in the plots of the elastic invariants. Furthermore, the particular role of the diffused interface with respect to the compression rates is indicated by the color coding. As first illustrative examples, we computed the average of different 2D objects as

Variational Methods in Image Matching and Motion Extraction

Fig. 22 Twenty shapes "device7" from the MPEG7 database and their average phase field. The *bottom line* shows $|\mathcal{D}\phi_i|_2$ and $\det(\mathcal{D}\phi_i)$ for shape 2, 8, and 18, with ranges of $[0.8, 1.2]$ and $[0.8\sqrt{2}, 1.2\sqrt{2}]$ color-coded as ▓▓▓▓. (Resolution 513×513, $\gamma = 10^7$, $\mu = 10^{-2}$.)

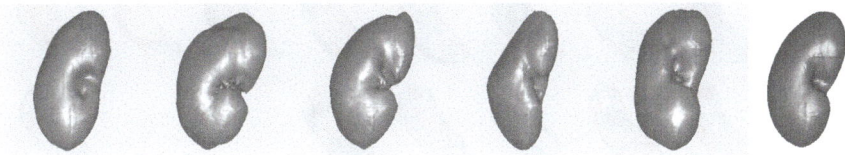

Fig. 23 Five segmented kidneys and their average (*right*). (Result obtained for resolution $257 \times 257 \times 257$ and parameter values $\gamma = 10^7$, $\mu = 1$, $(a_1, a_2, a_3) = (10^8, 0, 10^7)$.)

shown in Fig. 22. Furthermore, Fig. 20 has already shown that due to the invariance of the hyperelastic energy with respect to local rotations, the computed averages try to locally preserve isometries. Effectively, the different characteristics of the input shapes, both on the global and a local scale, are averaged in a physically intuitive way, and the scheme proves to perform fairly robust due to the diffusive approximation based on the phase field model and the multi scale relaxation. In what follows we will consider the averaging of 3D shapes originally given as triangulated surfaces and first converted to an implicit representation as binary images. A set of 48 kidneys and a set of 24 feet will serve as input data. We will employ a hyperelastic energy (8) with a density (9) with $p = q = 4$ and $s = r = 2$. The first five original kidneys and their computed average are shown in Fig. 23. In fact, since the average 3D phase field itself cannot be properly displayed, we instead depict one of the original kidneys, deformed to the average configuration. This is allowed, for all deformed kidneys look alike and constitute each just one representative of

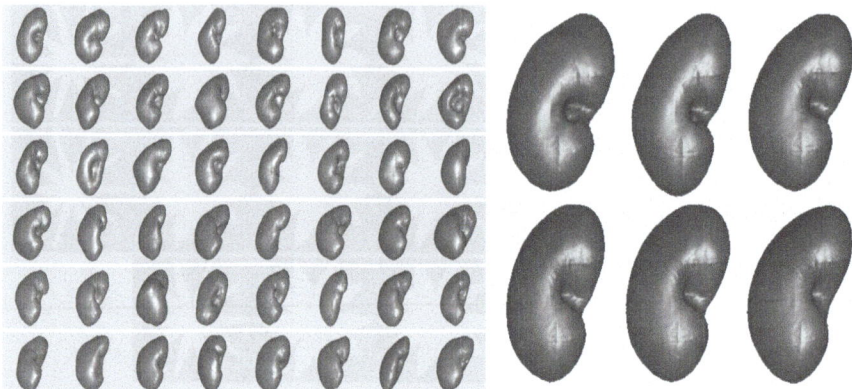

Fig. 24 On the *left*, 48 kidney shapes are shown. On the *right*, from *top left* to *bottom down* the averaged shape of the first 2, 4, 5, 6, 8 and of all 48 kidneys are depicted. The parameter values are as for Fig. 23

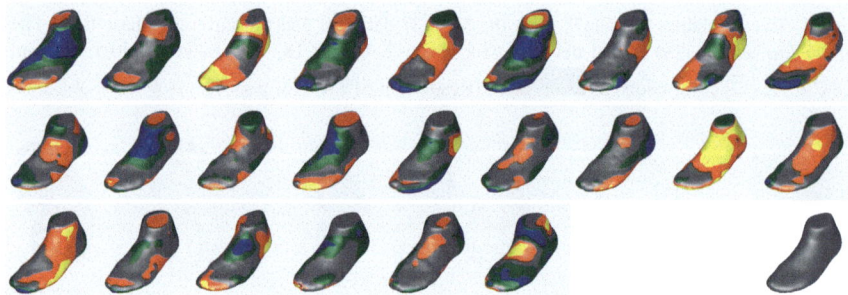

Fig. 25 Twenty-four given foot shapes, textured with the distance to the surface of the average foot (*bottom right*). Values range from 6 mm inside the average foot to 6 mm outside, color-coded as ▇▇▇▇. For that purpose, the shapes have been aligned to the average. The front of the instep can be identified as a region of comparatively low variation. (Result obtained for resolution 257×257 and parameter values $\gamma = 10^7$, $\mu = 1$, $(a_1, a_2, a_3) = (10^8, 0, 10^8)$.)

the average equivalence class in the quotient space of images relative to the edge equivalence relation (Fig. 24). (It goes without saying that the deformed kidney boundaries only coincide up to the width of the phase field.)

The next example consists of a set of feet, where the average may help to design an optimal shoe. The 24 original feet are displayed in Fig. 25. Their surface is colored according to the local distance to the surface of the computed average shape, which helps to identify regions of strong variation. Furthermore, to allow a better comparison, the foot shapes, have been aligned with the average for the final visualization. Let us emphasize that the algorithm itself robustly deals with even quite large rigid body motions. Apparently, the instep differs comparatively little between the given feet, whereas the toes show a rather strong variation. Note

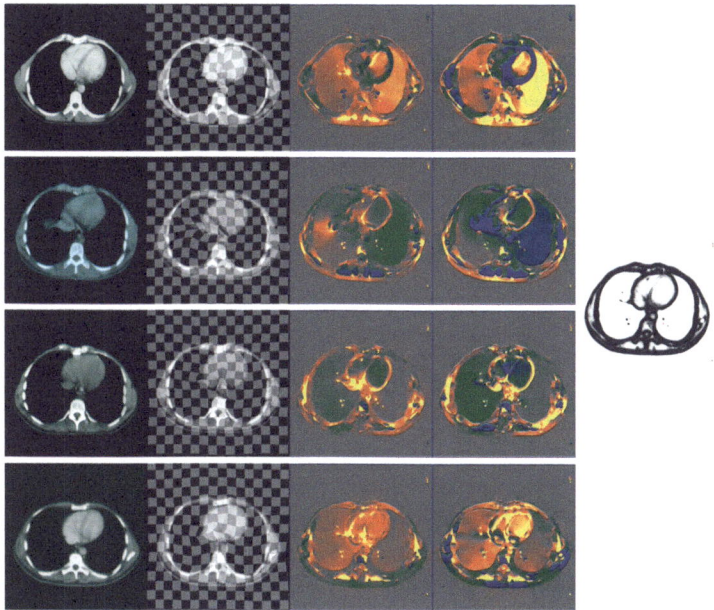

Fig. 26 Averaging of four CT scans of the thorax. From *left* to *right*: Original images, deformations ϕ_i, $|\mathcal{D}\phi_i|_2$ and $\det(D\phi_i)$ (color-coded as ▬▬▬ with ranges $[0.8, 1.2]$ and $[0.8\sqrt{2}, 1.2\sqrt{2}]$), and average phase field. (Result obtained for resolution 257×257 and parameter values $\gamma = 10^7$, $\mu = 0.1$.)

that—since we only display normal distance to the surface of the average foot—any potential tangential displacement is not visible, but could of course also be visualized when examining shape variation.

To illustrate that the approach can also be applied to average image morphologies, let us consider four two-dimensional, transversal CT scans of the thorax as input (Fig. 26, left). Unlike the previous examples, these images do not encode volumetric shapes homeomorphic to the unit ball, but contain far more complicated structures. Also, the quality of contrast differs between the images, and—even more problematic—the images do not show a one-to-one correspondence, i.e. several structures (the scapula, ribs, parts of the liver) are only visible in some images, but not in others. Nevertheless, the algorithm manages to segment and align the main features (the heart, the spine, the aorta, the sternum, the ribs, the back muscles, the skin), yielding sensible average contours (Fig. 26, right). In order to achieve this, we this time jointly segmented and averaged the original CT scans, i.e. we augmented our objective function (39) by the original Ambrosio–Tortorelli energy for each image and then alternatingly optimized for the v_i, v, and ϕ_i. The second to fourth column of Fig. 26 depict the corresponding deformations ϕ_i and the deformation invariants. Obviously, the deformation behaves quite regularly: Not only is it homeomorphic, but also too large and distorting deformations are prevented by the hyperelastic regularization. This enables the method to be applied to images containing also distinct structures, whereas for viscous flow regularization as in [11, 17] such individual structures are at risk of being matched with anything nearby (a frequently used example for viscous fluid regularization even matches a disc to a C-shape). The deformation energy is quite evenly distributed over the images and only peaks at pronounced features, where a local exact fit can be achieved (e.g. at the back muscles). Outside the thorax, the energy rapidly decreases to zero, justifying that in this example we did not weight the elastic energy differently inside and outside the body.

4.3 Joint Image Segmentation and Shape Averaging

In the derivation of our shape averaging model we have assumed that the shapes S_i to be averaged can be robustly extracted from a set of images u_i with $i = 1, \ldots, n$ and are a priori given. However, if we consider shapes being defined as the morphology of images represented by edges, some of these edges will be characterized by significant noise or low contrast and hence will be difficult to extract. Here, it might help to take into account the corresponding edges in the other images, which all refer to the same edge of the average shape. Indeed, in this case a joint approach of shape segmentation and registration with an averaged shape is particularly promising: On the one hand the quality of shape averaging highly depends on the robustness of the edge detection in the input images. On the other hand, a reliable average shape can be used to improve edge detection in case of a poor image quality. Incorporating the classical Mumford Shah segmentation approach in the above shape averaging functional (38) we end up with the joint functional

$$\mathcal{E}^{\gamma}_{\text{joint}}[\mathcal{S}, (u_i, \mathcal{S}_i, \phi_i)_{i=1,\ldots,n}] = \frac{1}{n} \sum_{i=1}^{n} (\mathcal{E}_{\text{MS}}[u_i, \mathcal{S}_i] + \mathcal{W}[\Omega, \phi_i] + \gamma \mathcal{F}[\mathcal{S}_i, \phi_i, \mathcal{S}])$$
$$+ \mu \mathcal{H}^{d-1}[\mathcal{S}],$$

which has to be relaxed simultaneously in u_i, $\mathcal{S}_i$, ϕ_i for $i = 1, \cdots, n$ and $\mathcal{S}$ for a given set of input images $(u_i^0)_{i=1,\cdots,n}$. Let us remark that we use the parameter μ for the weighting of the area of the shape $\mathcal{S}$, because an η already weight the area of the input shape $\mathcal{S}_i$. Finally, let us consider the phase field approximation for this joint model given by the functional

$$\mathcal{E}^{\gamma,\epsilon}_{\text{joint}}[v, (u_i, v_i, \phi_i)_{i=1,\ldots,n}] = \frac{1}{n} \sum_{i=1}^{n} (\mathcal{E}^{\epsilon}_{\text{AT}}[u_i, v_i] + \mathcal{W}^{\epsilon}[\mathcal{O}_i, \phi_i] + \gamma \mathcal{F}^{\epsilon}[v_i, \phi_i, v_i])$$
$$+ \mu \mathcal{H}^{d-1}_{\epsilon}(v).$$

Fig. 27 Blurred edges can be restored based on a joint approach for image segmentation and averaging. The three input images u_i^0 are depicted along with their phase field v_i as computed by the joint segmentation and averaging. The computed average shape is also shown (*right*). Apparently, the strongly blurred edges in the first input image were reconstructed based on the corresponding edges in the other images. (Resolution 513×513, $\gamma = 10^7$, $\mu = 10^{-2}$, $(a_1, a_2, a_3) = (10^8, 0, 10^8)$, $\alpha = 2 \cdot 10^{10}$, $\beta = 2 \cdot 10^5$, $\eta = 2 \cdot 10^6$.)

where $\mathcal{E}_{\text{AT}}^\epsilon[u_i, v_i]$ denotes the classical Ambrosio Tortorelli functional for given image intensities u_i^0 and for $i = 1, \ldots, n$. Figure 27 demonstrates that in a joint approach blurry edges in the input images can be segmented, if sufficiently strong evidence for this edge from other input images is integrated into the averaged shape.

Acknowledgements The author is grateful to Werner Bautz, radiology department at the university hospital Erlangen, Germany, for providing CT data of kidneys, as well as to Heiko Schlarb from Adidas, Herzogenaurach, Germany, for providing 3D scans of feet, and to Bruno Wirth, urology department at the Hospital zum Hl. Geist, Kempen, Germany, for providing thorax CT scans. Furthermore, the author thanks Stan Osher for pointing to the issue of elastic shape averaging and Marc Droske for discussion about the phase field approach. Finally, he acknowledges Helene Horn for her help in the careful preparation of the manuscript.

References

1. L. Alvarez, F. Guichard, P.-L. Lions, J.-M. Morel, Axioms and fundamental equations of image processing. Arch. Ration. Mech. Anal. **123**(3), 199–257 (1993)
2. L. Ambrosio, V.M. Tortorelli, Approximation of functionals depending on jumps by elliptic functionals via Γ-convergence. Comm. Pure Appl. Math. **43**, 999–1036 (1990)
3. L. Ambrosio, V.M. Tortorelli, On the approximation of free discontinuity problems. Boll. dell'Unione Matematica Ital. Sez. B **6**(7), 105–123 (1992)
4. L. Ambrosio, N. Fusco, D. Pallara, *Functions of Bounded Variation and Free Discontinuity Problems*. Oxford Mathematical Monographs (Oxford University Press, New York, 2000)
5. T.J. Baker, Three dimensional mesh generation by triangulation of arbitrary point sets, in *8th Computational Fluid Dynamics Conference*, vol. 1124-CP, Honolulu, 9–11 June 1987, pp. 255–271
6. J.M. Ball, Convexity conditions and existence theorems in nonlinear elasticity. Arch. Ration. Mech. Anal. **63**, 337–403 (1977)
7. J.M. Ball, Global invertibility of Sobolev functions and the interpenetration of matter. Proc. R. Soc. Edinb. **88A**, 315–328 (1988)
8. L. Bar, B. Berkels, M. Rumpf, G. Sapiro, A variational framework for simultaneous motion estimation and restoration of motion-blurred video, in *Eleventh IEEE International Conference on Computer Vision (ICCV'07)* (2007)
9. K.K. Bhatia, J.V. Hajnal, B.K. Puri, A.D. Edwards, D. Rueckert, Consistent groupwise non–rigid registration for atlas construction, in *IEEE International Symposium on Biomedical Imaging: Nano to Macro*, vol. 1 (2004), pp. 908–911

10. B. Bourdin, Image segmentation with a finite element method. Technical Report Nr. 00-14, Institut Galilée Université Paris Nord, UCLA CAM, 2000; Article: Math. Model. Numer. Anal. **33**(2), 229–244 (1999)
11. M. Bro-Nielsen, C. Gramkow, *Fast Fluid Registration of Medical Images* (Springer, Berlin, 1996), pp. 267–276
12. L.G. Brown, A survey of image registration techniques. ACM Comput. Surv. **24**(4), 325–376 (1992)
13. T. Chan, L. Vese, An active contour model without edges, in *Scale-Space Theories in Computer Vision. Second International Conference, Scale-Space '99*, Corfu, Greece, September 1999, ed. by M. Nielsen, P. Johansen, O.F. Olsen, J. Weickert. Lecture Notes in Computer Science, vol. 1682 (Springer, New York, 1999), pp. 141–151
14. G. Charpiat, O. Faugeras, R. Keriven, P. Maurel, Distance-based shape statistics, in *Acoustics, Speech and Signal Processing, 2006 (ICASSP 2006)*, vol. 5 (2006). doi: 10.1109/ICASSP.2006.1661428
15. G. Charpiat, O. Faugeras, R. Keriven, Approximations of shape metrics and application to shape warping and empirical shape statistics. Found. Comut. Math. **5**(1), 1–58 (2005)
16. G.E. Christensen, S.C. Joshi, M.I. Miller, Volumetric transformations of brain anatomy. IEEE Trans. Med. Imag. **16**(6), 864–877 (1997)
17. G.E. Christensen, R.D. Rabbitt, M.I. Miller, A deformable neuroanatomy textbook based on viscous fluid mechanics, in *Proceedings of the 27th Annual Conference Information Science and Systems*, ed. by J. Prince, T. Runolfsson (1993), pp. 211–216. doi: 10.1109/ICASSP.2006.1661428
18. G.E. Christensen, M.I. Miller, R.D. Rabbitt, Deformable templates using large deformation kinematics. IEEE Trans. Image Process. **5**(10), 1435–1447 (1996)
19. P.G. Ciarlet, *Three-Dimensional Elasticity* (Elsevier, Amsterdam, 1988)
20. P.G. Ciarlet, J. Nečas, Injectivity and self-contact in nonlinear elasticity. Arch. Ration. Mech. Anal. **97**(3), 171–188 (1987)
21. U. Clarenz, M. Droske, M. Rumpf, Towards fast non–rigid registration, in *Inverse Problems, Image Analysis and Medical Imaging, AMS Special Session Interaction of Inverse Problems and Image Analysis*, vol. 313 (American Mathematical Society, Providence, 2002), pp. 67–84
22. U. Clarenz, N. Litke, M. Rumpf, Axioms and variational problems in surface parameterization. Comput. Aided Geom. Des. **21**(8), 727–749 (2004)
23. T.F. Cootes, C.J. Taylor, D.H. Cooper, J. Graham, Active shape models—their training and application. Comput. Vis. Image Underst. **61**(1), 38–59 (1995)
24. D. Cremers, S. Soatto, Motion competition: a variational approach to piecewise parametric motion segmentation. Int. J. Comput. Vis. **85**(1), 115–132 (2004)
25. B. Dacorogna, *Direct Methods in the Calculus of Variations* (Springer, New York, 1989)
26. C.A. Davatzikos, R.N. Bryan, J.L. Prince, Image registration based on boundary mapping. IEEE Trans. Med. Imag. **15**(1), 112–115 (1996)
27. M.C. Delfour, J.P. Zolésio, *Geometries and Shapes: Analysis, Differential Calculus and Optimization*. Advances in Design and Control, vol. 4 (SIAM, Philadelphia, 2001)
28. M. Droske, W. Ring, A Mumford-Shah level-set approach for geometric image registration. SIAM Appl. Math. (2008, to appear)
29. M. Droske, M. Rumpf, A variational approach to non-rigid morphological registration. SIAM J. Appl. Math. 66, 2127–2148 (electronic) (2006). doi:10.1137/050630209. http://dx.doi.org/10.1137/050630209
30. M. Droske, M. Rumpf, Multi scale joint segmentation and registration of image morphology. IEEE Trans. Pattern Anal. Mach. Intell. **29**(12), 2181–2194 (2007)
31. M. Droske, W. Ring, M. Rumpf, Mumford-Shah based registration: a comparison of a level set and a phase field approach. Comput. Vis. Sci. Online First (2008)
32. P. Favaro, S. Soatto, A variational approach to scene reconstruction and image segmentation from motion-blur cues, in *2004 IEEE Computer Society Conference on Computer Vision and Pattern Recognition (CVPR'04)*, vol. 1 (2004), pp. 631–637

33. O. Féron, A. Mohammad-Djafari, Image fusion and unsupervised joint segmentation using a HMM and MCMC algorithms. J. Electron. Imag. 023014-1, 023014-12 (2005)
34. T. Fletcher, S. Venkatasubramanian, S. Joshi, Robust statistics on Riemannian manifolds via the geometric median, in *IEEE Conference on Computer Vision and Pattern Recognition (CVPR)* (IEEE, New York, 2008)
35. I. Fonseca, S. Müller, Quasi-convex integrands and lower semicontinuity in L^1. SIAM J. Math. Anal. **23**(5), 1081–1098 (1992)
36. M. Fuchs, B. Jüttler, O. Scherzer, H. Yang, Shape metrics based on elastic deformations. Forschungsschwerpunkt S92, Industrial Geometry 71, Universität Innsbruck, Innsbruck, 2008
37. E. Grinspun, A.H. Hirani, M. Desbrun, P. Schröder, Discrete shells, in *Eurographics/SIGGRAPH Symposium on Computer Animation* (Eurographics Association, 2003)
38. X. Gu, B.C. Vemuri, Matching 3D shapes using 2D conformal representations, in *Medical Image Computing and Computer-Assisted Intervention 2004*. Lecture Notes in Computer Science, vol. 3216 (2004), pp.771–780
39. W. Hackbusch, S.A. Sauter, Composite finite elements for problems containing small geometric details. Part II: implementation and numerical results. Comput. Vis. Sci. **1**(1), 15–25 (1997)
40. W. Hackbusch, S.A. Sauter, Composite finite elements for the approximation of PDEs on domains with complicated micro-structures. Numer. Math. **75**, 447–472 (1997)
41. W. Hackbusch, *Iterative Solution of Large Sparse Systems of Equations*. Applied Mathematical Sciences, vol. 95 (Springer, New York, 1994)
42. M. Hintermüller, W. Ring, An inexact Newton-cg-type active contour approach for the minimization of the Mumford-Shah functional. J. Math. Imag. Vis. **20**(1–2), 19–42 (2004)
43. S. Joshi, B. Davis, M. Jomier, G. Gerig, Unbiased diffeomorphic atlas construction for computational anatomy. NeuroImage **23**(Suppl. 1), S151–S160 (2004)
44. T. Kapur, L. Yezzi, L. Zöllei, A variational framework for joint segmentation and registration. *IEEE CVPR - Mathematical Methods in Biomedical Image Analysis* (IEEE, New York, 2001), pp. 44–51
45. S.L. Keeling, W. Ring, Medical image registration and interpolation by optical flow with maximal rigidity. J. Math. Imag. Vis. **23**(1), 47–65 (2005)
46. D.G. Kendall, Shape manifolds, procrustean metrics, and complex projective spaces. Bull. Lond. Math. Soc. **16**, 81–121 (1984)
47. V. Kraevoy, A. Sheffer, Cross-parameterization and compatible remeshing of 3D models. ACM Trans. Graph. **23**(3), 861–869 (2004)
48. A. Lee, D. Dobkin, W. Sweldens, P. Schröder, Multiresolution mesh morphing, in *Proceedings of SIGGRAPH 99*. Computer Graphics Proceedings, Annual Conference Series, August 1999, pp. 343–350
49. N. Litke, M. Droske, M. Rumpf, P. Schröder, An image processing approach to surface matching, in *Third Eurographics Symposium on Geometry Processing*, ed. by M. Desbrun, H. Pottmann (2005), pp. 207–216
50. J.E. Marsden, T.J.R. Hughes, *Mathematical Foundations of Elasticity* (Prentice-Hall, Englewood Cliffs, 1983)
51. M.I. Miller, L. Younes, Group actions, homeomorphisms and matching: a general framework. Int. J. Comput. Vis. **41**(1–2), 61–84 (2001)
52. M.I. Miller, A. Trouvé, L. Younes, On the metrics and Euler-Lagrange equations of computational anatomy. Annu. Rev. Biomed. Eng. **4**, 375–405 (2002)
53. J.M. Morel, S. Solimini, *Variational Models in Image Segmentation* (Birkäuser, Basel, 1994)
54. C. Morrey, Quasi-convexity and lower semicontinuity of multiple integrals. Pac. J. Math. **2**, 25–53 (1952)
55. C. Morrey, *Multiple Integrals in the Calculus of Variations*. Grundlehren der mathematischen Wissenschaften, vol.130 (Springer, New-York, 1966)
56. D. Mumford, J. Shah, Optimal approximation by piecewise smooth functions and associated variational problems. Comm. Pure Appl. Math. **42**, 577–685 (1989)
57. R.W. Ogden, *Non-Linear Elastic Deformations* (Wiley, London, 1984)

58. S.J. Osher, N. Paragios, *Geometric Level Set Methods in Imaging, Vision and Graphics* (Springer, New York, 2003)
59. S.J. Osher, J.A. Sethian, Fronts propagating with curvature dependent speed: algorithms based on Hamilton–Jacobi formulations. J. Comput. Phys. **79**, 12–49 (1988)
60. E. Praun, W. Sweldens, P. Schröder, Consistent mesh parameterizations. In *Proceedings of ACM SIGGRAPH 2001*. Computer Graphics Proceedings, Annual Conference Series, August 2001, pp. 179–184
61. L. Rudin, S. Osher, E. Fatemi, Nonlinear total variation based noise removal algorithms. Physica D **60**, 259–268 (1992)
62. D. Rueckert, A.F. Frangi, J.A. Schnabel, Automatic construction of 3D statistical deformation models using nonrigid registration, in *Medical Image Computing and Computer-Assisted Intervention (MICCAI)*, ed. by W. Niessen, M. Viergever. Lecture Notes in Computer Science, vol. 2208 (2001), pp. 77–84
63. M. Rumpf, B. Wirth, A nonlinear elastic shape averaging approach. SIAM J. Imag. Sci. **2**, 800–833. doi:10.1137/080738337
64. G. Sapiro, *Geometric Partial Differential Equations and Image Analysis* (Cambridge University Press, Cambridge, 2001)
65. J. Schreiner, A. Asirvatham, E. Praun, H. Hoppe, Inter-surface mapping. ACM Trans. Graph. **23**(3), 870–877 (2004)
66. J. Sokołowski, J.-P. Zolésio, *Introduction to Shape Optimization. Shape Sensitivity Analysis* (Springer, Berlin, 1992)
67. P.A. van den Elsen, E.-J.J. Pol, M.A. Viergever, Medical image matching: a review with classification. IEEE Eng. Med. Biol. **12**, 26–39 (1993)
68. V. Šverák, Regularity properties of deformations with finite energy. Arch. Ration. Mech. Anal. **100**, 105–127 (1988)
69. P.P. Wyatt, J.A. Noble, MAP MRF joint segmentation and registration, in *Medical Image Computing and Computer–Assisted Intervention (MICCAI)*, ed. by T. Dohi, R. Kikinis. Lecture Notes in Computer Science, vol. 2488 (2002), pp. 580–587
70. A. Yezzi, L. Zöllei, T. Kapur, A variational framework for integrating segmentation and registration through active contours. Med. Image Anal. **7**(2), 171–185 (2003)
71. Y.-N. Young, D. Levy, Registration-based morphing of active contours for segmentation of ct scans. Math. Biosci. Eng. **2**(1), 79–96 (2005)

Metrics of Curves in Shape Optimization and Analysis

Andrea C.G. Mennucci

Abstract In these lecture notes we will explore the mathematics of the space of immersed curves, as is nowadays used in applications in computer vision. In this field, the space of curves is employed as a "shape space"; for this reason, we will also define and study the space of geometric curves, which are immersed curves up to reparameterizations. To develop the usages of this space, we will consider the space of curves as an infinite dimensional differentiable manifold; we will then deploy an effective form of calculus and analysis, comprising tools such as a Riemannian metric, so as to be able to perform standard operations such as minimizing a goal functional by gradient descent, or computing the distance between two curves. Along this path of mathematics, we will also present some current literature results (and a few examples of different "shape spaces", including more than only curves).

Introduction

In these lecture notes we will explore the mathematics of the space of immersed curves, as is nowadays used in applications in computer vision. In this field, the space of curves is employed as a *shape space*; for this reason, we will also define and study the space of *geometric curves*, that are *immersed curves up to reparameterizations*. To develop the usages of this space, we will consider the *space of curves* as an *infinite dimensional differentiable manifold*; we will then deploy an effective form of calculus and analysis, comprising tools such as a *Riemannian metric*, so as to be able to perform standard operations: minimizing a goal functional by gradient descent, or computing the distance between two curves. Along this path of mathematics, we will also present some current literature results. (Another

---

A.C.G. Mennucci (✉)
Scuola Normale Superiore, Pisa, Italy
e-mail: mennucci@sns.it

common and interesting example of "shape spaces" is the space of all compact subsets of $\mathbb{R}^n$—we will briefly discuss this option as well, and relate it to the aforementioned theory).

These lecture notes aim to be as self-contained as possible, so as to be accessible to young mathematicians and non-mathematicians as well. For this reason, many examples are intermixed with the definitions and proofs; in presenting advanced and complex mathematical ideas, the rigorous mathematical definitions and proofs were sometimes sacrificed and replaced with an intuitive description.

These lecture notes are organized as follows. Section 1 introduces the definitions and some basilar concepts related to immersed and geometric curves. Section 2 overviews the realm of applications for a shape space in computer vision, that we divide in the fields of "shape optimization" and "shape analysis"; and highlights features and problems of those theories as were studied up to a few years ago, so as to identify the needs and obstacles to further developments. Section 3 contains a summary of all mathematical concepts that are needed for the rest of the notes. Section 4 coalesces all the above in more precise definitions of spaces of curves to be used as "shape spaces", and sets mathematical requirements and goals for applications in computer vision. Section 5 indexes examples of "shape spaces" from the current literature, inserting it in a common paradigm of "representation of shape"; some of this literature is then elaborated upon in the following Sects. 6, 7, 8, 9, containing two examples of metrics of compact subsets of $\mathbb{R}^n$, two examples of Finsler metrics of curves, two examples of Riemannian metrics of curves "up to pose", and four examples of Riemannian metrics of immersed curves. The last such example is the family of Sobolev-type Riemannian metrics of immersed curves, whose properties are studied in Sect. 10, with applications and numerical examples. Section 11 presents advanced mathematical topics regarding the Riemannian spaces of immersed and geometric curves.

I gratefully acknowledge that a part of the theory and many numerical experiments exposited were developed in joint work with Prof. Yezzi (GaTech) and Prof. Sundaramoorthi (UCLA); other numerical experiments were by A. Duci and myself. I also deeply thank the organizers for inviting me to Cetraro to give the lectures that were the basis for these lecture notes.

1 Shapes and Curves

What is this course about? In the first two sections we begin by summarizing in a simpler form the definitions, reviewing the goals, and presenting some useful mathematical tools.

Metrics of Curves in Shape Optimization and Analysis

1.1 Shapes

A wide interest for the study of *shape spaces* arose in recent years, in particular inside the *computer vision* community. Some examples of shape spaces are as follows.

- The family of all collections of k points in $\mathbb{R}^n$.
- The family of all non empty compact subsets of $\mathbb{R}^n$.

- The family of all closed curves in $\mathbb{R}^n$.

There are two different (but interconnected) fields of applications for a good shape space in *computer vision*:

Shape optimization. where we want to find the shape that best satisfies a design goal; a topic of interest in engineering at large;

Shape analysis. where we study a family of shapes for purposes of statistics, (automatic) cataloging, probabilistic modeling, among others, and possibly create an a-priori model for a better shape optimization.

1.2 Curves

We will use formulæ of the form $A := B$ to mean that the formula A is defined by the formula B.

$S^1 = \{x \in \mathbb{R}^2 \mid |x| = 1\}$ is the circle in the plane. S^1 is the template for all possible **closed curves**. (*Open curves* will be called **paths**, to avoid confusion).

Definition 1.1 (Classes of curves).

- A C^1 **curve** is a continuously differentiable map $c : S^1 \to \mathbb{R}^n$ such that the derivative $c'(\theta) := \partial_\theta c(\theta)$ exists at all points $\theta \in S^1$.
- An **immersed curve** is a C^1 curve c such that $c'(\theta) \neq 0$ at all points $\theta \in S^1$.

$$c : S^1 \to c(S^1)$$

Note that, in our terminology, the "curve" is the function c, and not just the image $c(S^1)$ inside $\mathbb{R}^n$.

Most of the theory following will be developed for curves in $\mathbb{R}^n$, when this does not complicate the math. We will call **planar curves** those whose image is in $\mathbb{R}^2$.

The class of immersed curves is a differentiable manifold. For the purposes of this introduction, we present a simple, intuitive definition.

Definition 1.2. The **manifold of (parametric) curves** M is the set of all closed immersed curves. Suppose that $c \in M$, $c : S^1 \to \mathbb{R}^n$ is a closed immersed curve.

- A deformation of c is a function $h : S^1 \to \mathbb{R}^n$.
- The set of all such h is the **tangent space** $T_c M$ of M at c.
- An infinitesimal deformation of the curve c_0 in "direction" h will yield (on first order) the curve $c_0(u) + \varepsilon h(u)$.
- A **homotopy** C connecting c_0 to c_1 is a continuously differentiable function $C : [0, 1] \times S^1 \to \mathbb{R}^n$ such that $c_0(\theta) = C(0, \theta)$ and $c_1(\theta) = C(1, \theta)$.
  By defining $\gamma(t) = C(t, \cdot)$ we can associate C to a path $\gamma : [0, 1] \to M$ in the space of curves M, connecting c_0 to c_1.

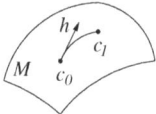

When dealing with homotopies $C = C(t, \theta)$, we will use the "prime notation" C' for the derivative $\partial_\theta C$ in the "curve parameter" θ, and the "dot notation" $\dot{C}$ for the derivative $\partial_t C$ in the "time parameter" t.

1.3 Geometric Curves and Functionals

Shapes are usually considered to be *geometric objects*. Representing a curve using $c : S^1 \to \mathbb{R}^n$ forces a choice of parameterization, that is not really part of the concept of "shape". To get rid of this, we first summarily present what *reparameterizations* are.

Definition 1.3. Let $\text{Diff}(S^1)$ be the family of **diffeomorphism**s of S^1: all the maps $\phi : S^1 \to S^1$ that are C^1 and invertible, and the inverse ϕ^{-1} is C^1.

$\text{Diff}(S^1)$ enjoys some important mathematical properties.

- $\text{Diff}(S^1)$ is a group, and its **group operation** is the composition $\phi, \psi \mapsto \phi \circ \psi$.
- $\text{Diff}(S^1)$ is divided in two connected components, the family $\text{Diff}^+(S^1)$ of diffeomorphisms with derivative $\phi' > 0$ at all points; and the family $\text{Diff}^-(S^1)$ of diffeomorphisms with derivative $\phi' < 0$ at all points.
  $\text{Diff}^+(S^1)$ is a subgroup of $\text{Diff}(S^1)$.

- Diff(S^1) acts on M, the **action**[1] is the right function composition $c \circ \phi$. The resulting curve $c \circ \phi$ is a **reparameterization** of c.

  The action of the subgroup $\text{Diff}^+(S^1)$ moreover does not change the orientation of the curve.

Definition 1.4. The **quotient space**

$$B := M/\text{Diff}(S^1)$$

is the space of *curves up to reparameterization*, also called **geometric curves** in the following. Two parametric curves $c_1, c_2 \in M$ such that $c_1 = c_2 \circ \phi$ are the same geometric curve inside B.

For some applications we may choose instead to consider the quotient w.r.to $\text{Diff}^+(S^1)$; the quotient space $M/\text{Diff}^+(S^1)$ is the space of **geometric oriented curves**.

B is mathematically defined as the set $B = \{[c]\}$ of all **equivalence classes** $[c]$ of curves that are equal but for reparameterization,

$$[c] := \{c \circ \phi \text{ for } \phi \in \text{Diff}(S^1)\} \, ;$$

and similarly for the quotient $M/\text{Diff}^+(S^1)$. More properties of these quotients will be presented in Sect. 4.5.

We can now define the **geometric functional**s (that are, loosely speaking invariant w.r.to reparameterization of the curve).

Definition 1.5. A functional $F(c)$ defined on curves will be called **geometric** if one of the two following alternative definitions holds.

- $F(c) = F(c \circ \phi)$ for all curves c and for all $\phi \in \text{Diff}(S^1)$.
- For all curves c and all ϕ, if $\phi \in \text{Diff}^+(S^1)$ then $F(c) = F(c \circ \phi)$, whereas if $\phi \in \text{Diff}^-(S^1)$ then $F(c) = -F(c \circ \phi)$

In the first case, F can be "projected" to B, that is, it may be considered as a function $F : B \to \mathbb{R}$. In the second case, F can be "projected" to $M/\text{Diff}^+(S^1)$.

It is important to remark that "geometric" theories have often provided the best results in *computer vision*.

1.4 Curve-Related Quantities

A good way to specify the *design goal* for shape optimization is to define an **objective function** (a.k.a. **energy**) $F : M \to \mathbb{R}$ that is minimum in the curve that is most fit for the task.

---

[1] We will provide more detailed definitions and properties of the "actions" in Sect. 3.8.

When designing our F, we will want it to be *geometric*; this is easily accomplished if we use geometric quantities to start from. We now list the most important such quantities.

In the following, given $v, w \in \mathbb{R}^n$ we will write $|v|$ for the standard **Euclidean norm**, and $\langle v, w \rangle$ or $(v \cdot w)$ for the standard **scalar product**. We will again write $c'(\theta) := \partial_\theta c(\theta)$.

Definition 1.6 (Derivations). If the curve c is *immersed*, we can define the **derivation with respect to the arc parameter**

$$\partial_s = \frac{1}{|c'|} \partial_\theta .$$

(We will sometimes also write D_s instead of ∂_s.)

Definition 1.7 (Tangent). At all points where $c'(\theta) \neq 0$, we define the **tangent vector**

$$T(\theta) = \frac{c'(\theta)}{|c'(\theta)|} = \partial_s c(\theta) .$$

(At the points where $c' = 0$ we may define $T = 0$).

It is easy to prove (and quite natural for our geometric intuition) that D_s and T are geometric quantities (according to the second Definition 1.5).

Definition 1.8 (Length). The **length** of the curve c is

$$\operatorname{len}(c) := \int_{S^1} |c'(\theta)| \, d\theta . \tag{1}$$

We can define formally the **arc parameter** s by

$$ds := |c'(\theta)| \, d\theta ;$$

we use it only in integration, as follows.

Definition 1.9 (Integration by arc parameter). We define the **integration by arc parameter** of a function $g : S^1 \to \mathbb{R}^k$ along the curve c by

$$\int_c g(s) \, ds := \int_{S^1} g(\theta) |c'(\theta)| \, d\theta .$$

and the **average integral**

$$\fint_c g(s) \, ds := \frac{1}{\operatorname{len}(c)} \int_c g(s) \, ds$$

and we will sometimes shorten this as **avg**$_c(g)$.

The above integrations are geometric quantities (according to the first Definition 1.5).

Fig. 1 Example of a regular curve and its curvature

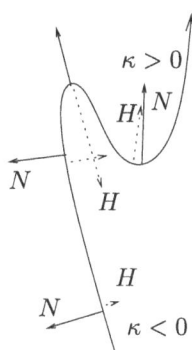

Curvature

Suppose moreover that the curve is immersed and is C^2 regular (that means that it is twice differentiable, and the second derivative is continuous); in this case we may define *curvatures*, that indicate how much the curve *bends*. There are two different definitions of *curvature* of an immersed curve: *mean curvature* H and *signed curvature* κ, that is defined when c is planar. H and k are *extrinsic curvatures*, they are properties of the *embedding* of c into $\mathbb{R}^n$.

Definition 1.10 (H). If c is C^2 regular and immersed, we can define the **mean curvature** H of c as

$$H = \partial_s \partial_s c = \partial_s T$$

where $\partial_s = \frac{1}{|c'|}\partial_\theta$ is the *derivation with respect to the arc parameter*. It enjoys the following properties.

Properties 1.11.
- It is easy to prove that $H \perp T$.
- H is a geometric quantity. If $\tilde{c} = c \circ \phi$ and $\tilde{H}$ is its curvature, then $\tilde{H} = H \circ \phi$.
- $1/|H|$ is the radius of a tangent circle that best approximates the curve to second order.

Definition 1.12 (N). When the curve c is immersed and planar, we can define a **normal vector** N to the curve, by requiring that $|N| = 1$, $N \perp T$ and N is rotated $\pi/2$ radians anticlockwise with respect to T.

Definition 1.13 (κ). If c is planar and C^2 regular, then we can define a **signed scalar curvature** $\kappa = \langle H, N \rangle$, so that

$$\partial_s T = \kappa N = H \text{ and } \partial_s N = -\kappa T.$$

See Fig. 1. The above equality implies that κ is a geometric quantity, according to the second Definition 1.5.

2 Shapes in Applications

A number of methods have been proposed in shape analysis to define distances between shapes, averages of shapes and statistical models of shapes. At the same time, there has been much previous work in shape optimization, for example image segmentation via active contours, 3D stereo reconstruction via deformable surfaces; in these later methods, many authors have defined energy functionals $F(c)$ on curves (or on surfaces), whose minima represent the desired segmentation/reconstruction; and then utilized the calculus of variations to derive curve evolutions to search minima of $F(c)$, often referring to these evolutions as gradient flows. The reference to these flows as gradient flows *implies a certain Riemannian metric on the space of curves; but this fact has been largely overlooked. We call this metric H^0, and properly define it in (5).*

2.1 Shape Analysis

Many method and tools comprise the **shape analysis**. We may list

- distances between shapes,
- averages for shapes,
- principle component analysis for shapes and
- probabilistic models of shapes.

We will present a short overview of the above, in theory and in applications. We begin by defining the *distance function* and *signed distance function*, two tools that we will use often in this theory.

Definition 2.1. Let $A, B \subset \mathbb{R}^n$ be closed sets.

- $u_A(x) := \inf_{y \in A} |x - y|$ is the **distance function**,

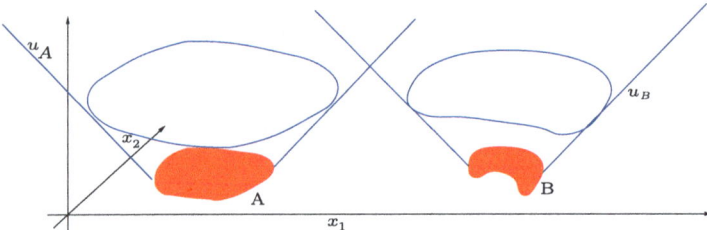

- $b_A(x) := u_A(x) - u_{\mathbb{R}^n \setminus A}(x)$ is the **signed distance function**.

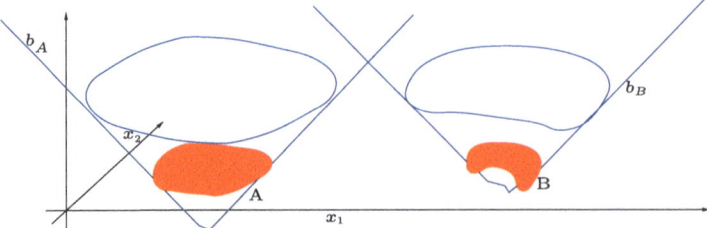

Metrics of Curves in Shape Optimization and Analysis 213

Shape Distances

A variety of distances have been proposed for measuring the difference between two given shapes. Two examples follows:

Definition 2.2. The **Hausdorff distance**

$$d_H(A, B) := \left(\sup_{x \in A} u_B(x) \right) \vee \left(\sup_{x \in B} u_A(x) \right) \quad (2)$$

where $A, B \subset \mathbb{R}^n$ are compact, and $u_A(x)$ is the *distance function*.

The above can be used for curves; in this case a distance may be defined by associating to the two curves c_1, c_2 the Hausdorff distance of the image of the curves

$$d_H(c_1(S^1), c_2(S^1)) .$$

If c_1, c_2 are immersed and planar, we may otherwise use $d_H(c_1, c_2)$ where c_1, c_2 denote the closed region of the plane that is enclosed by c_1, c_2.

Definition 2.3. Let $A, B \subset \mathbb{R}^n$ be two measurable sets, let

$$A \Delta B := (A \setminus B) \cup (B \setminus A)$$

be the **set symmetric difference**; let $|A|$ be the Lebesgue measure of A. We define the **set symmetric distance** as

$$d(A, B) = |A \Delta B| ,$$

(where it is to be intended that two sets are considered "equal" if they differ by a measure zero set).

In the case of planar curves c_1, c_2, we can apply to above idea to define

$$d(c_1, c_2) = |c_1(S^1) \Delta c_2(S^1)| .$$

Shape Averages

Many definitions have also been proposed for the average $\bar{c}$ of a finite set of shapes $c_1, \ldots, c_N$. There are methods based on the **arithmetic mean** (that are representation dependent).

- One such case is when the shapes are defined by a finite family of M parameters; so we can define the **parametric or landmark averaging**

$$\bar{c}(p) = \frac{1}{N} \sum_{n=1}^{N} c_n(p)$$

where p is a common parameter/landmark.

- More in general, we can define the **signed distance level set averaging** by using as a representative of a shape its signed distance function, and computing the average shape by

$$\bar{c} = \{x \mid \bar{b}(x) = 0\}, \text{ where } \bar{b}(x) = \frac{1}{N}\sum_{n=1}^{N} b_{c_n}(x)$$

where b_{c_n} is the signed distance function of c_n; or in case of planar curves, of the part of plane enclosed by c_n.

Then there are *non parametric, representation independent*, methods. The (possibly) most famous one is the

Definition 2.4. The **distance-based average**.[2] Given a distance d_M between shapes, an average shape $\bar{c}$ may be found by minimizing the sum of its squared distances.

$$\bar{c} = \arg\min_{c} \sum_{n=1}^{N} d_M(c, c_n)^2$$

It is interesting to note that in Euclidean spaces there is an unique minimum, that coincides with the arithmetic mean; while in general there may be no minimum, or more than one.

Principal Component Analysis (PCA)

Definition 2.5. Suppose that X is a *random vector* taking values in $\mathbb{R}^n$; let $\overline{X} = \mathbb{E}(X)$ be the mean of X. The **principal component analysis** is the representation of X as

$$X = \overline{X} + \sum_{i=1}^{n} Y_i S_i$$

where S_i are constant vectors, and Y_i are uncorrelated real random variables with zero mean, and with decreasing variance. S_i is known as the i-**th mode of principal variation**.

The PCA is possible in general in any finite dimensional linear space equipped with an inner product. In infinite dimensional spaces, or equivalently in case of *random processes*, the PCA is obtained by the *Karhunen-Loève theorem*.

---

[2] Due to Fréchet, 1948; but also attributed to Karcher [26].

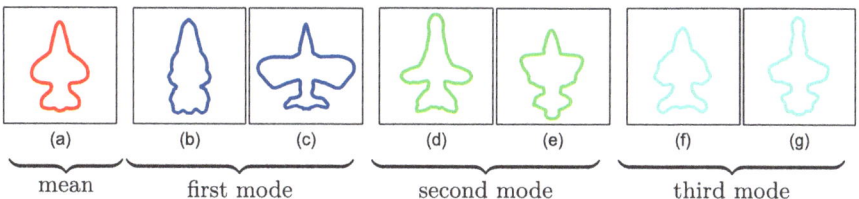

Fig. 2 PCA of plane shapes (From [60] ©2001 IEEE. Reproduced with permission)

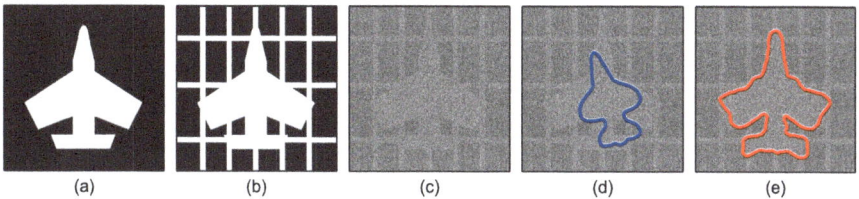

Fig. 3 Segmentation of occluded plane shape (From [60] ©2001 IEEE. Reproduced with permission)

Given a finite number of samples, it is also possible to define an *empirical principal component analysis*, by estimating the expectation and variance of the data.

In many practical cases, the variances of Y_i decrease quite fast: it is then sensible to replace X by a simplified variable

$$\tilde{X} := \overline{X} + \sum_{i=1}^{k} Y_i S_i$$

with $k < n$.

So, if shapes are represented by some *linear* representation, the PCA is a valuable tool to study the statistics, and it has then become popular for imposing *shape priors* in various shape optimization problems. For more general manifold representations of shapes, we may use the *exponential map* (defined in 3.29), replacing S_n by *tangent vectors* and following *geodesics* to perform the summation.

We present here an example in applications.

Example 2.6 (PCA by signed distance representation). In these pictures we see an example of synergy between analysis and optimization, from Tsai et al. [60]. Figure 2 contains a row of pictures that represent the (empirical) PCA of signed distance function of a family of plane shapes. Figure 3 contains a second row of images that represent: an image of a plane shape *(a)* occluded *(b)* and noise degraded *(c)*, an initialization for the shape optimizer *(d)* and the final optimal segmentation *(e)*.

2.2 Shape Optimization and Active Contours

A Short History of Active Contours

In the late twentieth century, the most common approach to *image segmentation* was a combination of **edge detection** and **contour continuation**. With edge detection [5], small **edge elements** would be identified by a local analysis of the image; then a **continuation** method would be employed to reconstruct full contours. The methods were suffering from two drawbacks, edge detection being too sensitive to noise and to photometric features (such as sharp shadows, reflections) that were not related to the physical structure of the image; and continuation was in essence a NP-complete algorithm.

Active contours were introduced by Kass et al. [27] as a simpler approach to the segmentation problem. The idea is to minimize an energy $F(c)$ (where the variable c is a contour i.e. a curve), that contains an **edge-based** attraction term and a smoothness term, which becomes large when the curve is irregular. To search for the minimum of $F(c)$, a curve evolution method was derived from the calculus of variations. The evolving curve would then be attracted to the features of interest, while mantaining the property of being a closed smooth contour.

An unjustified feature of the model of [27] is that the evolution is dependent on the way the contour is parameterized. Hence Caselles et al. [6], Malladi et al. [34] revisited the idea of [27] in the new form of a **geometric evolution**, which could be implemented by the **level set method** invented by Osher and Sethian [44]. Thereafter, Kichenassamy et al. [28] and Caselles et al. [7] presented the active contour approach to the edge detection problem in the form of the minimization of a geometric energy that is a generalization of Euclidean arc length. Again, the authors derived the **gradient descent flow** in order to minimize the geometric energy.

In contrast to the edge-based approaches for active contours (mentioned above), **region-based** energies for active contours have been proposed in [8, 46, 66, 70]. In these approaches, an energy is designed to be minimized when the curve partitions the image into statistically distinct regions. This kind of energy has provided many desirable features; for example, it provides less sensitivity to noise, better ability to capture concavities of objects, more dependence on global features of the image, and less sensitivity to the initial contour placement.

In [42, 43], the authors introduced and rigorously studied a region-based energy that was designed to both extract the boundary of distinct regions while also smoothing the image within these regions. Subsequently, Tsai et al. [61], Vese and Chan [62] implemented a **level set method** for the curve evolution minimizing the energy functional considered by Mumford and Shah.

Metrics of Curves in Shape Optimization and Analysis 217

Energies in *Computer Vision*

A variational approach to solving shape optimization problems is to define and consequently minimize geometric energy functionals $F(c)$ where c is a planar curve. We may identify two main families of energies,

- the **region based energies**:

$$F(c) = \int_R f_{in}(x)\,dx + \int_{R^c} f_{out}(x)\,dx$$

where $\int \cdots dx$ is area element integral, and R and R^c are the interior and exterior areas outlined by c; and

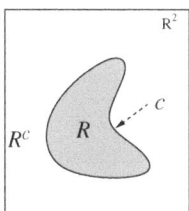

- the **boundary based energies**:

$$F(c) = \int_c \phi(c(s))\,ds$$

where ϕ is designed to be small on the salient features of the image, and $\int_c \cdots ds$ is the *integration by arc parameter* defined in Definition 1.9. Note that $\phi(x)\,ds$ may be interpreted as a conformal metric on $\mathbb{R}^2$, so minima curves are geodesics w.r.to the conformal metric.

Examples of Geometric Energy Functionals for Segmentation

Suppose $I : \Omega \to [0, 1]$ is the image (in black and white). We propose two examples of energies taken from the above families.

- The **Chan-Vese** segmentation energy [62]

$$F(c) = \int_R \left| I(x) - \mathrm{avg}_R(I) \right|^2 dx + \int_{R^c} \left| I(x) - \mathrm{avg}_{R^c}(I) \right|^2 dx$$

where $\mathrm{avg}_R I = \dfrac{1}{|R|} \int_R I(x)dx$ is the average of I on the region R.

- The **geodesic active contour** [7, 28]

$$F(c) = \int_c \phi(c(s))\,\mathrm{d}s \qquad (3)$$

where ϕ may be chosen to be

$$\phi(x) = \frac{1}{1 + |\nabla I(x)|^2}$$

that is small on sharp discontinuities of I (∇I is the gradient of I w.r.to x). Since real images are noisy, the function ϕ in practice would be more influenced by the noise than by the image features; for this reason, usually the function ϕ is actually defined by

$$\phi(x) = \frac{1}{1 + |\nabla G \star I(x)|^2}$$

where G is a smoothing kernel, such as the *Gaussian*.

2.3 Geodesic Active Contour Method

The "Geodesic Active Contour" Paradigm

The general procedure for geodesic active contours goes as follows.

1. Choose an appropriate *geometric* energy functional, E.
2. Compute the **directional derivative**

$$DE(c; h) = \frac{\mathrm{d}}{\mathrm{d}t} E(c + th) \Big|_{t=0}$$

where c is a curve and h is an arbitrary perturbation of c.
3. Manipulate $DE(c; h)$ into the form

$$\int_c h(s) \cdot v(s)\,\mathrm{d}s\ .$$

4. Consider v to be the *gradient*, the direction which increases E *fastest*.
5. Evolve $c = c(t, \theta)$ by the differential equation $\partial_t c = -v$; this is called the **gradient descent flow**.

(Note that the *directional derivative* is the basis for the Definition 3.13 of the *Gâteaux differential*, that will be then discussed in Sects. 3.7 and 4.2.)

Example: Geodesic Active Contour Edge Model

We propose an explicit computation starting from the classical active contour model.

1. Start from an energy that is minimal along image contours:

$$E(c) = \int_c \phi(c(s))\,ds$$

   where ϕ is defined to extract relevant features; for examples as discussed after (3).

2. Calculate the directional derivative:

$$\begin{aligned}DE(c;h) &= \int_c \nabla\phi(c)\cdot h + \phi(c)(D_s h \cdot D_s c)\,ds \\ &= \int_c \nabla\phi(c)\cdot h - \bigl(\nabla\phi(c)\cdot D_s c\bigr)(h \cdot D_s c) - \phi(c)(h \cdot D_{ss}c)\,ds \\ &= \int_c h \cdot \bigl((\nabla\phi(c)\cdot N)N - \phi(c)\kappa N\bigr)\,ds\,. \end{aligned} \quad (4)$$

3. Deduce the "gradient":

$$\nabla E = -\phi\kappa N + (\nabla\phi \cdot N)N\,.$$

4. Write the gradient flow

$$\frac{\partial c}{\partial t} = \phi\kappa N - (\nabla\phi \cdot N)N\,.$$

(Note that the flow of geometric energies moves only in orthogonal direction w.r.t the curve—this phenomenon will be explained in in Sect. 11.10.)

Implicit Assumption of H^0 Inner Product

We have made a critical assumption in going from the directional derivative

$$DE(c;h) = \int_c h(s) \cdot v(s)\,ds$$

to deducing that the gradient of E is $\nabla E(c) = v$. Namely, the definition of the gradient[3] is based on the following equality

---

[3]The precise definition of what the gradient is is in Sect. 3.7.

$$\langle h(s), \nabla E \rangle = \int_c \underbrace{h(s)}_{h_1=h} \cdot \underbrace{v(s)}_{h_2=\nabla E} \, ds,$$

that needs an inner-product structure.

This implies that we have been presuming all along that curves are equipped with a H^0-**type inner-product** defined as follows

$$\langle h_1, h_2 \rangle_{H^0} := \int_c h_1(s) \cdot h_2(s) \, ds \, . \tag{5}$$

2.4 Problems and Goals

We will now briefly list some examples that show some limits of the usual *active contour* method.

Example: Geometric Heat Flow

We first review one of the most mathematically studied gradient descent flows of curves. By direct computation, the *Gâteaux differential* of the length of a closed curve is

$$D \operatorname{len}(c; h) = \int_c \partial_s h \cdot T \, ds = - \int_c h \cdot H \, ds \tag{6}$$

Let $C = C(t, \theta)$ be an evolving family of curves. The **geometric heat flow** (also known as **motion by mean curvature**) is

$$\frac{\partial C}{\partial t} = H$$

where $H := \partial_s \partial_s C$ is the mean curvature. In the H^0 inner-product, this flow is the *gradient descent for curve length*.[4]

This flow has been studied deeply by the mathematical community, and is known to enjoy the following properties.

Properties 2.7. • *Embedded planar curves remain embedded.*
• *Embedded planar curves converge to a circular point.*

---

[4] A different gradient descent flow for curve length will be discussed in Remark 10.32.

Metrics of Curves in Shape Optimization and Analysis

- *Comparison principle: if two embedded curves are used as initialization, and the first contains the second, then it will contain the second for all time of evolution. This is important for* level set methods.
- *The flow is well posed only for positive times, that is, for increasing t (similarly to the usual* heat equation).

For the proofs, see [21, 23].

Short Length Bias

The usual edge-based active contour energy is of the form

$$E(c) = \int_c g(c(s))\,ds$$

supposing that g is reasonably constant in the region where c is running, then $E(c) \simeq \bar{g}\,\text{len}(c)$, due to the integral being performed w.r.to the arc parameter ds. So one way to reduce E is to shrink the curve. Consequently, when the image is smooth and featureless (as in medical imaging), the usual edge based energies would often drive the curve to a point.

To avoid it, an **inflationary term** νN (with $\nu > 0$) was usually added to the curve evolution, to obtain

$$\frac{\partial c}{\partial t} = \phi \kappa N - \nabla \phi + \nu N\,,$$

see [7, 28]. Unfortunately, this adds a parameter that has to be tuned to match the characteristics of each specific image.

Average Weighted Length

As an alternative approach we may normalize the energy to obtain a more length-independent energy, the **average weighted length**

$$\text{avg}_c(g) := \frac{1}{\text{len}(c)} \int_c g(c(s))\,ds \tag{7}$$

but this generates an **ill-posed** H^0-flow, as we will now show.

Flow Computations

Here we write $L := \text{len}(c)$. Let $g : \mathbb{R}^n \to \mathbb{R}^k$; let

$$\mathrm{avg}_c(g(c)) := \frac{1}{L}\int_c g(c(s))\,\mathrm{d}s = \frac{1}{L}\int_{S^1} g(c(\theta))|c'(\theta)|\,\mathrm{d}\theta\;;$$

then the *Gâteaux differential* is

$$D[\mathrm{avg}_c(g(c))](c;h) = \frac{1}{L}\int_c \nabla g(c)\cdot h + g(c)(\partial_s h\cdot T)\,\mathrm{d}s - \qquad(8)$$
$$-\frac{1}{L^2}\int_c g(c)\,\mathrm{d}s \int_c \partial_s h\cdot T\,\mathrm{d}s =$$
$$= \fint_c \nabla g(c)\cdot h + \bigl(g(c) - \mathrm{avg}_c(g(c))\bigr)(\partial_s h\cdot T)\,\mathrm{d}s\;.$$

In the above, we omit the argument (s) for brevity; ∇g is the gradient of g.

If the curve is planar, we define the normal and tangent vectors N and T as by Definitions 1.12 and 1.13; then, integrating by parts, the above becomes

$$D[\mathrm{avg}_c(g(c))](c;h) = \frac{1}{L}\int_c \nabla g(c)\cdot h - (\nabla g(c)\cdot T)(h\cdot T) - \qquad(9)$$
$$-\bigl(g(c) - \mathrm{avg}_c(g(c))\bigr)(h\cdot D_s^2 c)\,\mathrm{d}s =$$
$$= \fint_c \Bigl(\nabla g(c)\cdot N - \kappa\bigl(g(c) - \mathrm{avg}_c(g(c))\bigr)\Bigr)(h\cdot N)\,\mathrm{d}s\;.$$

Suppose now that we have a shape optimization functional E including a term of the form $\mathrm{avg}_c(g(c))$; let $C = C(t,\theta)$ be an evolving family of curves trying to minimize E; this flow would contain a term of the form

$$\frac{\partial C}{\partial t} = \ldots \bigl(g(c(s)) - \mathrm{avg}_c(g)\bigr)\kappa N \ldots$$

unfortunately the above flow is ill defined: it is a backward-running *geometric heat flow* on roughly half of the curve. We will come back to this example in Sect. 10.8; more examples and methods for computing Gâteaux differentials are in Sect. 4.2.

Centroid Energy

We will now propose another simple example where the above phenomenon is again evident.

Example 2.8 (Centroid-based energy). Let us fix a target point $v \in \mathbb{R}^n$. We recall that $\mathrm{avg}_c(c)$ is the **center of mass** of the curve. The energy

$$E(c) := \frac{1}{2}|\mathrm{avg}_c(c) - v|^2 \qquad(10)$$

Metrics of Curves in Shape Optimization and Analysis   223

penalizes the distance from the center of mass to v. Let in the following $\bar{c} = \text{avg}_c(c)$ for simplicity. The *directional derivative* of E in direction h is

$$DE(c;h) = \left\langle \bar{c} - v, D(\bar{c})(c;h) \right\rangle$$

where in turn (by (8) and (9))

$$D(\bar{c})(c;h) = \oint_c h + (c - \bar{c})(D_s h \cdot D_s c) \, ds = \tag{11}$$
$$= \oint_c h - D_s c \, (h \cdot D_s c) - (c - \bar{c})(h \cdot D_s^2 c) \, ds$$

supposing that the curve is planar, then

$$h - D_s c \, (h \cdot D_s c) = N(h \cdot N)$$

so

$$DE(c;h) = \oint \langle \bar{c} - v, N \rangle (h \cdot N) - \langle \bar{c} - v, c - \bar{c} \rangle \kappa (h \cdot N) \, ds .$$

The H^0 gradient descent flow is then

$$\frac{\partial c}{\partial t} = -\nabla_{H^0} E(c) = \langle (v - \bar{c}), N \rangle N - \kappa N \langle (c - \bar{c}), (v - \bar{c}) \rangle .$$

The first term $\langle (v - \bar{c}) \cdot N \rangle N$ in this gradient descent flow moves the whole curve towards v.

Let $P := \{w : \langle (w - \bar{c}) \cdot (v - \bar{c}) \rangle \geq 0\}$ be the half plane "on the v side". The second term $-\kappa N \langle (c - \bar{c}) \cdot (v - \bar{c}) \rangle$ in this gradient descent flow tries to decrease the curve length out of P and increase the curve length in P, **and this is ill posed**.

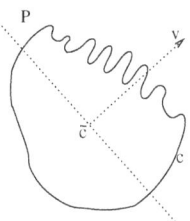

We will come back to this example in Proposition 10.20.

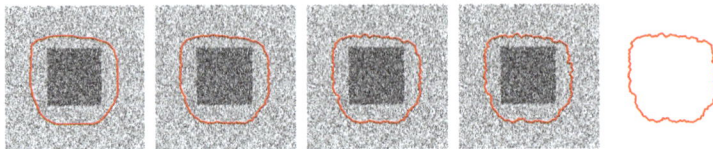

Fig. 4 H^0 gradient descent flow of Chan–Vese energy, to segment a noisy square

Conclusions

More recent works that use active contours for segmentation are not only based on information from the image to be segmented (edge-based or region-based), but also **prior information**, that is information known about the *shape* of the desired object to be segmented. The work of Leventon et al. [32] showed how to incorporate *prior information* into the active contour paradigm. Subsequently, there have been a number of works, for example Chen et al. [12], Cremers and Soatto [13], Raviv et al. [45], Rousson and Paragios [47], Tsai et al. [60], which design energy functionals that incorporate prior shape information of the desired object. In these works, the main idea is designing a novel term of the energy that is small when the curve is *close*, in some sense, to a pre-specified shape.

The need for *prior information* terms arose from several factors such as

- the fact that some images contain limited information,
- the energy functions considered previously could not incorporate complex information,
- the energy had too many extraneous local minima, and *the gradient flows* to minimize these energies allowed for arbitrary deformations that gave rise to unlikely shapes; as in the example in Fig. 4 of segmentation using the *Chan-Vese* energy, where the flowing contour gets trapped into noise.

Works on incorporating prior shape knowledge into active contours have led to a fundamental question on how to define distance between two curves or shapes. Many works, for example Charpiat et al. [9], Michor and Mumford [38], Mio and Srivastava [40], Soatto and Yezzi [51], Yezzi and Mennucci [64,65], Younes [67], in the *shape analysis* literature have proposed different ways of defining this distance.

However, Michor–Mumford and Yezzi–Mennucci [38, 65] observed that all previous works on geometric active contours that derive *gradient flows* to minimize energies, which were described earlier, imply a natural notion of Riemannian metric, given by the geometric inner product H^0 (that we just "discovered" in (5)). Subsequently, Michor–Mumford and Yezzi–Mennucci [38, 64] have shown a surprising property: the H^0 Riemannian metric on the space of curves is **pathological**, since the "distance" between *any* two curves is zero.

In addition to the pathologies of the Riemannian structure induced by H^0, there are also undesirable features of H^0 gradient flows. Some of these features are listed below.

1. There are no regularity terms in the definition of the H^0 inner product. That is, there is nothing in the definition of H^0 that discourages flows that are not smooth in the space of curves. Thus, when energies are designed to depend on the image that is to be segmented, the H^0 gradient is very sensitive to noise in the image.

   Therefore, in *geometric active contour* models, a penalty on the curve's length is added to keep the curve smooth. However, this changes the energy that is being optimized and ends up solving a different problem.
2. H^0 gradients, evaluated at a particular point on the curve, depend locally on derivatives of the curve. Therefore, as the curve becomes non-smooth, as mentioned above, the derivative estimates become inaccurate, and thus, the curve evolution becomes inaccurate. Moreover, for region-based and edge-based active contours, the H^0 gradient at a particular point on the curve depends locally on image data at the particular point. Although region-based energies may depend on global statistics, such as mean values of regions, the H^0 gradient still depends on local data. These facts imply that the H^0 gradient descent flow is sensitive to noise and local features.
3. The H^0 norm gives *non-preferential* treatment to arbitrary deformations regardless of whether the deformations are global motions (not changing the shape of the curve) such as translations, rotations and scales; or whether they are more local deformations.
4. Many *geometric active contour*s (such as edge and region-based active contours) require that the unit normal to the evolving curve be defined. As such, the evolution does not make sense for polygons. Moreover, since in general, an H^0 active contour does not remain smooth, one needs special numerical schemes based on viscosity theory in a level set framework to define the flow.
5. Many simple and apparently sound energies cannot be implemented for *shape optimization* tasks;

   - some energies generate *ill-posed* H^0 flows;
   - if an energy integrand uses derivatives of the curve of degree up to d, then the PDE driving the flow has degree $2d$; but derivatives of the curves of high degree are noise sensitive and are difficult to implement numerically, and
   - the active contours method works best when implemented using the *level set method*, but this is difficult for flow PDEs of degree higher than two.

In conclusion, if one wishes to have a consistent view of the geometry of the space of curves in both shape optimization and shape analysis, then one should use the H^0 metric when computing distances, averages and morphs between shapes. Unfortunately, H^0 does not yield a well define metric structure, since the associated distance is identically zero.

So to achieve our goal, we will need to devise new metrics.

3 Basic Mathematical Notions

In this section we provide the mathematical theory that will be needed in the rest of the course. (Some of the definitions are usually known to mathematics' students; we will present them nonetheless as a chance to remark less known facts.)We will though avoid technicalities in the definitions, and for the most part just provide a base intuition of the concepts. The interested reader may obtain more details from a books in analysis, such as [2], in functional analysis, such as [48], and in differential and Riemannian geometry, such as [14, 30] or [31].

We start with a basic notion, in a quite simplified form.

Definition 3.1 (Topological spaces). A **topological space** is a set M with associated a **topology** τ of subsets, that are the **open sets** in M.

The topology is the simplest and most general way to define what are the "convergent sequences of points" and the "continuous functions". We will not provide details regarding topological spaces, since in the following we will mostly deal with *normed spaces* and *metric spaces*, where the topology is induced by a norm or a metric. We just recall this definition.

Definition 3.2. A **homeomorphism** is an invertible continuous function $\phi : M \to N$ between two topological spaces M, N, whose inverse ϕ^{-1} is again continuous.

3.1 Distance and Metric Space

Definition 3.3. Given a set M, a **distance** $d = d_M$ is a function

$$d : M \times M \to [0, \infty]$$

such that

1. $d(x, x) = 0$,
2. if $d(x, y) = 0$ then $x = y$,
3. $d(x, y) = d(y, x)$ *(d is symmetric)*
4. $d(x, z) \leq d(x, y) + d(y, z)$ *(the triangular inequality)*.

There are some possible generalizations.

- The second request may be waived, in this case d is a **semidistance**.
- The third request may be waived: then d would be an **asymmetric distance**. Most theorems we will see can be generalized to asymmetric distances; see [36].

Metric Space

The pair (M, d) is a **metric space**. A metric space has a distinguished topology, generated by balls of the form $B(x, r) := \{y \mid d(x, y) < r\}$; according to this topology, $x_n \to_n x$ iff $d(x_n, x) \to_n 0$; and functions are continuous if they map convergent sequences to convergent sequences. We will assume in the following that the reader is acquainted with the concepts of *open, closed, compact sets* and *continuous functions* in metric spaces. We recall though the definition of *completeness*.

Definition 3.4. A metric space (M, d) is **complete** iff, for any given sequence $(c_n)_n \subset M$, the fact that

$$\lim_{m,n \to \infty} d(c_m, c_n) = 0$$

implies that c_n converges.

Example 3.5. • $\mathbb{R}^n$ is usually equipped with the **Euclidean distance**

$$d(x, y) = |x - y| = \sqrt{\sum_{i=1}^{n} (x_i - y_i)^2} \; ;$$

and $(\mathbb{R}^n, d)$ is a complete metric space.
- Any closed subset of a complete space is complete.
- If we cut away an accumulation point out of a space, the resulting space is not complete.

A complete metric space is a space without "missing points". This is important in optimization: if a space is not complete, any optimization method that moves the variable and searches for the optimal solution may fail since the solution may, in a sense, be "missing" from the space.

3.2 Banach, Hilbert and Fréchet Spaces

Definition 3.6. Given a vector space E, a **norm** $\|\cdot\|$ is a function

$$\|\cdot\| : E \to [0, \infty]$$

such that

1. $\|\cdot\|$ is convex
2. if $\|x\| = 0$ then $x = 0$.
3. $\|\lambda x\| = |\lambda| \|x\|$ for $\lambda \in \mathbb{R}$

Again, there are some possible generalizations.

- If the second request is waived, then $\|\cdot\|$ is a **seminorm**.
- If the third request holds only for $\lambda \geq 0$, then the norm is **asymmetric**; in this case, it may happen that $\|x\| \neq \|-x\|$.

Each (semi/asymmetric)norm $\|\cdot\|$ defines a (semi/asymmetric)distance

$$d(x, y) := \|x - y\|.$$

So a norm induces a topology.

Examples of Spaces of Functions

We present some examples and definitions.

Definition 3.7. A **locally-convex topological vector space** E (shortened as **l.c.t.v.s.** in the following) is a vector space equipped with a collection of seminorms $\|\cdot\|_k$ (with $k \in K$ an index set); the seminorms induce a topology, such that $c_n \to c$ iff $\|c_n - c\|_k \to_n 0$ for all k; and all vector space operations are continuous w.r.to this topology.

The simplest example of l.c.t.v.s. is obtained when there is only one norm; this gives raise to two renowned examples of spaces.

Definition 3.8 (Banach and Hilbert spaces).

- A **Banach space** is a vector space E with a norm $\|\cdot\|$ defining a distance $d(x, y) := \|x - y\|$ such that E is metrically complete.
- A **Hilbert space** is a space with an inner product $\langle f, g \rangle$, that defines a norm $\|f\| := \sqrt{\langle f, f \rangle}$ such that E is metrically complete.

(Note that a Hilbert space is also a Banach space).

Example 3.9. Let $I \subset \mathbb{R}^k$ be open; let $p \in [1, \infty]$. A standard example of Banach space is the L^p space of functions $f : I \to \mathbb{R}^n$. If $p \in [1, \infty)$, the L^p space contains all functions such that $|f|^p$ is Lebesgue integrable, and is equipped with the norm

$$\|f\|_{L^p} := \sqrt[p]{\int_I |f(x)|^p \, dx}.$$

For the case $p = \infty$, L^∞ contains all Lebesgue measurable functions $f : I \to \mathbb{R}^n$ such that there is a constant $c \geq 0$ for which $|f(x)| \leq c$ for all $x \in I \setminus N$, where N is a set of measure zero; and L^∞ is equipped with the norm

$$\|f\|_{L^\infty} := \mathrm{supess}_I |f(x)|$$

that is the lowest possible constant c.

If $p = 2$, L^2 is a Hilbert space by inner product

$$\langle f, g \rangle := \int_I f(x) \cdot g(x) \, dx \, .$$

Note that, in these spaces, by definition, $f = g$ iff the set $\{f \neq g\}$ has *Lebesgue measure* zero.

Fréchet Space

The following citations [24] are referred to the first part of Hamilton's 1982 survey on the Nash and Moser theorem.

Definition 3.10. A **Fréchet space** E is a *complete Hausdorff metrizable* l.c.t.v.s.; where we define that the l.c.t.v.s. E is

complete when, for any sequence (c_n), the fact that

$$\lim_{m,n \to \infty} \|c_m - c_n\|_k = 0$$

for all k implies that c_n converges;

Hausdorff when, for any given c, if $\|c\|_k = 0$ for all k then $c = 0$;

metrizable when there are countably many seminorms associated to E.

The reason for the last definition is that, if E is metrizable, then we can define a distance

$$d(x, y) := \sum_{k=0}^{\infty} \frac{2^{-k} \|x - y\|_k}{1 + \|x - y\|_k}$$

that generates the same topology as the family of seminorms $\|\cdot\|_k$; and the vice versa is true as well, see [48].

More Examples of Spaces of Functions

Example 3.11. Let $I \subset \mathbb{R}^m$ be open and non empty.

- The **Banach space** $C^j(I \to \mathbb{R}^n)$, with associated norm

$$\|f\| := \sup_{i \leq j} \sup_{t \in I} |f^{(i)}(t)| \, ;$$

where $f^{(i)}$ is the i-th derivative. In this space $f_n \to f$ iff $f_n^{(i)} \to f^{(i)}$ uniformly for all $i \leq j$.

- The **Sobolev space** $H^j(I \to \mathbb{R}^n)$, with scalar product

$$\langle f,g \rangle_{H^j} := \int_I f(t) \cdot g(t) + \cdots + f^{(j)}(t) \cdot g^{(j)}(t)\, dt$$

  where $f^{(j)}$ is the j-th derivative.[5]
- The **Fréchet space** of **smooth functions** $C^\infty(I \to \mathbb{R}^n)$.
  The seminorms are

$$\|f\|_k = \sup_{x \in I} |f^{(k)}(x)|$$

  where $f^{(k)}$ is the k-th derivative. In this space, $f_n \to f$ iff all the derivatives $f_n^{(k)}(x)$ converge to derivatives $f^{(k)}(x)$, and for any fixed k the convergence is uniform w.r.to $x \in I$.

This last is the strongest topology between the topologies usually associated to spaces of functions.

Dual Spaces

Definition 3.12. Given a *l.c.t.v.s.* E, the **dual space** E^* is the space of all linear functions $L : E \to \mathbb{R}$.

If E is a Banach space, it is easy to see that E^* is again a Banach space, with norm

$$\|L\|_{E^*} := \sup_{\|x\|_E \le 1} |Lx|\ .$$

The biggest problem when dealing with Fréchet spaces is that the dual of a Fréchet space is not in general a Fréchet space, since it often fails to be metrizable. (In most cases, the duals of Fréchet spaces are "quite wide" spaces; a classical example being the dual elements of smooth functions, that are the **distribution**s). So given F, G Fréchet spaces, we cannot easily work with *the space $\mathcal{L}(F, G)$ of linear functions between F and G*.

As a workaround, given an auxiliary space H, we will consider *indexed families of linear maps* $L : F \times H \to G$, where $L(\cdot, h)$ is linear, and L is jointly continuous; but we will *not* consider L as a map

$$\begin{aligned} h &\mapsto (f \mapsto L(f,h)) \\ H &\to \mathcal{L}(F,G) \end{aligned} \tag{12}$$

---

[5] The derivatives are computed in distributional sense, and must exists as Lebesgue integrable functions.

Gâteaux Differential

An example is the *Gâteaux differential*.

Definition 3.13. We say that a continuous map $P : U \to G$, where F, G are Fréchet spaces and $U \subset F$ is open, is **Gâteaux differentiable** if for any $h \in F$ the limit

$$DP(f;h) := \lim_{t \to 0} \frac{P(f+th) - P(f)}{t} = [\partial_t P(f+th)]|_{t=0} \tag{13}$$

exists. The map $DP(f;\cdot) : F \to G$ is the **Gâteaux differential**.

Note that the Gâteaux differential may fail to be linear in h; indeed, any one-homogeneous function P is Gâteaux differentiable at $f = 0$.

Definition 3.14. We say that P is C^1 if $DP : U \times F \to G$ exists and is jointly continuous.

(This is weaker than what is usually required in Banach spaces).

The basics of the usual calculus hold.

Theorem 3.15 ([24, Thm. 3.2.5]). *If P is C^1 then $DP(f;h)$ is linear in h.*

Theorem 3.16 (Chain rule [24, Thm. 3.3.4]). *If P, Q are C^1 then $P \circ Q$ is C^1 and*

$$D(P \circ Q)(f;h) = DP(Q(f); DQ(f;h)).$$

Also, the **implicit function theorem** holds, in the form due to Nash and Moser: see again [24] for details.

Troubles in Calculus

But some important parts of calculus are instead lost.

Example 3.17. Suppose $P : U \subset F \to F$ is smooth, and consider the O.D.E.

$$\frac{d}{dt} f = Pf$$

to be solved for a solution $f : (-\varepsilon, \varepsilon) \to F$.

- if F is a Banach space, then, given initial condition $f(0) = x$, the solution will exist and be unique (for $\varepsilon > 0$ small enough);
- but if F is a Fréchet space, then f may fail to exist or be unique.

See [24, Sect. 5.6]

A consequence (that we will discuss more later on) is that the exponential map in Riemannian manifolds may fail to be locally surjective. See [24, Sect. 5.5.2].

3.3 Manifold

To build a manifold of curves, we have ahead two main definitions of "manifold" to choose from.

Definition 3.18 (Differentiable Manifold). An **abstract differentiable manifold** is a topological space M associated to a model *l.c.t.v.s.* U. It is equipped with an **atlas** of **charts** $\phi_k : U_k \to V_k$, where $U_k \subset U$ are open sets, and $V_k \subset M$ are open sets that cover M. The maps ϕ_k are *homeomorphisms*.
The composition $\phi_1^{-1} \circ \phi_2$ restricted to $\phi_2^{-1}(V_1 \cap V_2)$ is usually required to be a smooth diffeomorphism.
The **dimension** $\dim(M)$ of M is the dimension of U. When M is finite-dimensional, the model space (in this book) is always $U = \mathbb{R}^n$.

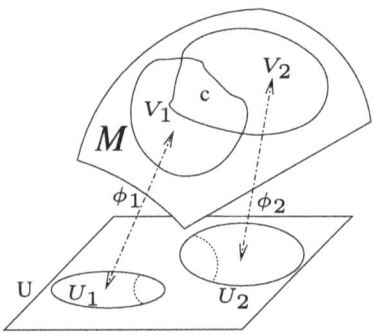

Since ϕ_k are homeomorphisms, then the topology of M is "identical" to the topology of U. The rôle of the charts is to define the differentiable structure of M mimicking that of U. See for example the definition of *directional derivative* in (15).

Submanifold

Definition 3.19. Suppose A, B are open subsets of two linear spaces. A **diffeomorphism** is an invertible C^1 function $\phi : A \to B$, whose inverse ϕ^{-1} is again C^1.

Let U be a fixed closed linear subspace of a *l.c.t.v.s.* X.

Definition 3.20. A **submanifold** is a subset M of X, such that, at any point $c \in M$ we may find a **chart** $\phi_k : U_k \to V_k$, with $V_k, U_k \subset X$ open sets, $c \in V_k$. The maps ϕ_k are diffeomorphisms, and ϕ_k maps $U \cap U_k$ onto $M \cap V_k$.

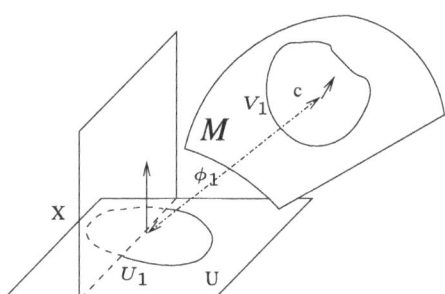

Most often, $M = \{\Phi(c) = 0\}$ where $\Phi : X \to Y$; so, to prove that M is a submanifold, we will use the *implicit function theorem*.

Note that M is itself an abstract manifold, and the model space is U; so $\dim(M) = \dim(U) \leq \dim(X)$. Vice versa any abstract manifold is a submanifold of some large X (by well known *embedding theorems*).

3.4 Tangent Space and Tangent Bundle

Let us fix a differentiable manifold M, and $c \in M$. We want to define the **tangent space** $T_c M$ of M at c.

Definition 3.21.
- The **tangent space** $T_c M$ is more easily described for submanifolds; in this case, we choose a chart ϕ_k and a point $x \in U_k$ s.t. $\phi_k(x) = c$; $T_c M$ is the image of the linear space U under the derivative $D_x \phi_k$. $T_c M$ is itself a linear subspace in X. In Fig. 5, we graphically represent $T_c M$ though as an affine subspace, by translating it to the point c.
- The **tangent bundle** TM is the collection of all tangent spaces. If M is a submanifold of the vector space X, then

$$TM := \{(c, h) \mid c \in M, h \in T_c M\} \subset X \times X .$$

The tangent bundle TM is itself a differentiable manifold; its charts are of the form $(\phi_k, D\phi_k)$, where ϕ_k are the charts for M.

3.5 Fréchet Manifold

When studying the space of all curves, we will deal with **Fréchet manifolds**, where the model space E will be a Fréchet space, and the composition of local charts $\phi_1^{-1} \circ \phi_2$ is smooth.

Fig. 5 Tangent space

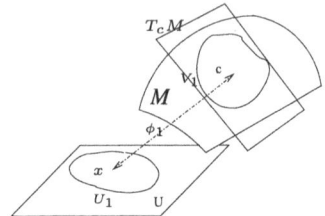

Some objects that we may find useful in the following are Fréchet manifolds.

Example 3.22 (Examples of Fréchet manifolds). Given two finite-dimensional manifolds S, R, with S compact and n-dimensional,

- [24, Example 4.1.3]. the space $C^\infty(S; R)$ of smooth maps $f : S \to R$ is a Fréchet manifold. It is modeled on $U = C^\infty(\mathbb{R}^n; R)$.
- [24, Example 4.4.6]. The space of smooth diffeomorphisms Diff(S) of S onto itself is a Fréchet manifold. The group operation $\phi, \psi \to \phi \circ \psi$ is smooth.
- But if we try to model Diff(S) on $C^k(S; S)$, then the group operation is not even differentiable.
- [24, Example 4.6.6]. The quotient of the two above

$$C^\infty(S; R)/\text{Diff}(S)$$

is a Fréchet manifold. It contains *all smooth maps from S to R, up to a diffeomorphism of S*.

So the theory of Fréchet spaces seems appropriate to define and operate on the *manifold of geometric curves*.

3.6 Riemann and Finsler Geometry

We first define *Riemannian geometries*, and then we generalize to *Finsler geometries*.

Riemann Metric, Length

Definition 3.23.
- A **Riemannian metric** G on a differentiable manifold M defines a **scalar product** $\langle h_1, h_2 \rangle_{G_{|c}}$ on $h_1, h_2 \in T_c M$, dependent on the point $c \in M$. We assume that the scalar product varies smoothly w.r.to c.
- The scalar product defines the **norm** $|h|_c = |h|_{G_{|c}} = \sqrt{\langle h, h \rangle_{G_{|c}}}$.

  Suppose $\gamma : [0, 1] \to M$ is a path connecting c_0 to c_1.

- The **length** is
$$\mathrm{Len}(\gamma) := \int_0^1 |\dot\gamma(v)|_{\gamma(v)} dv$$
where $\dot\gamma(v) := \partial_v \gamma(v)$.

- The **energy** (or **action**) is
$$\mathbb{E}(\gamma) := \int_0^1 |\dot\gamma(v)|^2_{\gamma(v)} dv$$

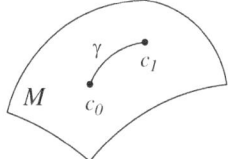

Finsler Metric, Length

Definition 3.24. We define a **Finsler metric** to be a function $F : TM \to \mathbb{R}^+$, such that

- F is continuous and,
- for all $c \in M$, $v \mapsto F(c, v)$ is a norm on $T_c M$.

We will sometimes write $|v|_c := F(c, v)$.

(Sometimes F is called a "Minkowski norm").

As for the case of norms, a *Finsler metric* may be asymmetric; but, for sake of simplicity, we will only consider the symmetric case.

Using the norm $|v|_c$ we can then again define the length of paths by

$$\mathrm{Len}^F(\gamma) = \int_0^1 |\dot\gamma(t)|_{\gamma(t)} dt = \int_0^1 F(\gamma(t), \dot\gamma(t)) dt$$

and similarly the action.

Distance

Definition 3.25. The **distance** $d(c_0, c_1)$ is the infimum of $\mathrm{Len}(\gamma)$ between all C^1 paths γ connecting $c_0, c_1 \in M$.

Remark 3.26. In the following chapter, we will define some differentiable manifolds M of curves, and add a Riemann (or Finsler) metric G on those; there are two different choices for the model space,

- suppose we model the differentiable manifold M on a Hilbert space U, with scalar product $\langle,\rangle_U$; this implies that M has a topology τ associated to it, and this topology, through the charts ϕ, is the same as that of U. Let now

G be a Riemannian metric; since the derivative of a chart $D\phi(c)$ maps U onto $T_c M$, one natural hypothesis will be to assume that $\langle,\rangle_U$ and $\langle,\rangle_{G,c}$ be locally equivalent (uniformly w.r.to small movements of c); as a consequence, the topology generated by the Riemannian distance d will coincide with the original topology τ. A similar request will hold for the case of a Finsler metric G, in this case U will be a Banach space with a norm equivalent to that defined by G on $T_c M$.

- We will though find out that, for technical reasons, we will initially model the spaces of curves on the Fréchet space C^∞; but in this case there cannot be a norm on $T_c M$ that generates the same original topology (for the proof, see I.1.46 in [48]).

Minimal Geodesics

Definition 3.27. If there is a path γ^* providing the minimum of $\text{Len}(\gamma)$ between all paths connecting $c_0, c_1 \in M$, then γ^* is called a **minimal geodesic**.

The *minimal geodesic* is also the minimum of the action (up to reparameterization).

Proposition 3.28. *Let ξ^* provide the minimum $\min_\gamma \mathbb{E}(\gamma)$ in the class of all paths γ in M connecting x to y. Then ξ^* is a minimal geodesic and its speed $|\dot\xi^*|$ is constant.*

Vice versa, let γ^* be a minimal geodesic, then there is a reparameterization $\xi^* = \gamma^* \circ \phi$ s.t. ξ^* provides the minimum $\min_\gamma \mathbb{E}(\gamma)$.

A proof may be found in [36].

Exponential Map

The *action* $\mathbb{E}$ is a smooth integral, quadratic in $\dot\gamma$, and we can compute the Euler-Lagrange equations; its minima are more regular, since they are guaranteed to have constant speed; consequently, when trying to find geodesics, we will try to minimize the *action* and not the *length*. This also related to the idea of the exponential map.

Definition 3.29. Let $\ddot\gamma = \Gamma(\dot\gamma, \gamma)$ be the Euler-Lagrange ODE of the action $\mathbb{E}(\gamma) = \int_0^1 |\dot\gamma(v)|^2 \, dv$. Any solution of this ODE is a **critical geodesic**. Note that any *minimal geodesic* is a *critical geodesic*.

Define the **exponential map** $\exp_c : T_c M \to M$ as $\exp_c(\eta) = \gamma(1)$, where γ is the solution of

$$\ddot\gamma(v) = \Gamma(\dot\gamma(v), \gamma(v)), \quad \gamma(0) = c, \quad \dot\gamma(0) = \eta \tag{14}$$

Solving the above ODE (14) is informally known as **shooting geodesics**.

The exponential map is often used as a chart, since it is the least possibly deforming map at the origin.

Theorem 3.30. *Suppose that M is a smooth differentiable manifold modeled on a Hilbert space with a smooth Riemannian metric. The derivative of the exponential map $\exp_c : T_c M \to M$ at the origin is an isometry, hence $\exp_c$ is a local diffeomorphism between a neighborhood of $0 \in T_c M$ and a neighborhood of $c \in M$.*

(See [31], VIII Sect. 5). The exponential map can then "linearize" small portions of M, and so it will enable us to use linear methods such as the *principal component analysis*. Unfortunately, the above result does not hold if M is modeled on a Fréchet space.

Hopf–Rinow Theorem

Theorem 3.31 (Hopf–Rinow). *Suppose M is a **finite** dimensional Riemannian or Finsler manifold. The following statements are equivalent:*

- *(M, d) is metrically complete;*
- *the O.D.E. (14) can be solved for all $c \in M$, $\eta \in T_c M$ and $v \in \mathbb{R}$;*
- *the map $\exp_c$ is surjective;*

and all those imply that, $\forall x, y \in M$ there exists a minimal geodesic connecting x to y.

Drawbacks in Infinite Dimensions

In a certain sense, infinite dimensional manifolds are simpler than their corresponding finite-dimensional counterparts: indeed, by Eells and Elworthy [18]

Theorem 3.32 (Eells–Elworthy). *Any smooth differentiable manifold M modeled on an infinite dimensional separable Hilbert space H may be embedded as an open subset of that Hilbert space.*

In other words, it is always possible to express M using one single chart. (But note that this may not be the best way for computations/applications).

When M is an infinite dimensional Riemannian manifold, though, only a small part of the Hopf–Rinow theorem still holds.

Proposition 3.33. *Suppose M is **infinite** dimensional, modeled on a Hilbert space, and (M, d) is complete, then the O.D.E. (14) of a critical geodesic can be solved for all $v \in \mathbb{R}$.*

But other implications fails.

Example 3.34 (Atkin [3]). There exists an infinite dimensional complete Hilbert smooth manifold M and $x, y \in M$ such that there is no *critical geodesic* connecting x to y.

That is,

- (M, d) is complete $\not\Rightarrow$ $\exp_c$ is surjective,
- and (M, d) is complete $\not\Rightarrow$ minimal geodesics exist.

It is then, in general, quite difficult to prove that an infinite dimensional manifold admits minimal geodesics (even when it is known to be metrically complete). There are though some positive results, such as Ekeland [19] (that we cannot properly discuss for lack of space); or the following.

Theorem 3.35 (Cartan–Hadamard). *Suppose that M is connected, simply connected and has seminegative sectional curvature; then these are equivalent:*

- (M, d) *is complete*
- *there exists a $c \in M$ such that the map $\eta \to \exp_c(\eta)$ is well defined*

and then there exists an unique minimal geodesic connecting any two points.

For the proof, see Corollaries 3.9 and 3.11 in Sect. IX.Sect. 3 in [31].

3.7 Gradient in Abstract Differentiable Manifold

Let M be a differentiable manifold, and U the model l.c.t.v.s.. Choose $c \in M$, and $\phi_1 : U_1 \to V_1$ a chart with $c \in V_1$, where V_1 is an open subset of V.

Definition 3.36. Let $E : M \to \mathbb{R}$ be a functional; we say that it is **Gâteaux differentiable** at c if for all $h \in T_c M$ there exists the **directional derivative** at c in direction h

$$DE(c; h) := \lim_{t \to 0} \frac{E(\phi_1(x + tk)) - E(c)}{t}. \tag{15}$$

where $c = \phi_1(x)$, $h = D\phi_1(x; k)$.

We suppose that E is C^1 near c; this is expressed using the chart, imposing that $DE(\phi_1(x_1), D\phi_1(x_1; h))$ is continuous, for $x_1 \in U_1$ and $h \in U$. By theorem 3.15, this implies that $DE(c; \cdot)$ is a linear function from $T_c M$ into $\mathbb{R}$; for this reason it is considered an element of the **cotangent space** $T_c^* M$ (that is the dual space of $T_c M$—and we will not discuss here further).

Suppose now that we add a Riemannian geometry to M; this defines an inner product $\langle \cdot, \cdot \rangle_c$ on $T_c M$, so we can then define the gradient.

Definition 3.37 (Gradient). The **gradient** $\nabla E(c)$ of E at c is the unique vector $v \in T_c M$ such that

$$\langle v, h \rangle_c = DE(c; h) \quad \forall h \in T_c M .$$

If M is modeled on a Hilbert space H, and the inner product $\langle \cdot, \cdot \rangle_c$ used on $T_c M$ is equivalent to the inner product in H (as we discussed in 3.26), then the above equation uniquely defines what the gradient is. When M is modeled on a Fréchet space, though, there is no choice of inner product that is "compatible" with M; and there are pathological situations where the gradient does not exist.

Example 3.38. Let $M = C^\infty([-1, 1] \to \mathbb{R})$; since M is a vector space, then it is trivially a manifold, with a single identity chart; and $T_c M = M$. Let $E : M \to \mathbb{R}$ be the **evaluation functional** $E(f) = f(0)$, then $DE(f; h) = h(0)$; let

$$\langle f, g \rangle = \int_{-1}^{1} f(t) g(t) \, dt ;$$

then the gradient of E would be δ_0 (the *Dirac's delta*), that is a *distribution* (or more simply a *probability measure*) but not an element of M.

3.8 Group Actions

Definition 3.39. Let $\mathcal{G}$ be a group, M a set. We say that $\mathcal{G}$ **acts on** M if there is a map

$$\begin{aligned} \mathcal{G} \times M &\to M \\ g, m &\mapsto g \cdot m \end{aligned}$$

that respects the group operations:

- if e is the identity element then $e \cdot m = m$, and
- for any $g, h \in \mathcal{G}, m \in M$

$$h \cdot (g \cdot m) = (hg) \cdot m . \tag{16}$$

(where hg is the product of the two elements in the group G).

If $\mathcal{G}$ is topological group, that is, a group with a topology such that the group operation is continuous, and M has a topology as well, we will require the action to be continuous. Similarly for smooth actions between smooth manifolds.

The action of a group $\mathcal{G}$ on M induces an **equivalence relation** $\sim$, defined by

$$m_1 \sim m_2 \text{ iff there exists } g \text{ such that } m_1 = g \cdot m_2.$$

We can then define the following objects.

Definition 3.40.
- The **orbit** $[m]$ of m is the family of all m_1 equivalent to m.
- The **quotient** $M/\mathcal{G}$ is the space $M/\sim$ of equivalence classes.
- There is a **projection** $\pi : M \to M/\mathcal{G}$, sending each m to its orbit $\pi(m) = [m]$.

We will see many examples of actions in Example 4.8.

Distances and Groups

Let $d_M(x, y)$ be a distance on a space M, and $\mathcal{G}$ a group acting on M; we may think of defining a distance on $B = M/\mathcal{G}$ by

$$d_B([x],[y]) := \inf_{x \in [x], y \in [y]} d_M(x,y) = \inf_{g,h \in \mathcal{G}} d_M(g \cdot x, h \cdot y) \quad (17)$$

that is the lowest distance between two orbits. Unfortunately, this definition does not in general satisfy the triangle inequality.

Proposition 3.41. *If d_M is invariant with respect to the action of the group $\mathcal{G}$, that is*

$$d_M(g \cdot x, g \cdot y) = d_M(x, y) \quad \forall g \in \mathcal{G},$$

then the above (17) can be simplified to

$$d_B([x],[y]) = \inf_{g \in \mathcal{G}} d_M(g \cdot x, y). \quad (18)$$

and d_B is a semidistance.

Proof. We write $d_B(x, y)$ instead of $d_B([x], [y])$ for simplicity. It is easy to check that

$$d_B(x,y) = \inf_{g \in \mathcal{G}} d_M(g \cdot x, y) = \inf_{g \in \mathcal{G}} d_M(y, g \cdot x) = \inf_{g \in \mathcal{G}} d_M(g^{-1} \cdot y, x) = d_B(y, x).$$

For the triangle inequality,

$$d_B(x,z) + d_B(z,y) = \inf_{g \in \mathcal{G}} d_M(g \cdot x, z) + \inf_{h \in \mathcal{G}} d_M(z, h \cdot y) =$$

$$= \inf_{g,h \in \mathcal{G}} d_M(g \cdot x, z) + d_M(z, h \cdot y) \geq \inf_{g,h \in \mathcal{G}} d_M(g \cdot x, h \cdot y) = d_B(x,y)$$

Remark 3.42. There is a main problem: Does $d_B([x], [y]) > 0$ hold when $[x] \neq [y]$? That is, there is no guarantee, in line of principle, that the above definition (18) will not simply result in $d_B \equiv 0$.

4 Spaces and Metrics of Curves

In this section we review the mathematical definitions regarding the space of curves in full detail, and express a set of mathematical goals for the theory.

4.1 Classes of Curves

Remember that $S^1 = \{x \in \mathbb{R}^2 \mid |x| = 1\}$ is the circle in the plane.

We will often associate $\mathbb{R}^2 = \mathbb{C}$, for convenience. Consequently,

- sometimes S^1 will be identified with $\mathbb{R}/(2\pi)$ (that is $\mathbb{R}$ modulus 2π translations).
- but other times (and in particular if the curve is planar) we will associate $S^1 = \{e^{it}, t \in \mathbb{R}\} = \{z, |z| = 1\} \subset \mathbb{C}$.

We recall that a **curve** is a map $c : S^1 \to \mathbb{R}^n$. The **image**, a.k.a. **trace**, of the curve is $c(S^1) = \{c(\theta), \theta \in S^1\}$.

Definition 4.1 (Classes of Curves).

- $\mathrm{Imm}(S^1, \mathbb{R}^n)$ is the class of **immersed curves** c, such that $c' \neq 0$ at all points.

- $\mathrm{Imm}_f(S^1, \mathbb{R}^n)$ is the class of **freely immersed curve**, the immersed curves c such that, moreover, if $\phi : S^1 \to S^1$ is a diffeomorphism and $c(\phi(t)) = c(t)$ for all t, then $\phi = \mathrm{Id}$. So, in a sense, the curve is "completely" characterized by its image.

- $\mathrm{Emb}(S^1, \mathbb{R}^n)$ are the **embedded curves**, maps c that are diffeomorphic onto their image $c(S^1)$; and the image is an embedded submanifold of $\mathbb{R}^n$ of dimension 1.

Each class contains the one following it. The following is an example of a non-freely immersed curve.

Example 4.2. We define the **doubly traversed circle** using the complex notation $c(z) = z^2$ for $z \in S^1 \subset \mathbb{C}$; or otherwise identifying $S^1 = \mathbb{R}/(2\pi)$, and in this case

$$c(\theta) = (\cos(2\theta), \sin(2\theta))$$

for $\theta \in \mathbb{R}/(2\pi)$. Setting $\phi(t) = t + \pi$, we have that $c = c \circ \phi$, so c is not freely immersed.

Vice versa, the following result is a sufficient condition to assert that a curve is freely immersed.

Proposition 4.3 (Michor and Mumford [38]). *If c is immersed and there is a $x \in \mathbb{R}^n$ s.t. $c(t) = x$ for one and only one t, then c is freely immersed.*

Remark 4.4. $\text{Imm}(S^1, \mathbb{R}^n)$, $\text{Imm}_f(S^1, \mathbb{R}^n)$, $\text{Emb}(S^1, \mathbb{R}^n)$, are open subset of the Banach space $C^r(S^1 \to \mathbb{R}^n)$ when $r \geq 1$, so they are trivially submanifolds (the charts are identity maps); and similarly for the Fréchet space $C^\infty(S^1 \to \mathbb{R}^n)$.

$\text{Imm}(S^1, \mathbb{R}^n)$, $\text{Imm}_f(S^1, \mathbb{R}^n)$, $\text{Emb}(S^1, \mathbb{R}^n)$ are connected iff $n \geq 3$; whereas in the case $n = 2$ of planar curves, they are divided in connected components each containing curves with the same *rotation index* (see Proposition 8.1 for the definition).

4.2 Gâteaux Differentials in the Space of Immersed Curves

Let once again M be the manifold of smooth immersed curves. We will mainly be interested in the Gâteaux derivation of operators O and functions $E : M \to \mathbb{R}^k$. By **operator** we mean any one of these possible operations.

- An operator

$$O : M \to T_c M$$

such that $O(c)$ is a smooth a vector field; examples are the operations of computing the tangent field T, and the curvature H.

- An operator

$$O : M \to M$$

such that $O(c)$ is an immersed curve; examples are all the group actions listed in Sect. 4.3

- An operator

$$O : M \to (T_c M \to T_c M)$$

such that $O(c)$ is itself an operator on vector fields along c; an example is the derivative D_s;

- or sometimes

$$O : M \to (T_c M \to \mathbb{R}^n)$$

such as the average along the curve $O(c) = \text{avg}_c(\cdot)$.

In all above cases, the evaluated operator $O(c)$ is itself a function; for this reason, to avoid confusion in the notation, in this section we write the Gâteaux derivation as $D_{c,h}O$ instead of $DO(c;h)$.

The charts in M are trivial (as noted in 4.4), so the definition (15) of *Gâteaux differential* simplifies to

$$D_{c,h}E := \partial_t E(c+th)|_{t=0} \qquad (19)$$

and similarly for an operator.

The following rules are the building blocks that are used (by means of standard calculus rules for derivatives) to compute the derivation of all other geometric operators usually found in computer vision applications.

Proposition 4.5.

$$D_{c,h}\text{len}(c) = \int_c (D_s h \cdot D_s c)\,ds = -\int_c (h \cdot D_s^2 c)\,ds, \qquad (20)$$

$$D_{c,h}(D_s O) = -(D_s h \cdot D_s c)(D_s O) + D_s(D_{c,h}O), \qquad (21)$$

$$D_{c,h}\int_c O\,ds = \int_c D_{c,h}O + O.(D_s h \cdot D_s c)\,ds, \qquad (22)$$

$$D_{c,h}\fint_c O\,ds = \fint_c D_{c,h}O + O.(D_s h \cdot D_s c)\,ds - \fint_c O\,ds \fint_c (D_s h \cdot D_s c)\,ds$$
$$\qquad (23)$$

The proof is by direct computation.

For example, from (23) we easily obtain (8), that is

$$D_{c,h}\text{avg}_c(c) = \fint_c h + (c - \text{avg}_c(c))(D_s h \cdot D_s c)\,ds.$$

If $C(t,\theta)$ is a homotopy, we can obtain a different interpretation of all previous equalities substituting formally C for O, ∂_t for $D_{c,h}$ and eventually $\partial_t C$ for h.

The above proposition helps in formalizing and speeding up calculus on geometric quantities, as in this example.

Example 4.6. We can Gâteaux-derive the "second arc parameter derivative operator D_{ss}" by repeated applications of (21), as follows

$$D_{c,h}(D_{ss}g) = -(D_s h \cdot D_s c)D_{ss}g + D_s[D_{c,h}(D_s g)] =$$
$$= -(D_s h \cdot D_s c)D_{ss}g + D_s[-(D_s h \cdot D_s c)(D_s g) + D_s(D_{c,h}g)] =$$
$$= D_{ss}(D_{c,h}g) - 2(D_s h \cdot D_s c)D_{ss}g - [D_s(D_s h \cdot D_s c)](D_s g)$$

and hence obtain **the Gâteaux differential of the curvature**

$$D_{c,h}D_{ss}c = D_{ss}h - 2(D_s h \cdot D_s c)D_{ss}c - [D_s(D_s h \cdot D_s c)]D_s c$$

and eventually obtain **the Gâteaux differential of the elastica energy** by substituting this last equality in (22)

$$D_{c,h}\int_c |D_{ss}c|^2 \, ds = \int_c 2D_{ss}c \cdot (D_{c,h}D_{ss}c) + |D_{ss}c|^2 \cdot (D_s h \cdot D_s c) \, ds =$$

$$= \int_c 2D_{ss}c \cdot D_{ss}h - 3(D_s h \cdot D_s c)|D_{ss}c|^2 \, ds$$

since $(D_{ss}c \cdot D_s c) = 0$.

Remark 4.7. Most objects we deal with share an important property: they are *reparameterization invariant*. In the case of operators, in all cases in which $O(c)$ is a curve or a vector field, then this means that for any diffeomorphism $\varphi \in \text{Diff}^+(S^1)$ there holds

$$O(c \circ \phi)(\theta) = O(c)(\phi(\theta)) . \tag{24}$$

(whereas for $\varphi \in \text{Diff}^-(S^1)$ there is a possible sign choice as explained in Definition 1.5).

By considering an infinitesimal perturbation $a : S^1 \to \mathbb{R}$ of the identity in $\text{Diff}(S^1)$, we obtain the formula

$$D_{c,h}O(c)(\theta) = a(\theta) \cdot \partial_\theta O(c)(\theta) \text{ for any } h(\theta) = a(\theta).\partial_\theta c(\theta) \tag{25}$$

and this last may be rewritten in arc parameter

$$D_{c,h}O(c)(s) = a(s) \cdot D_s O(c)(s) \text{ for any } h(s) = a(s).D_s c(s) . \tag{26}$$

Similarly if $E : M \to \mathbb{R}$ is a *reparameterization invariant* energy, then

$$D_{c,h}E(c) = 0 \text{ for any } h = a.D_s c . \tag{27}$$

It is also possible to prove that the condition (26) is equivalent to saying that the operator is reparameterization invariant; and similarly for E.

4.3 Group Actions on Curves

Let M again be the manifold of immersed curves. An example of group action that we saw in Definition 1.3 is obtained when $\mathcal{G} = \text{Diff}(S^1)$; $\mathcal{G}$ acts on curves by right composition

$$c, \phi \mapsto c \circ \phi$$

and this action is the *reparameterization*.

In general, all groups that act on $\mathbb{R}^n$ also act on M; many are of interest in *computer vision*. The action of these groups on curves is always of the form $c, A \mapsto A \circ c$, where $A : \mathbb{R}^n \to \mathbb{R}^n$ is the group action on $\mathbb{R}^n$.

Example 4.8. • $O(n)$ is the group of **rotation**s in $\mathbb{R}^n$. It is represented by the group of orthogonal matrixes

$$O(n) := \{A \in \mathbb{R}^{n \times n} \mid AA^t = A^t A = \mathrm{Id}\}$$

The action on $v \in \mathbb{R}^n$ is the matrix·vector multiplication Av.
- $SO(n)$ is the group of **special rotation**s in $\mathbb{R}^n$ with $\det(A) = 1$.
- $\mathbb{R}^n$ is the group of **translation**s in $\mathbb{R}^n$. The action on $v \in \mathbb{R}^n$ is the vector sum $v \mapsto v + T$.
- $E(n)$ is the **Euclidean group**, generated by rotations and translations.
- $\mathbb{R}^+$ is the group of **rescaling**s.

The *quotient spaces* $M/\mathcal{G}$ are the spaces of *curves up to rotation and/or translations and/or scaling (...)*.

4.4 Two Categories of Shape Spaces

Let M be a manifold of curves. We should distinguish two different ideas of shape spaces of curves.

	Geometric curves	Geometric curves up to pose
The space of shapes	Curves up to reparameterization,	Curves up to reparameterization, rotation, translation, scaling,
is modeled as	$M/\mathrm{Diff}(S^1)$,	$M/\mathrm{Diff}(S^1)/E(n)/\mathbb{R}^+$,
is well-suited to	shape optimization.	shape analysis.
References:	[22, 35, 38, 53–59]	[29, 41, 52, 67, 68]

The term *preshape space* is sometimes used for the leftmost space, when both spaces are studied in the same paper.

4.5 Geometric Curves

Unfortunately the quotient

$$B_i = \mathrm{Imm}(S^1, \mathbb{R}^n)/\mathrm{Diff}(S^1)$$

of *immersed curves up to reparameterization* is not a Fréchet manifold.

We (re)define the space of **geometric curves**.

Definition 4.9.

$$B_{i,f}(S^1, \mathbb{R}^n) = \text{Imm}_f(S^1, \mathbb{R}^n)/\text{Diff}(S^1)$$

is the quotient of $\text{Imm}_f(S^1, \mathbb{R}^n)$ (the free immersions) by the diffeomorphisms $\text{Diff}(S^1)$ (that act as reparameterizations).

The good news is that

Proposition 4.10 (Sect. 2.4.3 in [38]). *If Imm_f and $\text{Diff}(S^1)$ have the topology of the Fréchet space of C^∞ functions, then $B_{i,f}$ is a Fréchet manifold modeled on C^∞.*

The bad news is that

- when we add a simple Riemannian metric to $B_{i,f}$, the resulting metric space is not metrically complete; indeed, there cannot be any norm on C^∞ that generates the same topology of the Fréchet space C^∞ (as we discussed in 3.26);
- by modeling $B_{i,f}$ as a Fréchet manifold, some calculus is lost, as we saw in Sect. 3.2.

Remark 4.11. It seems that this is the only way to properly define the manifold. If otherwise we choose $M = C^k(S^1 \to \mathbb{R}^n)$ to be the manifold of curves, then if $c \in M$, $c' \notin T_c M$. Hence we (must?) model M on C^∞ functions.

Research Path

Following [38] we so obtained a possible program of math research:

- define

$$B = B_{i,f}(S^1, \mathbb{R}^n) = \text{Imm}_f(S^1, \mathbb{R}^n)/\text{Diff}(S^1)$$

and consider B as a Fréchet manifold modeled on C^∞,
- define a Riemann/Finsler geometry on it, study its properties,
- metrically complete the space.

In the last step, we would hope to obtain a differentiable manifold; unfortunately, this is sometimes not true, as we will see in the overview of the literature.

4.6 Goals (Revisited)

We formulate an abstract set of goal properties on a metric $\langle h_1, h_2 \rangle_{G_{|c}}$ on spaces of curves.

Metrics of Curves in Shape Optimization and Analysis   247

1. [**rescaling invariance**] For any $\lambda > 0$, if we rescale c to λc, then

$$\langle h_1, h_2 \rangle_{G_{|\lambda c}} = \lambda^a \langle h_1, h_2 \rangle_{G_{|c}}$$

(where $a \in \mathbb{R}$ is an universal constant);

2. [**Euclidean invariance**] Suppose that A is an Euclidean transformation, and R is its rotational part; if we apply A to c and R to h_1, h_2, then

$$\langle Rh_1, Rh_2 \rangle_{G_{|Ac}} = \langle h_1, h_2 \rangle_{G_{|c}} ;$$

3. [**parameterization invariance**]
   the metric does not depend on the parameterization of the curve, that is $\|\tilde{h}\|_{\tilde{c}} = \|h\|_c$ when $\tilde{c}(t) = c(\varphi(t))$ and $\tilde{h}(t) = h(\varphi(t))$.

If a metric satisfies the above three properties, we say that it is a **geometric metric**.

Well Posedness

We also add a more basic set of requirements.

0. [**well-posedness and existence of minimal geodesics**]

- The metric induces a *good* distance d: that is, the distance between different curves is positive, and d generates the same topology that the atlas of the manifold M induces;
- (M, d) is complete;
- for any two curves in M, there exists a minimal geodesic connecting them.

Are the Goal Properties Consistent?

So we state the abstract problem:

Problem 4.12. Consider the space of curves M, and the family of all Riemannian (or Finsler) Geometries F on M.
   Does there exist a metric F satisfying the above properties 1,2,3?
   Consider metrics F that may be written in integral form

$$F(c, h) = \int_c f\left(c(s), \partial_s c(s), \ldots, \partial^j_{s^j} c(s), h(s), \ldots, \partial^i_{s^i} h(s)\right) ds.$$

What is the relationship between the degrees i, j and the properties in this section?

All this boils down to a fundamental question: *can we design metrics to satisfy our needs?*

5 Representation/Embedding/Quotient in the Current Literature

5.1 The Representation/Embedding/Quotient Paradigm

A common way to model shapes is by **representation/embedding**:

- we **represent** the shape A by a function u_A
- and then we **embed** this representation in a space E, so that we can operate on the shapes A by operating on the representations u_A.

Most often, representation/embedding alone do not directly provide a satisfactory *shape space*. In particular, in many cases it happens that the representation is "redundant", that is, the same shape has many different possible representations. An appropriate **quotient** is then introduced.

There are many examples of *shape spaces* in the literature that are studied by means of the *representation/embedding/quotient* scheme. Understanding the basic math properties of these three operations is then a key step in understanding *shape spaces* and designing/improving them.

We now present a rapid overview of how this scheme is exploited in the current literature of shape spaces; then we will come back to some of them to explain more in depth.

5.2 Current Literature

Example 5.1 (The family of all non empty compact subsets of $\mathbb{R}^N$). A standard representation is obtained by associating each closed subset A to the *distance function* u_A or the *signed distance function* b_A (that were defined in Definition 2.1). We may then define a *topology of shapes* by deciding that $A_n \to A$ when $u_{A_n} \to u_A$ uniformly on compact sets. This convergence coincides with the Kuratowski topology of closed sets; if we limit shapes to be compact sets, the Kuratowski topology is induced by the Hausdorff distance. See Sect. 6.2.

Example 5.2. Trouvé-Younes et al. (see [22, 59] and references therein) modeled the motion of shapes by studying a left invariant Riemannian metric on the family $\mathcal{G}$ of diffeomorphisms of the space $\mathbb{R}^N$; to recover a true metric of shapes, a quotient is then performed w.r.to all diffeomorphisms $\mathcal{G}_0$ that do not move a template shape.

But the *representation/embedding/quotient* scheme is also found when dealing with spaces of curves:

Example 5.3 (Representation by angle function). In the work of Klassen et al. [29], Mio and Srivastava [41], Srivastava et al. [52], smooth planar closed curves c :

$S^1 \to \mathbb{R}^2$ of length 2π are represented by a pair of angle-velocity functions $c'(u) = \exp(\phi(u) + i\alpha(u))$ (identifying $\mathbb{R}^2 = \mathbb{C}$) then (ϕ, α) are embedded as a subset N in $L^2(0, 2\pi)$ or $W^{1,2}(0, 2\pi)$. Since the goal is to obtain a *shape space* representation for shape analysis purposes, a quotient is then introduced on N. See Sect. 8.2.

Example 5.4. Another representation of planar curves for shape analysis is found in [67]. In this case the angle function is considered mod(π). This representation is both simple and very powerful at the same time. Indeed, it is possible to prove that **geodesics do exist and to explicitly show examples of geodesics**. See Sect. 8.3.

Example 5.5 (Harmonic representation). A. Duci et al. (see [16, 17]) represent a closed planar contour as the zero level of a harmonic function. This novel representation for contours is explicitly designed to possess a linear structure, which greatly simplifies linear operations such as averaging, principal component analysis or differentiation in the space of shapes.

And, of course, we have in this list the spaces of embedded curves.

Example 5.6. When studying embedded curves, usually, for the sake of mathematical analysis, the curves are modeled as immersed parametric curves; a quotient w.r.to the group of possible reparameterizations of the curve c (that coincides with the group of diffeomorphisms $\mathrm{Diff}(S^1)$) is applied afterward to all the mathematical structures that are defined (such as the manifold of curves, the Riemannian metric, the induced distance, etc.).

It is interesting to note this fact.

Remark 5.7 (A remark on the quotienting order). When we started talking of geometric curves, we proposed the quotient $B = M/\mathrm{Diff}(S^1)$ (the space of *curves up to reparameterization*); and this had to be followed by other quotients with respect to Euclidean motions and/or scaling, to obtain $M/\mathrm{Diff}(S^1)/E(n)/\mathbb{R}^+$. In practice, though, the space B happens to be more difficult to study; hence most *shape space* theories that deal with curves prefer a different order: the quotient $M/E(n)/\mathbb{R}^+$ is modeled and studied first; then the quotient $M/E(n)/\mathbb{R}^+/\mathrm{Diff}(S^1)$ is performed (often, only numerically).

6 Metrics of Sets

We now present two examples of Shape Theories where a shape may be a generic subset of the plane; with particular attention to how they behave with respect to curves.

6.1 Some More Math on Distance and Geodesics

We start by reviewing some basic results in abstract metric spaces theory.

Length Induced by a Distance

In this subsection (M, d) is a generic metric space.

Definition 6.1 (Length by total variation). Define the **length** of a continuous curve $\gamma : [\alpha, \beta] \to M$, by using the total variation

$$\operatorname{Len}^d \gamma := \sup_T \sum_{i=1}^n d\big(\gamma(t_{i-1}), \gamma(t_i)\big).$$

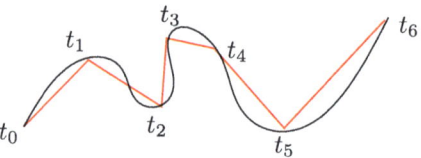

where the supremum is carried out over all finite subsets $T = \{t_0, \cdots, t_n\} \subset [\alpha, \beta]$ and $t_0 \le \cdots \le t_n$.

Definition 6.2 (The induced geodesic distance).

$$d^g(x, y) := \inf_\gamma \operatorname{Len}^d \gamma \tag{28}$$

the infimum is in the class of all continuous γ in M connecting x to y.

Note that $d^g(x, y) < \infty$ iff x, y may be connected by a Lipschitz path.

Minimal Geodesics

Definition 6.3. If in the definition (28) there is a curve γ^* providing the minimum, then γ^* is called a **minimal geodesic** connecting x to y.

Proposition 6.4. *In Riemann and Finsler manifolds, the integral length* $\operatorname{Len}(\gamma)$ *that we defined in Sect. 3.6 coincides with the* total variation length $\operatorname{Len}^d \gamma$ *that we defined in Definition 6.1. As a consequence,* $d = d^g$: *we say that* d *is* **path-metric**.

In general, though, it is easy to devise examples where $d \ne d^g$.

Example 6.5. The set $M = S^1 \subset \mathbb{R}^2$ is a metric space, if associated with $d(x, y) = |x - y|$ (that is represented as a dotted segment in the picture); in this case, $d^g(x, y) = |\arg(x) - \arg(y)|$ (that is represented as a dashed arc in the picture).

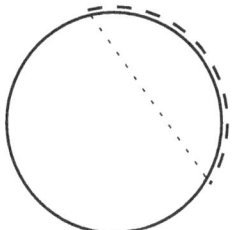

Existence of Geodesics and of Average

Proposition 6.6. *If for a $\rho > 0$ the closed ball*

$$\mathbb{D}^g(x, \rho) := \{y \mid d^g(x, y) \leq \rho\}$$

is compact, then x and any $y \in \mathbb{D}^g$ may be connected by a geodesic.

Proposition 6.7. *If for a $x \in M$ and all $\rho > 0$, $\mathbb{D}^g(x, \rho)$ is compact, then the distance-based average (that was defined in 2.4) exists.*

The proofs may be found in [15].

Geodesic Rays

More in general,

Definition 6.8. a continuous curve $\gamma : I \to M$ (where $I \subset \mathbb{R}$ is an interval) is a **geodesic ray** if for each $t \in I$ there is a $\varepsilon > 0$ s.t. $J = [t - \varepsilon, t + \varepsilon] \subset I$ and γ restricted to J is the geodesic between $\gamma(t - \varepsilon)$ and $\gamma(t + \varepsilon)$.

A critical geodesic in a Riemann smooth manifold (modeled on a Hilbert space) is always a *geodesic ray*, as in this example.

Example 6.9. The multiply traversed full circle $\gamma(t) = (0, \cos(t), \sin(t))$ with $t \in \mathbb{R}$ is a *geodesic ray* in the sphere S^2.

Hopf–Rinow, Cohn-Vossen Theorem

Definition 6.10. • We recall that (M, d) is **path-metric** if $d = d^g$.
• Moreover, (M, d) is **locally compact** if small closed balls are compact.

Theorem 6.11 (Hopf–Rinow, Cohn-Vossen). *Suppose (M, d) is locally compact and path-metric; the following statements are equivalent:*

- for all $x \in M, \rho > 0$, the closed ball

$$\mathbb{D}(x, \rho) := \{y \mid d(x, y) \leq \rho\}$$

is compact;
- (M, d) is complete;
- any geodesic ray $\gamma : [0, 1) \to M$ may be extended on $[0, 1]$;

and all imply that any two points may be connected by a minimal geodesic.

Note that the above theorem cannot be used in infinite dimensional differentiable manifolds, since the the closed balls are never compact in that case (see Theorems 1.21, 1.22 in Chap. I in [48]).

6.2 Hausdorff Metric

Let Ξ be the collection of all compact subsets of $\mathbb{R}^N$. (This is sometimes called the *topological hyperspace* of $\mathbb{R}^N$). Let $A, B \subset \mathbb{R}^N$ compact non-empty. We already defined the **distance function** $u_A(x) := \inf_{y \in A} |x - y|$, and the Hausdorff distance of two sets A, B as

$$d_H(A, B) := \left(\sup_{x \in A} u_B(x)\right) \vee \left(\sup_{x \in B} u_A(x)\right).$$

We now list some known properties.

Properties 6.12.
- $d_H(A, B) = \sup_{x \in \mathbb{R}^N} |u_A(x) - u_B(x)|$.
- (Ξ, d_H) is path-metric;
- each family of compact sets that are contained in a fixed large closed ball in $\mathbb{R}^N$ is compact in (Ξ, d_H); so
- any closed ball $\mathbb{D}(A, \rho) := \{B \mid d_H(A, B) \leq \rho\}$ is compact in (Ξ, d_H), and moreover
- by Theorem 6.11, minimal geodesics exist in (Ξ, d_H).

An Alternative Definition

Let $D_r(x)$ be the closed ball of center x and radius $r > 0$ in $\mathbb{R}^N$, and $D_r = D_r(0)$. We define the **fattened set** to be

$$A + D_r := \{x + y \mid x \in A, |y| \leq r\} = \bigcup_{x \in A} D_r(x) = \{y \mid u_A(y) \leq r\}.$$

Fig. 6 Fattening of a set

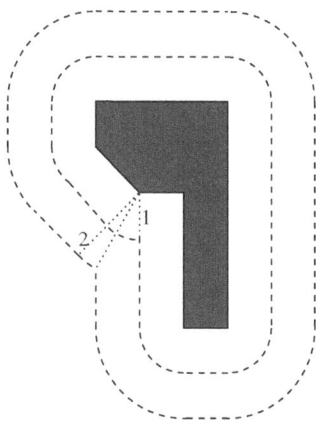

Note that the fattened set is always closed, (since the distance function $u_A(x)$ is continuous).

Example 6.13. In Fig. 6 we see an example of a set A fattened to $r = 1, 2$; the set A is the black polygon (and is filled in), whereas the dashed lines in the drawing are the contours of the fattened sets.[6]

We can then state the following equivalent definition of the *Hausdorff distance*:

$$d_H(A, B) = \min\{\delta > 0 \mid A \subset (B + D_\delta), B \subset (A + D_\delta)\}.$$

Uncountable Many Geodesics

Unfortunately d_H is quite "unsmooth". There are choices of A, B compact that may be joined by an uncountable number of geodesics. In fact we can consider this simple example.

Example 6.14 (Duci and Mennucci [15]). Let us set

$$A = \{x = 0, 0 \le y \le 2\}$$
$$B = \{x = 2, 0 \le y \le 1\}$$
$$C_t = \{x = 1, 0 \le y \le \tfrac{3}{2}\} \cup \{y = 0, 1 \le x \le t\}$$
$$\text{with } 1 \le t \le \sqrt{5}/2;$$

---

[6]The fattened sets are not drawn filled—otherwise they would cover A.

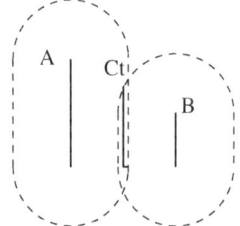

and in the picture we represent (dashed) the fattened sets $A + D_{\sqrt{5}/2}$ and $B + D_{\sqrt{5}/2}$. Note that $d_H(A, B) = \sqrt{5}$ while $d_H(A, C_t) = d_H(B, C_t) = \sqrt{5}/2$: so C_t are all midpoints that are on different geodesics between A and B.

Curves and Connected Sets

Let again Ξ be the collection of all compact subsets of $\mathbb{R}^N$. Let Ξ_c be the subclass of compact *connected* subsets of $\mathbb{R}^N$. We now relate the space of curves to this metric space, by listing these properties and remarks.

- Ξ_c is a closed subset of (Ξ, d_H);
- Ξ_c is the closure in (Ξ, d_H) of the class of (images of) all embedded curves.
- Ξ_c is Lipschitz-path-connected[7];
- for all above reasons, it is possible to connect any two $A, B \in \Xi_c$ by a minimal geodesic moving in Ξ_c.
- So, if we try to find a minimal geodesic connecting two curves using the metric d_H, we will end up finding a geodesic in (Ξ_c, d_H); and similarly if we try to optimize an energy defined on curves.
- But note that Ξ_c is not geodesically convex in Ξ, that is, there exist two points $A, B \in \Xi_c$ and a minimal geodesics ξ connecting A to B in the metric space (Ξ, d), such that the image of ξ is not contained inside Ξ_c.
- We do not know if (Ξ_c, d_H) is path-metric.

Applications in *Computer Vision*

Charpiat et al. [9] propose an approximation method to compute $\text{len}^H(\xi)$ by means of a family of energies defined using a smooth integrand; the approximation is mainly based on the property $\|f\|_{L^p} \to_p \|f\|_{L^\infty}$, for any measurable function f defined on a bounded domain; they successively devise a method to find approximation of geodesics.

---

[7]That is, any $A, B \in \Xi_c$ can be connected by a Lipschitz arc $\gamma : [0, 1] \to \Xi_c$.

6.3 A Hausdorff-Like Distance of Compact Sets

In [15] a L^p-like distance on the compact subsets of $\mathbb{R}^N$ was proposed. (Here $p \in [1, \infty)$.)

To this end, we fix $\varphi : [0, \infty) \to (0, \infty)$, decreasing, C^1, with $\varphi(|x|) \in L^p$. We then define $v_A(x) := \varphi(u_A(x))$, where u_A is the *distance function*. We eventually define the distance

$$d(A, B) := \|v_A - v_B\|_{L^p}. \qquad (29)$$

Remark 6.15. This *shape space* is a perfect example of the representation/embedding/quotient scheme. Indeed, this *shape space* is represented as $N_c = \{v_A \mid A \text{ compact}\}$ and embedded in L^p. Given $v \in N_c$, we recover the shape $A = \{v = \varphi^{-1}(0)\}$ (that is a *level set* of v).

Example 6.16. A simple example (that works for all p) is given by $\varphi(t) = e^{-t}$, so that $v_A(x) = \exp(-u_A(x))$; in this case, $A = \{v = 1\}$.

The distance d of (29) enjoys the following properties.

Properties 6.17.
- *It is Euclidean invariant;*
- *it is locally compact but not path-metric;*
- *the topology induced is the same as that induced by d_H;*
- *minimal geodesics do exist, since $\mathbb{D}^g := \{B \mid d^g(A, B) \leq \rho\}$ is compact.*

The proofs are in [15]. We present a numerical computation of a minimal geodesic (by A. Duci) in Fig. 7.

Analogy with the Hausdorff Metric, L^p vs L^∞

We recall that $d_H(A, B) = \|u_A(x) - u_B(x)\|_{L^\infty}$; whereas instead now we are proposing $d(A, B) := \|v_A - v_B\|_{L^p}$. The idea being that this distance of compact sets is modeled on L^p, whereas the Hausdorff distance is "modeled" on L^∞. (Note that the Hausdorff distance is not really obtained by embedding, since $u_A \notin L^\infty$). L^p is more regular than L^∞, as shown by this remark.

Remark 6.18. Given any $f, g \in L^p$ with $p \in (1, \infty)$, the segment connecting f to g is the unique minimal geodesic connecting them. Suppose now that the dimension of $L^\infty(\Omega, \mathcal{A}, \mu)$ is greater than 1. Given generically $f, g \in L^\infty$, there is an uncountable number of minimal geodesics connecting them.[8]

(For the proof see 2.11 and 2.13 in [15]). The above result suggests that it may be possible to shoot geodesics in the "Riemannian metric" associated to this distance.

---

[8]"Generically" is meant in the Baire sense: the set of exceptions is of first category.

Fig. 7 Example of a minimal geodesic

Contingent Cone

Let again $N_c = \{v_A \mid A \text{ compact}\}$ be the family of all representations. Given a $v \in N_c$, let $T_v N_c \subset L^p$ be the **contingent cone**

$$T_v N_c := \{\lim_n t_n(v_n - v) \mid t_n > 0, v_n \in N_c, v_n \to v\}$$

$$= \left\{ \lambda \lim_n \frac{v_n - v}{\|v_n - v\|_{L^p}} \mid \lambda \geq 0, v_n \to v \right\}.$$

where it is intended that the above limits are in the sense of strong convergence in L^p.

$T_v N_c$ contains all directions in which it is possible *in the L^p sense* to infinitesimally deform a compact set. For example, if $\Phi(x, t) : \mathbb{R}^N \times (-\varepsilon, \varepsilon) \to \mathbb{R}^N$ is smooth diffeomorphical motion of $\mathbb{R}^N$, and $A_t := \Phi(A, t)$, then v_{A_t} is (locally) Lipschitz, so it is differentiable for almost all t, and the derivative is in $T N_c$. This includes all perspective, affine, and Euclidean deformations. Unfortunately the contingent cone is not capable of expressing some shape deformations.

Example 6.19. We consider the **removing** motion; let A be compact, and suppose that x is in the internal part of A; let $A_t := A \setminus B(x, t)$ be the removal of a small ball from A. The motion v_{A_t} inside N_c is Lipschitz, but the limit

$$\lim_{t \to 0+} \frac{v_{A_t} - v_A}{\|v_{A_t} - v_A\|_{L^p}}$$

does not exist in L^p. (Morally, if $p = 1$, the limit would be the measure δ_x).

Fig. 8 Polar coordinates around a convex set

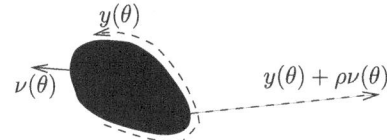

Riemannian Metric

Let now $p = 2$. The set N_c may fail to be a smooth submanifold of L^2; yet we will, as much as possible, pretend that it is, in order to induce a sort of "Riemannian metric" on N_c from the standard L^2 metric.

Definition 6.20. We define the "Riemannian metric" on N_c simply by

$$\langle h, k \rangle := \langle h, k \rangle_{L^2}$$

for $h, k \in T_v N_c$ and correspondingly a norm by

$$|h| := \sqrt{\langle h, h \rangle}$$

where $T_v N_c$ is the *contingent cone*.

Proposition 6.21. *The distance induced by this "Riemannian metric" coincides with the geodesically induce distance d^g.*

The proof is in 3.22 in [15].

To conclude, we propose an explicit computation of the Riemannian Metric for the case of compact sets in the plane with smooth boundaries; we then pull back the metric to obtain a metric of closed embedded planar curves. We start with the case of convex sets. We fix $p = 2$, $N = 2$.

Polar Coordinates of Smooth Convex Sets

Let $\Omega \subset \mathbb{R}^2$ be a convex set with smooth boundary; let $y(\theta) : [0, L] \to \partial \Omega$ be a parameterization of the boundary (by arc parameter), $\nu(\theta)$ the unit vector normal to $\partial \Omega$ and pointing external to Ω. The following *polar* **change of coordinates** ψ holds:

$$\psi : \mathbb{R}^+ \times [0, L] \to \mathbb{R}^2 \setminus \Omega \quad , \quad \psi(\rho, \theta) = y(\theta) + \rho \nu(\theta) \tag{30}$$

see Fig. 8. We suppose that $y(\theta)$ moves on $\partial \Omega$ in anticlockwise direction; so $\nu = J \partial_s y$, $\partial_{ss} y = -\kappa \nu$; where J is the rotation matrix (of angle $-\pi/2$), κ is the curvature, and $\partial_s y$ is the tangent vector.

We can then express a generic integral through this change of coordinates as

$$\int_{\mathbb{R}^2\setminus\Omega} f(x)\,\mathrm{d}x = \int_{\mathbb{R}+}\int_{\partial\Omega} f(\psi(\rho,s))|1+\rho\kappa(s)|\,\mathrm{d}\rho\,\mathrm{d}s$$

where s is the arc parameter, and ds is integration in arc parameter.

Smooth Deformations of a Convex Set

We want to study a smooth deformation of Ω, that we call Ω_t; then the border $y(\theta,t)$ depends on a time parameter t. Suppose also that $\kappa(\theta) > 0$, that is, the set is strictly convex: then for small smooth deformations, the set Ω_t will still be strictly convex. We suppose that the border of Ω_t moves with orthogonal speed α; more precisely, we assume that $(\partial_t y) \perp \partial_s y$, that is, $(\partial_t y) = \alpha \nu$ with $\alpha = \alpha(t,\theta) \in \mathbb{R}$. Since this deformation is smooth, we expect that it will be associated to a vector $h_\alpha \in T_\nu N_c$, defined by $h_\alpha := \partial_t v_{\Omega_t}$. We now show briefly how to explicitly compute it.

Suppose that x is a fixed point in the plane, $x \notin \Omega_t$, and express it using polar coordinates $x = \psi(\rho,\theta)$, with $\rho = \rho(t), \theta = \theta(t)$. With some computations, $\rho' = -\alpha$. Now, for $x \notin \Omega_t$, $u_{\Omega_t}(x) = \rho(t)$ hence we obtain the explicit formula for h_α

$$h_\alpha := \partial_t v_{\Omega_t}(x) = -\varphi'(u_{\Omega_t}(x))\alpha \;;$$

whereas $h_\alpha(x) = 0$ for $x \in \Omega_t$.

Pullback of the Metric on Convex Boundaries

Let us fix two orthogonal smooth vector fields $\alpha(s)\nu(s), \beta(s)\nu(s)$, that represent two possible deformations of $\partial\Omega$; those correspond to two vectors $h_\alpha, h_\beta \in T_\nu N_c$; so the Riemannian Metric that we defined in 6.20 can be pulled back on $\partial\Omega$, to provide the metric

$$\langle \alpha, \beta \rangle := \int_{\mathbb{R}^2} h_\alpha(x) h_\beta(x) dx = \int_{\mathbb{R}^2\setminus\Omega} h_\alpha(x) h_\beta(x) dx =$$
$$= \int_{\partial\Omega} \left[\int_{\mathbb{R}+} (\varphi'(\rho))^2 (1+\rho\kappa(s))\,\mathrm{d}\rho \right] \alpha(s)\beta(s) ds$$

that is,

$$\langle \alpha, \beta \rangle = \int_{\partial\Omega} (a + b\kappa(s))\alpha(s)\beta(s) ds \tag{31}$$

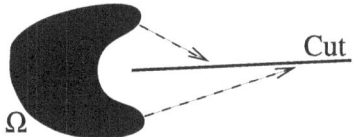

Fig. 9 Smooth contour and cutlocus (The *arrows* represent the distance $R(s)$ from $y(s) \in \partial\Omega$ to the cutlocus)

with

$$a := \int_{\mathbb{R}^+} (\varphi'(\rho))^2 \, d\rho \quad , \quad b := \int_{\mathbb{R}^+} (\varphi'(\rho))^2 \rho \, d\rho .$$

Pullback of the Metric on Smooth Contours

If Ω is smooth but not convex, then the above formula holds up to the cutlocus.

Definition 6.22. The **cutlocus** (*a.k.a.* the **external skeleton**) is the set of points $x \notin \Omega$ such that there are two (or more) different points $y_1, y_2 \in \Omega$ of minimum distance from x to Ω, that is,

$$|x - y_1| = |x - y_2| = u_\Omega(x) .$$

We define a function $R(s) : [0, L] \to [0, \infty]$ that measures the distance from $\partial\Omega$ to the cutlocus Cut; in turn, we can parameterize Cut as

$$\mathrm{Cut} = \{\psi(R(s), s) \mid s \in [0, L], R(s) < \infty\} ;$$

$R(s)$ is locally Lipschitz where it is finite (by results in [25, 33]). The *polar* change of coordinates ψ (defined in (30)) is a diffeomorphism between the sets

$$\{(\rho, s) : s \in [0, L], 0 < \rho < R(s)\} \leftrightarrow \mathbb{R}^2 \setminus (\Omega \cup \mathrm{Cut}) .$$

See Fig. 9.

The pullback metric for deformations of the contour $\partial\Omega$ is

$$\langle h, k \rangle = \int_{\partial\Omega} \left[\int_0^{R(s)} (\varphi'(\rho))^2 (1 + \rho\kappa(s)) \, d\rho \right] \alpha(s) \beta(s) \, ds$$

Conclusion

In this case we have found (a posteriori) a Riemannian metric of closed embedded planar curves, and we know the structure of the completion, and the completion admits minimal geodesics. On the down side, the completion is not really "a smooth Riemannian manifold". For example, it is difficult to study the minimal geodesics, and to prove any property about them. Anyway, the fact that N_c is locally compact and the regularity property 6.18 of L^p spaces suggest that it may be possible to "shoot geodesics" (in some weak form). Unfortunately the topology has too few open sets to be used for shape optimization: many simple energy functionals are not continuous.

7 Finsler Geometries of Curves

We present two examples of Finsler geometries of curves that have been used (sometimes covertly) in the literature.

7.1 Tangent and Normal

Let $v, T \in \mathbb{R}^N$, let

$$N = T^\perp = \{w \in \mathbb{R}^N : w \cdot T = 0\}$$

be the space orthogonal to T. Usually T will be the tangent to a curve c.[9]

Definition 7.1. We define the projection onto the normal space $N = T^\perp$

$$\pi_N : \mathbb{R}^N \to \mathbb{R}^N \quad , \quad \pi_N v = v - \langle v, T \rangle T$$

and the projection on the tangent T

$$\pi_T : \mathbb{R}^N \to \mathbb{R}^N \quad , \quad \pi_T v = \langle v, T \rangle T$$

so $\pi_N v + \pi_T v = v$ and $|\pi_N v|^2 + |\pi_T v|^2 = |v|^2$.

---

[9] There is a slight abuse of notation here, since in the definition $N = T^\perp$ given for planar curves in 1.12, we defined N to be a "vector" and not the "vector space orthogonal to T".

7.2 L^∞ Metric and Fréchet Distance

If we wish to define a norm on $T_c M$ that is modeled on the norm of the Banach space $L^\infty(S^1 \to \mathbb{R}^N)$, we define

$$F^\infty(c,h) := \|\pi_N h\|_{L^\infty} = \sup_\theta |\pi_N h(\theta)|$$

where $N = (D_s c)^\perp$. This metric weights the maximum normal motion of the curve. *(The rôle of π_N will be properly explained in Sect. 11.10)*. This Finsler metric is geometric. The length of a homotopy is

$$\mathrm{Len}_\infty(C) := \int_0^1 \sup_{\theta \in S^1} |\pi_N \partial_t C(t,\theta)| \, dt \ .$$

Definition 7.2 (Fréchet distance).

$$d_f(c_0, c_1) := \inf_\phi \sup_u |c_1(\phi(u)) - c_0(u)|$$

where $u \in S^1$ and ϕ is chosen in the class of diffeomorphisms of S^1.

This distance is induced by the Finsler metric $F^\infty(c,h)$.

Theorem 7.3. *$d_f(c_0, c_1)$ coincides with the infimum of the length $\mathrm{Len}_\infty(C)$ for C connecting c_0 to (a reparameterization of) c_1.*

For the proof, see Theorem 15 in [35].

7.3 L^1 Metric and Plateau Problem

If we wish to define a geometric norm on $T_c M$ that is modeled on the norm of the Banach space $L^1(S^1 \to \mathbb{R}^N)$, we may define the metric

$$F^1(c,h) = \|\pi_N h\|_{L^1} = \int |\pi_N h(\theta)||c'(\theta)| d\theta \ ;$$

the length of a homotopy is then

$$\mathrm{Len}(C) = \iint |\pi_N \partial_t C(t,\theta)||C'(t,\theta)| \, dt \, d\theta$$

which coincides with

$$\mathrm{Len}(C) = \iint |\partial_t C(t,\theta) \times \partial_\theta C(t,\theta)| \, d\theta \, dt$$

The last is easily recognizable as the surface area of the homotopy (up to multiplicity); the problem of finding a minimal geodesic connecting c_0 and c_1 in the F^1 metric may be reconducted to the Plateau problem of finding a surface that is an immersion of $I = S^1 \times [0, 1]$ and which has fixed borders to the curves c_0 and c_1. The Plateau problem is a wide and well studied subject upon which Fomenko expounds in the monograph [20].

8 Riemannian Metrics of Curves up to Pose

A particular approach to the study of shapes is to define a shape to be a *curve up to reparameterization, rotation, translation, scaling*; to abbreviate, we call this the *shape space of curves up to pose*. We present two examples of Riemannian metrics.

8.1 Representation by Angle Function

We consider in the following planar curves $\xi : S^1 \to \mathbb{R}^2$ of length 2π and parameterized by arc length.

Proposition 8.1 (Continuous lifting, rotation index). *If $\xi \in C^1$ and parameterized by arc parameter, then ξ' is continuous and $|\xi'| = 1$, so there exists a continuous function $\alpha : \mathbb{R} \to \mathbb{R}$ satisfying*

$$\xi'(s) = \big(\cos(\alpha(s)), \sin(\alpha(s))\big) \tag{32}$$

and $\alpha(s)$ is unique, up to adding the constant $k2\pi$ with $k \in \mathbb{Z}$.

Moreover $\alpha(s + 2\pi) - \alpha(s) = 2\pi I$, where I is an integer, known as **rotation index** *of ξ. This number is unaltered if ξ is homotopically deformed in a smooth way.*

See the examples in Fig. 10.

The addition of a real constant to $\alpha(s)$ is equivalent to a rotation of ξ. We then understand that we may represent arc parameterized curves $\xi(s)$, up to translation, scaling, and rotations, by considering a suitable class of liftings $\alpha(s)$ for $s \in [0, 2\pi]$.

The use of the angle function for representing (possibly non-closed) planar curves dates back to [69].

8.2 Shape Representation Using Direction Functions

Klassen, Mio, Srivastava et al. in [29, 41] represent a closed planar curve c by a pair of velocity-angle functions (ϕ, α) through the identity

Fig. 10 Examples of curves of different rotation index

$$c'(u) = \exp(\phi(u) + i\alpha(u))$$

(identifying $\mathbb{R}^2 = \mathbb{C}$), and then defining a metric on the velocity-angle function space. They propose models of spaces of curves where the metrics involve higher order derivatives in [29].

We review here the simplest such model, where $\phi = 0$.

In [29] two spaces of closed planar curves are defined; we present the case of *Shape Representation using Direction Functions*:

Definition 8.2. Let $L^2 := L^2([0, 2\pi] \to \mathbb{R})$; $\Phi : L^2 \to \mathbb{R}^3$ is defined by

$$\Phi_1(\alpha) := \int_0^{2\pi} \alpha(s)\,ds\ , \quad \Phi_2(\alpha) := \int_0^{2\pi} \cos\alpha(s)\,ds\ , \quad \Phi_3(\alpha) := \int_0^{2\pi} \sin\alpha(s)\,ds\ .$$

and the space of (pre)-shapes is defined as the closed subset S of L^2,

$$S := \{\alpha \in L^2 \mid \Phi(\alpha) = (2\pi^2, 0, 0)\}\ ;$$

the condition $\Phi_1(\alpha) = 2\pi^2$ forces a choice of rotation, while $\Phi_2 = 0, \Phi_3 = 0$ ensure that α represents a closed curve.

For any $\alpha \in S$, it is possible to reconstruct the curve by integrating

$$\xi(s) = \int_0^s \Big(\cos(\alpha(t)), \sin(\alpha(t))\Big) dt \tag{33}$$

This means that α identifies an unique curve (of length 2π, and arc parameterized) up to rotations and translations, and to the choice of the base point $\xi(0)$; for this last reason, S is called in [29] a **preshape space**.

Inside the family of arc parameterized curves, the reparameterization[10] of ξ is restricted to the transformation $\hat{\xi}(u) = \xi(u - s_0)$, that is a different choice of base point for the parameterization along the curve. Inside the *preshape space* S, the same operation is encoded by the relation $\hat{\alpha}(s) = \alpha(s - s_0)$.

To obtain a model of *immersed curves up to reparameterization, rotation, translation, scaling* we need to quotient S by the relation $\alpha \sim \hat{\alpha}$, for all possible $s_0 \in \mathbb{R}$. We do not discuss this quotient here.

We now prove that $S \setminus Z$ is a smooth manifold. We first define Z.

---

[10]We are considering only reparameterizations in $\mathrm{Diff}^+(S^1)$.

Definition 8.3 (Flat curves). Let Z be the set of all $\alpha \in L^2([0, 2\pi])$ such that $\alpha(s) = a + k(s)\pi$ where $k(s) \in \mathbb{Z}$ and k is measurable, $a = 2\pi - \int k \in \mathbb{R}$, and

$$|\{k(s) = 0 \mod 2\}| = |\{k(s) = 1 \mod 2\}| = \pi$$

- Z is closed (by Theorem 4.9 in [4]).
- $S \setminus Z$ contains the (representation α by continuous lifting of) all smooth immersed curves.
- Z contains the (representations α of) flat curves ξ, that is, curves ξ whose image is contained in a line.

Example 8.4. An example of a flat curve is

$$\xi_1(s) = \xi_2(s) = \begin{cases} s/\sqrt{2} & s \in [0, \pi] \\ (2\pi - s)/\sqrt{2} & s \in (\pi, 2\pi] \end{cases}, \quad \alpha = \begin{cases} \pi/2 & s \in [0, \pi] \\ 3\pi/2 & s \in (\pi, 2\pi] \end{cases}$$

Proposition 8.5. *$S \setminus Z$ is a smooth immersed submanifold of codimension 3 in L^2.*

Proof. By the implicit function theorem. Indeed, suppose by contradiction that $\nabla \Phi_1, \nabla \Phi_2, \nabla \Phi_3$ are linearly dependent at $\alpha \in S$, that is, there exists $a \in \mathbb{R}^3, a \neq 0$ s.t.

$$a_1 \cos(\alpha(s)) + a_2 \sin(\alpha(s)) + a_3 = 0$$

for almost all s; then, by integrating, $a_3 = 0$, therefore $a_1 \cos(\alpha(s)) + a_2 \sin(\alpha(s)) = 0$ that means that $\alpha \in Z$.

The manifold $S \setminus Z$ inherits a Riemannian structure, induced by the scalar product of L^2; (critical) geodesics may be prolonged smoothly as long as they do not meet Z.

Even if S may not be a manifold at Z, we may define the **geodesic distance** $d^g(x, y)$ in S as the infimum of the length of Lipschitz paths $\gamma : [0, 1] \to L^2$ going from x to y and whose image is contained in S[11]; since $d^g(x, y) \geq \|x - y\|_{L^2}$, and S is closed in L^2, then the metric space (S, d^g) is metrically complete.

We do not know if (S, d^g) admits minimal geodesics, or if it falls in the same category as the Atkin example 3.34.

Multiple Measurable Representations

We may represent any Lipschitz closed arc parameterized curve ξ using a measurable $\alpha \in S$.

---

[11] It seems that S is Lipschitz-arc-connected, so $d^g(x, y) < \infty$; but we did not carry on a detailed proof.

Definition 8.6. A **measurable lifting** is a measurable function $\alpha : \mathbb{R} \to \mathbb{R}$ satisfying (32). Consequently, a **measurable representation** is a measurable lifting α satisfying conditions $\Phi(\alpha) = (2\pi^2, 0, 0)$ (from Definition 8.2).

We remark that

- a measurable representation always exists; for example, let

$$\text{arc} : S^1 \to [0, 2\pi)$$

be the inverse of

$$\alpha \mapsto (\cos(\alpha), \sin(\alpha))$$

when $\alpha \in [0, 2\pi)$. arc() is a Borel function; then $((\text{arc} \circ \xi')(s) + a) \in S$, for a choice of $a \in \mathbb{R}$.
- The measurable representation is never unique: for example, given any measurable $A, B \subset [0, 2\pi]$ with $|A| = |B|$,

$$\alpha(s) + 2\pi \mathbf{1}_A(s) - 2\pi \mathbf{1}_B(s)$$

will as well represent ξ.

This implies that the family A_ξ of measurable $\alpha \in S$ that represent the same curve ξ is infinite. It may be then advisable to define a **quotient distance** $\hat{d}$ as follows:

$$\hat{d}(\xi_1, \xi_2) := \inf_{\alpha_1 \in A_{\xi_1}, \alpha_2 \in A_{\xi_2}} d(\alpha_1, \alpha_2) \qquad (34)$$

where $d(\alpha_1, \alpha_2) = \|\alpha_1 - \alpha_2\|_{L^2}$, or alternatively $d = d^g$ is the geodesic distance on S.

Continuous vs Measurable Lifting: No Rotation Index

If $\xi \in C^1$, we have an unique[12] continuous representation $\alpha \in S$; but note that, even if $\xi_1, \xi_2 \in C^1$, the infimum (34) may not be given by the continuous representations α_1, α_2 of ξ_1, ξ_2. Moreover there are rectifiable curves ξ that do not admit a continuous representation α, as for example the polygons.

A problem similar to the above is expressed by this proposition.

---

[12] Indeed, the continuous lifting is unique up to addition of a constant to $\alpha(s)$, which is equivalent to a rotation of ξ; and the constant is decided by $\Phi_1(\alpha) = 2\pi^2$.

Proposition 8.7. *For any $h \in \mathbb{Z}$, the set of closed smooth curves ξ with rotation index h, when represented in S using the continuous lifting, is dense in $S \setminus Z$.*

It implies that we cannot properly extend the concept of *rotation index* to S.

The proof is based on this lemma.

Lemma 8.8. *Suppose that ξ is not flat, let τ be one of the measurable liftings of ξ. There exists a smooth projection $\pi : V \to S$ defined in a neighborhood $V \subset L^2$ of τ such that, if $f \in L^2 \cap C^\infty$ then $\pi(f)$ is in $L^2 \cap C^\infty$.*

This is the proof of both the lemma and the proposition.

Proof. Fix $\alpha_0 \in S \setminus Z$. Let $T = T_{\alpha_0} S$ be the tangent at α_0. T is the vector space orthogonal to $\nabla \phi_i(\alpha_0)$ for $i = 1, 2, 3$. Let $e_i = e_i(s) \in L^2 \cap C_c^\infty$ be near $\nabla \phi_i(\alpha_0)$ in L^2, so that the map $(x, y) : T \times \mathbb{R}^3 \to L^2$

$$(x, y) \mapsto \alpha = \alpha_0 + x + \sum_{i=1}^{3} e_i y_i \tag{35}$$

is an isomorphism. Let S' be S in these coordinates; by the Implicit Function Theorem (5.9 in [31]), there exists an open set $U' \subset T, 0 \in U'$, an open $V' \subset \mathbb{R}^3$, $0 \in V'$, and a smooth function $F : U \to \mathbb{R}^3$ such that the local part $S' \cap (U' \times V')$ of the manifold S' is the graph of $y = F(x)$.

We immediately define a smooth projection $\pi : U' \times V' \to S'$ by setting $\pi'(x, y) = (x, F(x))$; this may be expressed in the original L^2 space; let $(x(\alpha), y(\alpha))$ be the inverse of (35) and $U = x^{-1}(U')$; we define the projection $\pi : U \to S$ by setting

$$\pi(\alpha) = \alpha_0 + x + \sum_{i=1}^{3} e_i F_i(x(\alpha))$$

Then

$$\pi(\alpha)(s) - \alpha(s) = \sum_{i=1}^{3} e_i(s) a_i , \ a_i := (F_i(x(\alpha)) - y_i(\alpha)) \in \mathbb{R} \tag{36}$$

so if $\alpha(s)$ is smooth, then $\pi(\alpha)(s)$ is smooth.

Let α_n be smooth functions such that $\alpha_n \to \alpha$ in L^2, then $\pi(\alpha_n) \to \alpha_0$; if we choose them to satisfy $\alpha_n(2\pi) - \alpha_n(0) = 2\pi h$, then, by the formula (36), $\pi(\alpha)(2\pi) - \pi(\alpha)(0) = 2\pi h$ so that $\pi(\alpha_n) \in S$ and it represents a smooth curve with the assigned rotation index h. $\square$

Metrics of Curves in Shape Optimization and Analysis

8.3 A Metric with Explicit Geodesics

A similar method has been proposed recently in [68], based on an idea originally in [67]. We consider immersed planar curves and again we identify $\mathbb{R}^2 = \mathbb{C}$.

Proposition 8.9 (Continuous lifting of square root). *If $\xi : [0, 2\pi] \to \mathbb{C}$ is an immersed planar curve, then ξ' is continuous and $\xi' \neq 0$, so there exists a continuous function $\alpha :\to \mathbb{C}$ satisfying*

$$\xi'(\theta) = \alpha(\theta)^2 \tag{37}$$

and α is uniquely identified up to multiplying by ± 1.

If the rotation index of ξ is even then $\alpha(0) = \alpha(2\pi)$, whereas if the rotation index of ξ is odd then $\alpha(0) = -\alpha(2\pi)$.

We will use this lifting to obtain, as a first step, a representation of *curves up to rotation, translation, scaling*. To this end, let ξ be an immersed planar closed curve of $\operatorname{len}(\xi) = 2$ (not necessarily parameterized by arc parameter); let α be the **square root lifting** of the derivative ξ'. Let e, f be the real and imaginary part of α, that is, $\alpha = e + if$. The condition that ξ is a closed curve translates into $\int_0^{2\pi} (e + if)^2 \, d\theta = 0$ (where equality is in $\mathbb{C}$), hence we have two equalities (for real and imaginary part)

$$\int_0^{2\pi} e^2 - f^2 \, d\theta = 0, \quad \int_0^{2\pi} ef \, d\theta = 0$$

while the condition that $\operatorname{len}(\xi) = 2$ translates into

$$\int_0^{2\pi} |\xi'| \, d\theta = \int_0^{2\pi} |e + if|^2 \, d\theta = \int_0^{2\pi} (e^2 + f^2) \, d\theta = 2 \ .$$

With some algebra we obtain that the above conditions are equivalent to

$$\int_0^{2\pi} e^2 \, d\theta = 1, \ \int_0^{2\pi} f^2 \, d\theta = 1, \ \int_0^{2\pi} ef \, d\theta = 0.$$

Let $L^2 = L^2([0, 2\pi] \to \mathbb{R})$; let $S \subset L^2 \times L^2$ be defined by

$$S := \left\{ (e, f) \mid \int_0^{2\pi} e^2 \, d\theta = 1 = \int_0^{2\pi} f^2 \, d\theta, \int_0^{2\pi} ef \, d\theta = 0 \right\} .$$

S is the **Stiefel manifold** of orthonormal pairs in L^2. S is a smooth manifold, and inherits the (flat) metric of $L^2 \times L^2$. What is most surprising is that

Theorem 8.10 (2.2 in [68]). *Let $\delta\xi$ be a small deformation of ξ, and $\delta e, \delta f$ be the corresponding small deformation of the representation e, f. Then*

$$\int_\xi |D_s \delta\xi|^2 \, ds = \int_0^{2\pi} (\delta e)^2 + (\delta f)^2 \, d\theta$$

that is, the (geometric) Riemannian metric in M is mapped into the (flat and parametric) metric in S.

It is then natural to "embed" closed planar curves (up to translation and scaling) into S.

Curves up to Rotation Represented as a Grassmanian

We note that rotation of ξ by an angle τ is equivalent to rotation of the frame (e, f) by an angle $\tau/2$. So the orbit of all rotations of ξ is associated to the plane generated by (e, f) in L^2: the space of *curves up to rotation/translation/scaling* is represented by the **Grassmanian manifold** of 2-planes in L^2. A theorem by Neretin then applies, which provides a closed form formula for critical and minimal geodesics. See Sect. 4 in [68].

Representation

The *Stiefel manifold* is a complete smooth Riemannian manifold; it contains the (representation of) all closed rectifiable parametric planar curves, up to scaling and translation. So with this choice of metric and representation, we obtain that the completion of the Fréchet manifold of smooth curves is a Riemannian smooth manifold. There are two problems left.

- The quotient w.r.to reparameterization. This is studied in [68], where it is proven that unfortunately geodesics in the quotient space may develop singularities in the reparameterizations at the end times of the geodesic.
- But there is also the quotient w.r.to representation.

Quotient w.r.to Representation

As in the previous section, we may define a *measurable square root lifting*; this lifting will not be unique. Indeed, let (e, f) be the square root representation of ξ, that is $\xi' = (e + if)^2$; choose any function $a : S^1 \to \{-1, 1\}$ arbitrarily (but measurable); then (ae, af) represents the same curve. So again we may define a quotient metric $\hat{d}(\xi_1, \xi_2)$ as we did in (34); similar comments to those at the end of Sect. 8.2 hold.

9 Riemannian Metrics of Immersed Curves

Metrics of "geometric" curves have been studied by Michor and Mumford [37, 38, 39] and Yezzi and Mennucci [63, 64, 65]; more recently, Yezzi-M.-Sundaramoorthi [35, 53–58] have studied Sobolev-like metrics of curves and shown many good properties for applications to shape optimization; similar results have also been shown independently by Charpiat et al. [9, 10, 11].

We now discuss some Riemannian metrics on immersed curves.

- The H^0 metric

$$\langle h_1, h_2 \rangle_{H^0} = \int_c \langle h_1(s), h_2(s) \rangle \, \mathrm{d}s \ .$$

- Michor and Mumford [38]'s metric

$$\langle h_1, h_2 \rangle_{H^A} = \int_c (1 + A|\kappa_c|^2)\langle h_1, h_2 \rangle \, \mathrm{d}s \ .$$

- Yezzi and Mennucci [64]'s *conformal metric*

$$\langle h_1, h_2 \rangle_{H^0_\phi} = \phi(c) \int_c \langle h_1, h_2 \rangle \, \mathrm{d}s \ .$$

- Charpiat et al. [10] *rigidified norms*
- Sundaramoorthi et al. [53] Sobolev type metrics

$$\langle h_1, h_2 \rangle_{H^n} = \fint_c \langle h_1, h_2 \rangle \, \mathrm{d}s + \mathrm{len}(c)^{2n} \fint_c \langle \partial_s^n h_1, \partial_s^n h_2 \rangle \, \mathrm{d}s$$

$$\langle h_1, h_2 \rangle_{\tilde{H}^n} = \left\langle \fint_c h_1 \, \mathrm{d}s, \fint_c h_2 \, \mathrm{d}s \right\rangle + \mathrm{len}(c)^{2n} \fint_c \langle \partial_s^n h_1, \partial_s^n h_2 \rangle \, \mathrm{d}s \ .$$

We will now present a quick overview of all metrics (but for the latter, that is discussed in the next section).

9.1 H^0

The H^0 inner product was defined in (5) as

$$\langle h_1, h_2 \rangle_{H^0} := \int_c h_1(s) \cdot h_2(s) \, \mathrm{d}s \ ; \tag{38}$$

it is possibly the simplest geometric inner product that we may imagine to apply to curves. We already noted that the minimizing flows implemented in traditional active contour methods are "gradient flows" only if we use this H^0 inner product.

We will show in Sect. 11.3 that the H^0-induced distance is identically zero. In [64] there is a result that shows that the distance is non degenerate, and minimal geodesics exist, when the shape space is restricted to curves with uniformly bounded curvature.

9.2 H^A

Michor and Mumford [38] propose the metric H^A

$$\langle h_1, h_2 \rangle_{H^A_{|c}} = \int_0^L (1 + A|\kappa_c|^2) \langle h_1, h_2 \rangle \, ds$$

where κ_c is the curvature of c, and $A > 0$ is fixed.

Properties 9.1. • *The induced distance is non degenerate.*
• *The completion $\overline{M}$ (intended in the metric sense) is*

$$BV^2 \subset \overline{M} \subset Lip$$

where BV^2 are the curves that admit curvature as a measure and Lip are the rectifiable curves.
• *There are compactness bounds.*

9.3 Conformal Metrics

Yezzi and Mennucci [64] proposed to change the metric, from H^0 to a **conformal metric**

$$\langle h_1, h_2 \rangle_{H^0_\psi} = \psi(c) \int_c \langle h_1, h_2 \rangle \, ds$$

where $\psi(c)$ associates to each curve c a positive number. Then the *gradient descent flow* of an energy E defined on curves $C(t, \cdot)$ is

$$\frac{\partial C}{\partial t} = -\nabla^\psi E(C) = -\frac{1}{\psi(C)} \nabla E(C)$$

where $\nabla E(C)$ is the gradient for the H^0 metric. This is equivalent to a change of time variable t in the gradient descent flow. So all properties of the flows are unaffected if we switch from a H^0 to a conformal-H^0 metric.

Properties 9.2. *Consider a conformal metric where $\psi(c)$ depends (monotonically) on the length $\mathrm{len}(c)$ of the curve.*

- *If $\psi(c) \geq \mathrm{len}(c)$, the induced metric is non degenerate;*
- *unfortunately, according to a result by Shah [50], when $\psi(c) \equiv L(c)$ very few (minimal) geodesics do exist (only "grassfire" geodesics, moving by constant normal speed).*

9.4 "Rigidified" Norms

Charpiat et al. [10] consider norms that favor pre-specified rigid motions. They decompose a motion h using the H^0 projection as

$$h = h_{\mathrm{rigid}} + h_{\mathrm{rest}}$$

where h_{rigid} contains the rigid part of the motion; then they choose λ large, and construct the norm

$$\|h\|_{\mathrm{rigid}}^2 = \lambda \|h_{\mathrm{rigid}}\|_{H^0}^2 + \|h_{\mathrm{rest}}\|_{H^0}^2.$$

Note that these norms are equivalent to the H^0-type norm; as a result the induced distance is (again) degenerate.

10 Sobolev Type Riemannian Metrics

In this part we discuss the Sobolev norms, with applications and experiments. What follows summarizes [35, 56–58].

10.1 Sobolev-Type Metrics

Recently in [35, 53–58] Yezzi-M.-Sundaramoorthi studied a family of Sobolev-type metrics. Let $D_s := \frac{1}{|c'|}\partial_\theta$ be the derivative with respect to arc parameter. Let $j \geq 1$ be an integer[13] and

---

[13] It is though possible to define Sobolev metrics for any $j \in \mathbb{R}, j > 0$; see Prop. 3.1 in [57].

$$\langle h_1, h_2 \rangle_{H_0^j} := \fint_c \langle D_s^j h_1, D_s^j h_2 \rangle \, ds \tag{39}$$

where $\fint_c \cdots ds$ was defined in 1.9. Let $\lambda > 0$ a fixed constant; we define the **Sobolev-type metrics**

$$\langle h_1, h_2 \rangle_{H^j} := \fint_c \langle h_1, h_2 \rangle \, ds + \lambda L^{2j} \langle h_1, h_2 \rangle_{H_0^j} \tag{40}$$

$$\langle h_1, h_2 \rangle_{\tilde{H}^j} := \left\langle \fint_c h_1 \, ds, \fint_c h_2 \, ds \right\rangle + \lambda L^{2j} \langle h_1, h_2 \rangle_{H_0^j} \tag{41}$$

where $L = \text{len}(c)$. Notice that these metrics are *geometric*:

- they are easily seen to be invariant with respect to rotations and translations, (in a stronger sense than in page 246, indeed in this case

$$\langle h_1, h_2 \rangle_{Ac} = \langle h_1, h_2 \rangle_c \,, \, \langle Rh_1, Rh_2 \rangle_c = \langle h_1, h_2 \rangle_c$$

for any Euclidean transformation A and rotation R);
- they are *reparameterization invariant* due to the usage of the arc parameter in derivatives and integrals;
- they are *scale invariant*, since the normalizing factors L^{2j} make them zero-homogeneous w.r.to rescaling of the curve.

For this last reason, we **redefine the H^0 metric** to be

$$\left\langle h_1, h_2 \right\rangle_{H^0} := \fint_c h_1 \cdot h_2 \, ds \tag{42}$$

so that it is again zero-homogeneous. This is a conformal version of the H^0 metric defined in (5), so a gradient descent flow is a time reparameterization of the flow for the original H^0 metric; and the induced distance is again degenerate.

When we will present a shape optimization energy E and we will minimize E using a gradient descent flow driven by the H^j or $\tilde{H}^j$ gradient of E, we will call the resulting algorithm a **Sobolev active contour** method (abbreviated as **SAC** in the following).

Related Works

A family of metrics similar to H^j above (but for the length dependent scale factors[14]) was studied (independently) in [39]: the Sobolev-type weak Riemannian

---

[14]Though, a scale-invariant Sobolev metric is proposed in [39] in Sect. 4.8 as a sensible generalization.

metric on $\mathrm{Imm}(S^1, \mathbb{R}^2)$

$$\langle h, k \rangle_{G_c^j} = \int_c \sum_{i=0}^{j} \langle D_s^i h, D_s^i k \rangle \, ds \; ;$$

in that paper the geodesic equation, horizontality, conserved momenta, lower and upper bounds on the induced distance, and scalar curvatures are computed. Note that this metric is locally equivalent to the above metrics defined in (40) and (41).

Charpiat et al. in [10, 11] studied (again independently) some generalized metrics and relative gradient flows; in particular they defined the Sobolev-type metric

$$\fint_c \langle h_1, h_2 \rangle + \langle D_s h_1, D_s h_2 \rangle \, ds \; .$$

Properties of H^j Metrics

This is a list of important properties that will be discussed in the following sections.

- Flow regularization: Sobolev gradient flows are smoother than H^0 flows.
- PDE order reducing property: Sobolev gradient flows are lower order than H^0 flows.
- SAC do not require derivatives of the curve to be defined for many commonly used region-based and edge-based energies.
- Coarse-to-fine deformation property: a SAC automatically favors coarse-scale deformations before moving to fine-scale deformations; this is ideal for visual tracking.
- Sobolev-type norms induce a well-defined distance on space of curves;
- moreover the structure of the completion of the space of immersed curves with respect to H^1 and H^2 norm is fairly well understood. So they offer a *consistent* theory of *shape optimization* and *shape analysis*.

10.2 Mathematical Properties

We start summarizing the main mathematical properties, that were presented in [35] mostly. We first of all cite this lemma.

Lemma 10.1 (Poincaré inequality). *Pick $h : [0, l] \to \mathbb{R}^n$, weakly differentiable, with $h(0) = h(l)$ (so h is periodically extensible); let $\hat{h} = \frac{1}{l} \int_0^l h(x) \, dx$; then*

$$\sup_u |h(u) - \hat{h}| \leq \frac{1}{2} \int_0^l |h'(x)| \, dx \; . \tag{43}$$

This is proved as Lemma 18 in [35]; it is one of the main ingredients for the following propositions, whose full proofs are given in [35].

Proposition 10.2. *The H^j and $\tilde{H}^j$ distances are equivalent:*

$$d_{\tilde{H}^j} \leq d_{H^j} \leq \sqrt{\frac{1 + (2\pi)^{2j}\lambda}{(2\pi)^{2j}\lambda}} d_{\tilde{H}^j}$$

whereas $d_{H^j} \leq d_{H^k}$ for $j < k$.

Proposition 10.3. *The H^j and $\tilde{H}^j$ distances are lower bounded (with appropriate constants depending on λ) by the* Fréchet distance *(defined in 7.2).*

Proposition 10.4. *$c \mapsto len(c)$ is Lipschitz in M with H^j metric, that is,*

$$|len(c_0) - len(c_1)| \leq d_{H^j}(c_0, c_1)$$

Theorem 10.5 (Completion of B_i with respect to H^1). *let d_{H^1} be the distance induced by H^1; the metric completion of the space of curves is equal to the space of all rectifiable curves.*

Theorem 10.6 (Completion of B_i with respect to H^2). *Let $E(c) := \int |D_s^2 c|^2 \, ds$ be defined on non-constant smooth curves; then E is locally Lipschitz in with respect to d_{H^2}. Moreover the completion of $C^\infty(S^1)$ according to the metric H^2 is the space of curves that admit curvature $D_s^2 c \in L^2(S^1)$.*

The hopeful consequence of the above theorem would be a possible solution to the problem 4.12; it implies that the space of geometric curves B with the H^2 Riemannian metric completes onto the usual Hilbert space H^2.[15] This result in turn would ease a (possible) proof of existence of geodesics.

Concluding, it seems that, to have a complete Riemannian manifold of geometric (freely) immersed curves, a metric should penalize derivatives of 2nd order (at least).

10.3 Sobolev Metrics in Shape Optimization

We first present a definition.

Definition 10.7 (Convolution). A *arc-parameterized convolutional kernel* K along the curve c of length L is a L-periodic function $K : \mathbb{R} \to \mathbb{R}$. Given a vector field $f : S^1 \to \mathbb{R}^n$ and a kernel K, we define the **convolution by arc parameter**[16] formally as

---

[15]The detailed proof has not yet been written...

[16]Note this definition is different from (13) in [56].

$$(f \star K)(s) := \int_c K(s - \hat{s}) f(\hat{s}) \, d\hat{s} \ . \tag{44}$$

By defining the **run-length function** $l : \mathbb{R} \to \mathbb{R}$

$$l(\tau) := \int_0^\tau |c'(x)| \, dx$$

we can rewrite the above equation (44) explicitly in θ parameter as

$$(f \star K)(\theta) := \int_0^{2\pi} K\big(l(\theta) - l(\tau)\big) f(\tau) |c'(\tau)| \, d\tau \ . \tag{45}$$

Recall the definition 3.37 of gradient ∇E by means of the identity

$$\langle \nabla E, h \rangle_c = DE(c; h) \quad \forall h \in T_c M \ .$$

Let $f = \nabla_{H^0} E$, $g = \nabla_{H^1} E$ be the gradients w.r.to the inner products H^0 and H^1; by the definition of gradient, we obtain that

$$\langle f, h \rangle_{H^0, c} = \langle g, h \rangle_{H^1, c} \quad \forall h \in T_c M$$

that is[17]

$$\oint_c h \cdot f \, ds = \oint_c h \cdot g + \lambda L^2 (D_s h \cdot D_s g) \, ds \quad \forall h$$

by integrating by parts this becomes

$$\int_c h \cdot (f - g + \lambda L^2 D_s^2 g) \, ds = 0 \quad \forall h,$$

then we conclude that

$$\nabla_{H^0} E = \nabla_{H^1} E - \lambda L^2 D_s^2 (\nabla_{H^1} E) \ . \tag{46}$$

With similar computation, for $\tilde{H}^1$ we obtain that

$$\nabla_{H^0} E = \oint_c \nabla_{\tilde{H}^1} E \, ds - \lambda L^2 D_s^2 (\nabla_{\tilde{H}^1} E) \ . \tag{47}$$

Given $\nabla_{H^0} E$, both equations can be solved for $\nabla_{H^1} E$ and $\nabla_{\tilde{H}^1} E$, by means of suitable convolution kernels $\tilde{K}_\lambda$, K_λ, that is, we have the formulas

---

[17] We use the definition (42) of H^0.

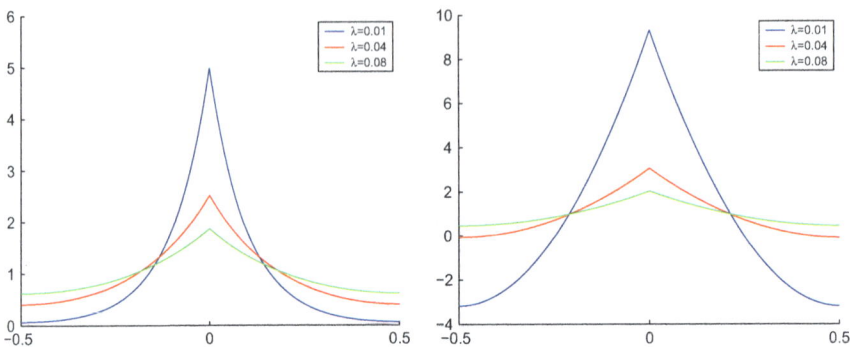

Fig. 11 Plots of K_λ (*left*) and $\tilde{K}_\lambda$ (*right*) for various λ with $L = 1$. The plots show the kernels over one period

$$\nabla_{H^1} E = \nabla_{H^0} E \star K_\lambda \quad , \quad \nabla_{\tilde{H}^1} E = \nabla_{H^0} E \star \tilde{K}_\lambda \; ;$$

The kernels $\tilde{K}_\lambda, K_\lambda$ are known in closed form:

$$K_\lambda(s) = \frac{\cosh\left(\frac{s-\frac{L}{2}}{\sqrt{\lambda}L}\right)}{2L\sqrt{\lambda}\sinh\left(\frac{1}{2\sqrt{\lambda}}\right)}, \quad \text{for } s \in [0, L], \tag{48}$$

$$\tilde{K}_\lambda(s) = \frac{1}{L}\left(1 + \frac{(s/L)^2 - (s/L) + 1/6}{2\lambda}\right), \quad s \in [0, L]. \tag{49}$$

and $K_\lambda, \tilde{K}_\lambda$ are periodically extended to all of $\mathbb{R}$. See Fig. 11.

The above idea extends to any j: it is possible to obtain $\nabla_{H^j} E$ and $\nabla_{\tilde{H}^j} E$ from $\nabla_{H^0} E$, by convolution. But there is also a way to compute $\nabla_{\tilde{H}^j} E$ without resolving to convolutions, see Proposition 10.10.

The bad news is that, when shape optimization is implemented using a *level set method* (as is usually the case), the curve must be traced out to compute gradients.[18] This in particular leads to some problems when a curve undergoes a topological change, since the Sobolev gradient depends on the *global* shape of the curve; but those problems are easily bypassed when the optimization energy is continuous across topological changes of the curves: see Sect. 5.1 in [56].

---

[18]Though, an alternate method that does not need the tracing of the curve is described in Sect. 5.3 in [56]—but it was not successively used, since it depends on some difficult-to-tune parameters.

Smoothing of Gradients, Coarse-to-Fine Flowing

In [57] it was noted that the regularizing properties may be explained in the Fourier domain: indeed, if we calculate Sobolev gradients $\nabla_{H^j} E$ of an arbitrary energy E in the frequency domain, then

$$\widehat{\nabla_{H^j} E}(l) = \frac{\widehat{\nabla_{H^0} E}(l)}{1 + \lambda(2\pi l)^{2j}} \quad \text{for } l \in \mathbb{Z} \tag{50}$$

and

$$\widehat{\nabla_{\tilde{H}^j} E}(0) = \widehat{\nabla_{H^0} E}(0), \quad \widehat{\nabla_{\tilde{H}^j} E}(l) = \frac{\widehat{\nabla_{H^0} E}(l)}{\lambda(2\pi l)^{2j}} \quad \text{for } l \in \mathbb{Z}\setminus\{0\}, \tag{51}$$

It is clear from the previous expressions that high frequency components of $\nabla_{H^0} E(c)$ are increasingly less pronounced in the various forms of the H^j gradients. So H^j and $\tilde{H}^j$ gradients are much smoother than H^0 gradients. Note that using a SAC method smooths the gradients, so it induces smoother minimization flows, but it does **not** smooth the curves themselves.

The above phenomenon may be also explained by the following argument. The gradient satisfies the following property.

Proposition 10.8. *If $DE(c) \neq 0$, the gradient $\nabla E(c)$ is the vector v in $T_c M$ that provides the maximum in*

$$\frac{DE(c)(v)}{\|v\|_c} = \sup_{h \in T_c M \setminus \{0\}} \frac{|DE(c)(h)|}{\|h\|_c}. \tag{52}$$

Thus, the gradient is the most *efficient* perturbation, i.e., it maximizes

$$\frac{\text{change in energy by moving in direction } h}{\text{cost of moving in direction } h}$$

By constructing $\|\cdot\|_c$ to penalize "unfavorable" directions, we have control over the path taken to minimize E *without changing the energy E*.[19]

Since higher frequencies are more penalized in H^1 than in H^0, this explains why a SAC prefers to move the curves in a coarser scale w.r.to a traditional active contour method.

---

[19] In a sense, the focus of research in active contours has been mostly on the numerator in (52)— whereas we now focus on the denominator.

Flow Regularization

It is also possible to show mathematically that the Sobolev-type gradients regularize the flows of well known energies, by reducing the degree of the P.D.E., as shown in this example.

Example 10.9 (Elastica). In the case of the elastic energy

$$E(c) = \int_c \kappa^2 \, ds = \int_c |D_s^2 c|^2 \, ds ,$$

the H^0 gradient is[20]

$$\nabla_{H^0} E = L D_s (2 D_s^3 c + 3 |D_s^2 c|^2 D_s c)$$

that includes fourth order derivatives; whereas the $\tilde{H}^1$-gradient is

$$-\frac{2}{\lambda L} D_s^2 c + 3 L (|D_s^2 c|^2 D_s c) \star (D_s \tilde{K}_\lambda) \tag{53}$$

that is an integro-differential second order P.D.E.

A practical positive outcome of this phenomenon will be shown in Sect. 10.9.

10.4 $\tilde{H}^j$ is Faster than H^j

We will now show that the metric $\tilde{H}^1$ (that is metrically equivalent to H^1, by Prop. 10.2) leads to a simpler calculus for gradients; with benefits in numerical applications as well.

We recall that $\mathrm{avg}_c(f) := \frac{1}{\mathrm{len}(c)} \int_c f(s) \, ds$.

Let $E(c)$ be an energy of curves. Gradients are implicitly defined by the following relations

$$DE(c; h) = \langle h, \nabla_{H^0} E \rangle_{H^0} = \langle h, \nabla_{H^1} E \rangle_{H^1} = \langle h, \nabla_{\tilde{H}^1} E \rangle_{\tilde{H}^1} \quad \forall h .$$

As we saw in (46) and (47), the above leads to the ODEs. The second ODE is much easier to solve.

Norm	ODE		
H^1	$\nabla_{H^0} E$	$=$	$\nabla_{H^1} E - \lambda L^2 D_s^2 (\nabla_{H^1} E)$
$\tilde{H}^1$	$\nabla_{H^0} E$	$=$	$\mathrm{avg}_c(\nabla_{\tilde{H}^1} E) - \lambda L^2 D_s^2(\nabla_{\tilde{H}^1} E)$

---

[20]We use the definition (42) of H^0; the directional derivative was computed in 4.6.

Proposition 10.10. *Let $f = \nabla_{H^0} E$, $g = \nabla_{\tilde{H}^1} E$, then g is derived from f by the following closed form formulas*

$$g(s) = g(0) + sg'(0) - \frac{1}{\lambda L^2} \int_0^s (s-\hat{s})(f(\hat{s}) - \text{avg}_c(f)) \, d\hat{s}$$

$$g'(0) = -\frac{1}{\lambda L^3} \int_0^L s(f(s) - \text{avg}_c(f)) \, ds$$

$$g(0) = \int_0^L f(s) \tilde{K}_\lambda(s) \, ds .$$

where $\tilde{K}_\lambda$ was defined in (49).

Proof. The ODE

$$f = \text{avg}_c(g) - \lambda L^2 D_s^2 g$$

immediately tells that $\text{avg}_c(f) = \text{avg}_c(g)$; so we rewrite it as

$$\lambda L^2 D_s^2 g = -f + \text{avg}_c(f)$$

and we simply integrate twice! Moreover, exploiting the identity

$$\int_0^s \left(\int_0^t h(r) \, dr \right) dt = \int_0^s (s-r) h(r) \, dr ,$$

and with some extra computations, we obtain the result.

In the end we obtain that the $\tilde{H}^1$ gradient need not be computed as a convolution; so the $\tilde{H}^1$ gradient enjoys nearly the same computational speed as the H^0 gradient; moreover the resulting gradient flow is more stable, so it works fine in numerical implementations for a larger choice of time step discretization. For these reasons, $\tilde{H}^1$ Sobolev active contours are very fast.

10.5 Analysis and Calculus of $\tilde{H}^1$ Gradients

We write $h \in T_c M$ as $h = \text{avg}_c(h) + \tilde{h}$; this decomposes

$$T_c M = \mathbb{R}^n \oplus D_c M \tag{54}$$

with

$$D_c M := \left\{ h : S^1 \to \mathbb{R}^n \mid \text{avg}_c(h) = 0 \right\} .$$

If we assign to $\mathbb{R}^n$ its usual Euclidean norm, and to $D_c M$ the H_0^j norm,[21] then

$$\|h\|_{\tilde{H}^j}^2 = |\text{avg}_c(h)|_{\mathbb{R}^n}^2 + \lambda L^{2j} \|\tilde{h}\|_{H_0^j}^2 \ .$$

This means that the two spaces $\mathbb{R}^n$ and $D_c M$ are **orthogonal** w.r.to $\tilde{H}^j$.

A remark on this decomposition is due.

Remark 10.12. In the above, $\mathbb{R}^n$ is *akin* to be the **space of translations** and $D_c M$ the **space of non-translating deformations**.

That labeling is not rigorous, though! Since the subspace of $T_c M$ of deformations that do not move the center of mass $\text{avg}_c(c)$ is not $D_c M$, but rather

$$\left\{ h : \int_c h + (c - \text{avg}_c(c))\langle D_s h \cdot T \rangle \, ds = 0 \right\} ,$$

as is easily deduced from (9).

The decomposition (54) is quite useful in the calculus when analytically computing $\tilde{H}^1$ gradients and proving existence of flows. Indeed, let $f = \nabla_{H^0} E$, $g = \nabla_{\tilde{H}^1} E$; decompose

$$f = \text{avg}_c(f) + \tilde{f} \ , \ g = \text{avg}_c(g) + \tilde{g} \ .$$

To solve

$$f = \text{avg}_c(g) - \lambda L^2 D_s^2 g$$

we have to solve two equations:

$$\text{avg}_c(g) = \text{avg}_c(f) \qquad \lambda L^2 D_s^2 \tilde{g} = -\tilde{f}$$
$$\mathbb{R}^n \quad \oplus \quad D_c M \ .$$

We will now show how to properly solve these coupled equations.

We will need to define some useful objects.

Definition 10.13. We define the **projection operator**

$$\Pi_c : T_c M \to D_c M \tag{55}$$
$$h \mapsto h - \fint_c h \, ds$$

---

[21] H_0^j was defined in (39). Note that H_0^j is a seminorm on $T_c M$, since its value is zero on constant fields; but H_0^j is a norm on $D_c M$, by Poincaré inequality (43).

Definition 10.14. When we consider the *derivation with respect to the arc parameter* as a linear operator

$$\mathcal{D}_c : D_c M \to D_c M \qquad (56)$$
$$h \mapsto D_s h$$

then it admits the inverse, that is the **primitive operator**

$$\mathcal{P}_c : D_c M \to D_c M \qquad (57)$$

Proof. The proof is just based on noting that, for any smooth $h \in T_c M$, $h \in D_c M$ iff there is a smooth $k \in T_c M$ with $h = D_s k$. Vice versa, two primitives differ by a constant, hence when $h \in D_c M$ there is exactly one primitive k in $D_c M$ such that $h = D_s k$.

Example 10.15. The tangent vector field $D_s c$ is in $D_c M$, and its primitive is

$$\mathcal{P}_c D_s c = c - \mathrm{avg}_c(c) \ .$$

Proposition 10.16. *Fix a curve c, let $L = \mathrm{len}(c)$. We can extend the primitive operator by composing it with the projection; the resulting composite operator can be expressed in convolutional form as*

$$\mathcal{P}_c \Pi_c h = K_c^{\mathcal{P}} \star h \qquad (58)$$

for any continuous $h \in T_c M$; where the kernel $K_c^{\mathcal{P}}$ is

$$K_c^{\mathcal{P}}(s) := -\frac{s}{L} + \frac{1}{2} \quad \text{for } s \in [0, L) \qquad (59)$$

and $K_c^{\mathcal{P}}(s)$ is extended periodically to $s \in \mathbb{R}$ (note that $K_c^{\mathcal{P}}(s)$ jumps at all points of the form $s = nL, n \in \mathbb{Z}$).

The above properties lead to this theorem, that can be used to shed a new light to what was expressed in 10.10 in the previous section.

Theorem 10.17. *Let $f = \nabla_{H^0} E$, $g = \nabla_{\tilde{H}^1} E$; the solution of*

$$f = \mathrm{avg}_c(g) - \lambda L^2 D_s^2 g$$

can be expressed as

$$g = \mathrm{avg}_c(f) - \frac{1}{\lambda L^2} \mathcal{P}_c \mathcal{P}_c \Pi_c f = \mathrm{avg}_c(f) - \frac{1}{\lambda L^2} K_c^{\mathcal{P}} \star K_c^{\mathcal{P}} \star f \ .$$

Corollary 10.18. *In particular, the kernels defined in* (48), (59) *are related by*

$$\tilde{K}_\lambda = \frac{1}{L} - \frac{1}{\lambda L^2} K_c^{\mathcal{P}} \star K_c^{\mathcal{P}}$$

By deriving,

$$D_s K_c^{\mathcal{P}} = \delta_0 - \frac{1}{L} \tag{61}$$

where δ_0 is the Dirac's delta; so we obtain the relations

$$D_s \tilde{K}_\lambda = -\frac{1}{\lambda L^2} K_c^{\mathcal{P}}$$

$$D_{ss} \tilde{K}_\lambda = -\frac{1}{\lambda L^2}\delta_0 + \frac{1}{\lambda L^3} \ .$$

We also recall this Lemma.

Lemma 10.19 (De la Vallée-Poussin). *Suppose that $f : S^1 \to \mathbb{R}^n$ is integrable and satisfies*

$$\int_c k \cdot f \, \mathrm{d}s = 0$$

for all $k \in D_c M$ smooth; then f is almost everywhere equal to a constant.

For the proof, see Chap. 13 in [1], or Lemma VIII.1 in [4].

We now compute the gradient of the centroid energy (10).

Proposition 10.20 (Gradient of the centroid-based energy). *Let in the following again, for simplicity of notation, $\overline{c} = \mathrm{avg}_c(c)$ be the center of mass of the curve. Let*

$$E(c) = \frac{1}{2}|\overline{c} - v|^2 \ .$$

The $\tilde{H}^1$ gradient of E is

$$\nabla_{\tilde{H}^1} E(c) = \overline{c} - v + \frac{1}{\lambda L^2} \mathcal{P}_c \Pi_c \big(D_s c \langle (\overline{c} - v), (c - \overline{c}) \rangle \big) \ . \tag{62}$$

Proof. We already computed the Gâteaux differential of E; but this time we prefer to use the first form of (11), so as to write

$$DE(c; h) = (\overline{c} - v) \cdot \overline{h} + \fint_c \langle \overline{c} - v, c - \overline{c} \rangle (D_s h, D_s c) \, \mathrm{d}s \ .$$

Let $f = \nabla_{\tilde{H}^1} E(c)$ be the $\tilde{H}^1$ gradient of E. The equality

$$DE(c;h) = \langle h, f \rangle_{\tilde{H}^1} \quad \forall h$$

becomes

$$\langle \overline{c} - v - \overline{f}, \overline{h} \rangle + \fint_c D_s h \cdot \left(\langle \overline{c} - v, c - \overline{c} \rangle D_s c - \lambda L^2 D_s f \right) ds = 0, \; \forall h \,.$$

Since h is arbitrary, using de la Vallée-Poussin Lemma, we obtain that

$$\overline{f} = \overline{c} - v \;, \quad \lambda L^2 D_s f = D_s c \langle (\overline{c} - v), (c - \overline{c}) \rangle + \alpha \,, \tag{63}$$

where the constant α is the unique $\alpha \in \mathbb{R}^n$ such that the rightmost term is in $D_c M$. In conclusion we obtain (62).

We similarly compute the gradient of the active contour energy.

Proposition 10.21 (Gradient of geodesic active contour model). *We consider once again the* geodesic active contour *model [7], [28] (that was presented in Sect. 3) where the energy is*

$$E(c) = \int_c \phi(c(s)) \, ds$$

with $\phi : \mathbb{R}^2 \to \mathbb{R}^+$ *appropriately designed. The gradient with respect to* $\tilde{H}^1$ *is*

$$\nabla_{\tilde{H}^1} E(c) = L \operatorname{avg}_c(\nabla \phi(c)) - \frac{1}{\lambda L} \mathcal{P}_c \mathcal{P}_c \Pi_c (\nabla \phi(c)) + \frac{1}{\lambda L} \mathcal{P}_c \Pi_c (\phi(c) D_s c) \,. \tag{64}$$

Proof. Let us note[22] (recalling (4)) that

$$\nabla_{H^0} E = L \nabla \phi(c) - L D_s(\phi(c) D_s c) \,, \tag{65}$$

so the above equation (64) can be obtained by the relation in Theorem 10.17. Alternatively, we know that

$$DE(c;h) = L \fint_c \nabla \phi(c) \cdot h + \phi(c)(D_s h \cdot D_s c) \, ds \, ;$$

let $f = \nabla_{\tilde{H}^1} E(c)$ be the $\tilde{H}^1$ gradient of E; the equality

$$DE(c;h) = \langle h, f \rangle_{\tilde{H}^1} \quad \forall h$$

---

[22] We use the definition (42) of H^0.

becomes

$$\int_c (L\nabla\phi(c) - avg_c(f)) \cdot h + D_s h \cdot (L\phi(c)D_s c - \lambda L^2 D_s f) \, ds = 0$$

imitating the proof of the previous proposition, we obtain (64). □

We remark a few things.

- Note that the first term in the formula (64) is in $\mathbb{R}^n$ while the other two are in $D_c M$.
- Using the kernel $\tilde{K}_\lambda$ that was defined in (48), we can rewrite

$$\nabla_{\tilde{H}^1} E(c) = L\tilde{K}_\lambda \star (\nabla\phi(c)) + \frac{1}{\lambda L} \mathcal{P}_c \Pi_c (\phi(c) D_s c) \qquad (66)$$

- The formula for the gradient does not require that the curve be twice differentiable: we will use this fact to prove an existence result for the gradient flow, in Theorem 10.31.

10.6 Existence of Gradient Flows

We recall this definition (that was already presented informally in the introduction).

Definition 10.22 (Gradient descent flow). Given a differentiable energy $E : M \to \mathbb{R}$, and a metric $\langle,\rangle_c$, let $\nabla E(c)$ be the gradient. Let us fix moreover $c_0 : S^1 \to \mathbb{R}^n, c_0 \in M$. The **gradient descent flow** of E is the solution $C = C(t, \theta)$ of the initial value P.D.E.

$$\begin{cases} \partial_t C = -\nabla E(C) \\ C(0, \theta) = c_0(\theta) \end{cases}$$

We present an example computation for the *geodesic active contour* model on a radially symmetric "image".

Example 10.23. Let

$$C(t, \theta) = r(t)(\cos\theta, \sin\theta), \quad \phi(x) = d(|x|)$$

with $r, d : \mathbb{R} \to \mathbb{R}^+$; the energy takes the form

$$E(C) = \int_C \phi(C(t, s)) \, ds = 2\pi \, r(t) \, d(r(t))$$

Then C is the $\tilde{H}^1$ gradient descent iff

Metrics of Curves in Shape Optimization and Analysis

$$r'(t) = -\frac{1}{\lambda 2\pi}\big(d(r(t)) + r(t)\,d'(r(t))\big)\,.$$

whereas C is the H^0 gradient descent iff

$$r'(t) = -2\pi\big(d(r(t)) + r(t)d'(r(t))\big)\,,$$

where we use the definition (42) of H^0.

Note also that

$$\partial_t E(C) = r'(t)\big(d(r(t)) + r(t)d'(r(t))\big)\,.$$

Proof. Indeed,

$$ds = r(t)\,d\theta$$
$$D_s = \frac{1}{r(t)} D_\theta$$
$$\text{len}(C) = 2\pi r(t)$$
$$\phi(C) = d(r(t))$$
$$\nabla\phi(x) = \frac{x}{|x|} d'(|x|) \text{ for } x \neq 0$$
$$\nabla\phi(C) = d'(r(t))(\cos\theta, \sin\theta)$$
$$\text{avg}_c(\nabla\phi(C)) = 0$$

$$\mathcal{P}_c \Pi_c \nabla\phi(C) = \mathcal{P}_c \nabla\phi(C) = r(t)\,d'(r(t))\,(\sin\theta, -\cos\theta)$$
$$\mathcal{P}_c \mathcal{P}_c \Pi_C \nabla\phi(C) = -r(t)^2\,d'(r(t))\,(\cos\theta, \sin\theta)$$
$$\phi(C) D_s C = d(r(t))(-\sin\theta, \cos\theta)$$
$$\mathcal{P}_c(\phi(C) D_s C) = r(t) d(r(t))\,(\cos\theta, \sin\theta)$$
$$\nabla_{\tilde{H}^1} E(C) = L\,\text{avg}_c(\nabla\phi(C)) - \frac{1}{\lambda L}\mathcal{P}_c \mathcal{P}_c \Pi_C(\nabla\phi(C))$$
$$+ \frac{1}{\lambda L}\mathcal{P}_c \Pi_C(\phi(C) D_s C) =$$
$$= \frac{1}{\lambda 2\pi}\big(d(r(t)) + r(t)\,d'(r(t))\big)(\cos\theta, \sin\theta)\,;$$

whereas for the $\tilde{H}^0$ gradient descent we use the formula (65)

$$\nabla_{H^0} E = L\nabla\phi(C) - L D_s(\phi(C) D_s C) =$$
$$= 2\pi\big(r(t)d'(r(t)) + d(r(t))\big)(\cos\theta, \sin\theta)$$

We will show in the following theorems how to prove that $\tilde{H}^1$ gradient flows of common energies are well defined; to this end, we will provide a detailed proof for the *centroid energy* E (10) discussed in the previous section, and for the *geodesic active contour* model [7, 28] (that was presented in Sect. 3); those proofs illustrates methods that may be used for many other energies.

Lemmas and Inequalities

Let us also prepare the proof by presenting some useful inequalities in three Lemma.

Definition 10.24. We recall from 3.11 that $C^0 = C^0(S^1 \to \mathbb{R}^n)$ is the space of continuous functions, that is a Banach space with norm

$$\|c\|_0 := \sup_{\theta \in [0, 2\pi]} |c(\theta)|.$$

Similarly, $C^1 = C^1(S^1 \to \mathbb{R}^n)$ is the space of continuously differentiable functions, that is a Banach space with norm

$$\|c\|_1 := \|c\|_0 + \|c'\|_0$$

where $c'(\theta) = \frac{\partial c}{\partial \theta}(\theta)$ is the usual parametric derivative of c.

Lemma 10.25. • *We will use repeatedly the two following inequalities. If $a_1, a_2, b_1, b_2 \in \mathbb{R}$ then*

$$|a_1 b_1 - a_2 b_2| \leq \tfrac{|b_1|+|b_2|}{2}|a_1 - a_2| + \tfrac{|a_1|+|a_2|}{2}|b_1 - b_2| \tag{i}$$

$$|\tfrac{a_1}{b_1} - \tfrac{a_2}{b_2}| \leq \tfrac{|b_1|+|b_2|}{2|b_1 b_2|}|a_1 - a_2| + \tfrac{|a_1|+|a_2|}{2|b_1 b_2|}|b_1 - b_2|$$

- *The length functional (from C^1 to $\mathbb{R}$) is Lipschitz, since*

$$\operatorname{len}(c_1) - \operatorname{len}(c_2) \leq 2\pi \|c_1' - c_2'\|_0 ; \tag{ii}$$

- *if $h_1, h_2 : S^1 \to \mathbb{R}^n$ are continuous fields, by* (i)

$$\left| \int_{c_1} h_1(s)\,ds - \int_{c_2} h_2(s)\,ds \right| \leq \pi(\|h_1\|_0 + \|h_2\|_0)\|c_1' - c_2'\|_0 +$$
$$+ \pi(\|c_1'\|_0 + \|c_2'\|_0)\|h_1 - h_2\|_0 ; \tag{iii}$$

- *similarly*

$$\left| \fint_{c_1} h_1(s)\,ds - \fint_{c_2} h_2(s)\,ds \right| \leq A \|c_1' - c_2'\|_0 + \|h_1 - h_2\|_0 \tag{iv}$$

where

$$A := \frac{2\pi^2(\|h_1\|_0 + \|h_2\|_0)(\|c_1'\|_0 + \|c_2'\|_0)}{\operatorname{len}(c_0)\operatorname{len}(c_1)} \ . \tag{v}$$

- Let then Π_c be the projection operator *(as in definition (55))*; note that $(\Pi_c h)' = h'$; *from all above inequalities we obtain that*

$$\|\Pi_{c_1}h_1 - \Pi_{c_2}h_2\|_0 \leq A\|c_1' - c_2'\|_0 + 2\|h_1 - h_2\|_0$$
$$\|\Pi_{c_1}h_1 - \Pi_{c_2}h_2\|_1 = \|\Pi_{c_1}h_1 - \Pi_{c_2}h_2\|_0 + \|(\Pi_{c_1}h_1)' - (\Pi_{c_2}h_2)'\|_0 \leq$$
$$\leq A\|c_1' - c_2'\|_0 + 2\|h_1 - h_2\|_0 + \|h_1' - h_2'\|_0$$
$$\leq A\|c_1 - c_2\|_1 + 2\|h_1 - h_2\|_1 \ . \tag{vi}$$

We will now prove the local Lipschitz regularity of the operators $\mathcal{P}_c(h)$, $\mathcal{D}_c(h)$ (that were defined in 10.14) in both the two variables h and c. Unfortunately to prove the Theorem 10.29 we will need to also compare the action of $\mathcal{P}_c$ for different values of c; since $\mathcal{P}_c h$ is properly defined only when $h \in D_c M$ (and in general $D_{c_1} M \neq D_{c_2} M$), we will actually need to study the composite operator $\mathcal{P}_c \Pi_c h$.

Lemma 10.26. *Let Π_c the projector from $T_c M$ to $D_c M$ (that we defined in (55)). Let c, c_1, c_2 be C^1 immersed curves, and h, h_1, h_2 be continuous fields; then*

$$\|\mathcal{P}_c \Pi_c h\|_1 \leq (2\pi + 2)\|c'\|_0 \|h\|_0 \tag{67}$$

and

$$\|\mathcal{P}_{c_1} \Pi_{c_1} h_1 - \mathcal{P}_{c_2} \Pi_{c_2} h_2\|_1 \leq P \|c_1' - c_2'\|_0 + Q \|h_2 - h_0\|_0 \tag{68}$$

where

$$Q := (1/2 + \pi)(\|c_2'\|_0 + \|c_1'\|_0) \ .$$

and similarly P is the evaluation

$$P = p(\|h_1\|_0, \|h_2\|_0, \|c_1'\|_0, \|c_2'\|_0, 1/(\operatorname{len}(c_1)\operatorname{len}(c_2)))$$

of the polynomial

$$p(x_1, x_2, x_3, x_4, x_5) = (\pi + 1/2)(x_1 + x_2) + 2\pi^3(x_1 x_3 + x_2 x_4)(x_3 + x_4)x_5 \tag{69}$$

(that has constant positive coefficients).

Proof. Fix c_i immersed, and $h_i \in T_{c_i} M$ for $i = 1, 2$; let $L_i = \operatorname{len}(c_i)$; let

$$k_i := \mathcal{P}_{c_i} \Pi_{c_i} h_i(s)$$

for simplicity. We rewrite this in the convolutional form

$$k_i(s) := \int_{c_i} K_i(s - \hat{s}) h_i(\hat{s}) \, d\hat{s} \tag{70}$$

when integrals are performed in arc parameter, and the kernel (following (59)) is

$$K_i(s) := -\frac{s}{L_i} + \frac{1}{2} \quad \text{for } s \in [0, L_i]$$

and K_i is extended periodically. By substituting and integrating on only one period of K_i,

$$k_i(s) = \int_{s-L_i}^{s} \left(\frac{1}{2} - \frac{s - \hat{s}}{L_i} \right) h_i(\hat{s}) \, d\hat{s} \, .$$

We can then prove easily (67): indeed by the convolutional representation (70)

$$|(\mathcal{P}_c \Pi_c h)(t)| \le \|h\|_0 \int_0^{\operatorname{len}(c)} \left| -\frac{s}{\operatorname{len}(c)} + \frac{1}{2} \right| ds \le \operatorname{len}(c) \|h\|_0$$

and instead, deriving and applying (iv) from Lemma 10.25,

$$|(\mathcal{P}_c \Pi_c h)'(\theta)| = |\Pi_c h(\theta)| \, |c'(\theta)| \le 2 \|h\|_0 \|c'\|_0 \, .$$

To prove (68), we write (70) in θ parameter, as was done in the Definition 10.7; we need the run-length functions $l_i : \mathbb{R} \to \mathbb{R}$

$$l_i(\tau) := \int_0^{\tau} |c_i'(x)| \, dx$$

so that, setting $s = l_i(\tau)$ and $\hat{s} = l_i(\theta)$ we can write

$$k_i(\tau) = \int_{\tau - 2\pi}^{\tau} \left(\frac{1}{2} - \frac{l_i(\tau) - l_i(\theta)}{L_i} \right) h_i(\theta) |c_i'(\theta)| \, d\theta$$

We can eventually estimate the difference $|k_1(\tau) - k_2(\tau)|$ by using e.g. the inequality

$$|A_2 B_2 C_2 - A_1 B_1 C_1| \le |A_1 B_1| \, |C_2 - C_1| + |A_1 C_2| \, |B_2 - B_1| + |B_2 C_2| \, |A_2 - A_1|$$

on the difference of the integrands

$$\underbrace{\left(\frac{1}{2} - \frac{l_2(\tau) - l_2(\theta)}{L_2}\right)}_{A_2} \underbrace{h_2(\theta)}_{B_2} \underbrace{|c_2'(\theta)|}_{C_2} - \underbrace{\left(\frac{1}{2} - \frac{l_1(\tau) - l_1(\theta)}{L_1}\right)}_{A_1} \underbrace{h_1(\theta)}_{B_1} \underbrace{|c_1'(\theta)|}_{C_1}$$

to obtain that the above term is less or equal than

$$\|h_1\|_0 \|c_2' - c_1'\|_0 + \|c_2'\|_0 \|h_2 - h_0\|_0 + \|h_2\|_0 \|c_2'\|_0 \left| \frac{l_2(\theta) - l_2(\tau)}{L_2} - \frac{l_1(\theta) - l_1(\tau)}{L_1} \right|$$

since $|A_i| \le 1/2$. In turn (since the formulas defining $k_1(\tau), k_2(\tau)$ the parameters are bound by $\tau - 2\pi \le \theta \le \tau$) then

$$|A_2 - A_1| = \left| \frac{l_2(\tau) - l_2(\theta)}{L_2} - \frac{l_1(\tau) - l_1(\theta)}{L_1} \right| = \left| \frac{\int_\theta^\tau |c_2'(x)| \, dx}{L_2} - \frac{\int_\theta^\tau |c_1'(x)| \, dx}{L_1} \right| \le$$

$$\le \pi \frac{L_1 + L_2}{L_1 L_2} \|c_1' - c_2'\|_0 + \pi \frac{\|c_1'\|_0 + \|c_2'\|_0}{L_1 L_2} |L_1 - L_2| \le$$

$$\le 4\pi^2 \frac{\|c_1'\|_0 + \|c_2'\|_0}{L_1 L_2} \|c_1' - c_2'\|_0$$

(by (ii) and (i) in Lemma 10.25). Summarizing

$$|k_1(\tau) - k_2(\tau)| \le 2\pi \left(\|h_1\|_0 + \|h_2\|_0 \|c_2'\|_0 4\pi^2 \frac{\|c_1'\|_0 + \|c_2'\|_0}{L_1 L_2} \right) \|c_1' - c_2'\|_0 +$$
$$+ 2\pi \|c_2'\|_0 \|h_2 - h_0\|_0 .$$

If we derive, $k_i'(\theta) = h_i(\theta) |c_i'(\theta)|$, so

$$|k_2'(\theta) - k_1'(\theta)| \le \frac{|h_2(\theta)| + |h_1(\theta)|}{2} |c_2'(\theta) - c_1'(\theta)| + \frac{|c_2'(\theta)| + |c_1'(\theta)|}{2} |h_2'(\theta) - h_1'(\theta)| .$$

Symmetrizing we prove (68).

Lemma 10.27. *Conversely, let c_1, c_2 be two C^1 immersed curves, and h_1, h_2 be two differentiable fields; then (by using once again (i) from the Lemma 10.25)*

$$\|\mathcal{D}_{c_1} h_1 - \mathcal{D}_{c_2} h_2\|_0 \le \frac{\|c_1\|_1 + \|c_2\|_1}{2\varepsilon^2} \|h_1 - h_2\|_1 + \frac{\|h_1\|_1 + \|h_2\|_1}{2\varepsilon^2} \|c_1 - c_2\|_1 \quad (73)$$

where $\varepsilon = \min(\inf_{S^1} |c_1'|, \inf_{S^1} |c_2'|)$.

Existence of Flow for the Centroid Energy (10)

Theorem 10.29. *Let us fix $v \in \mathbb{R}^n$, let $E(c) = \frac{1}{2}|\text{avg}_c(c) - v|^2$; let the initial curve $c_0 \in C^1$, then the $\tilde{H}^1$ gradient descent flow of E has an unique solution $C = C(t, \theta)$, for all $t \in \mathbb{R}$, and $C(t, \cdot) \in C^1$.*

Results of a numerical simulation of the above gradient descent flow are shown in Fig. 12.

Before proving the theorem, let us comment on an interesting property of the above gradient flow.

Remark 10.30. The length of the curves $\text{len}(C(t, \cdot))$ is constant during the evolution in t.

Proof. Indeed it is easy to prove that $\partial_t \text{len}(C(t, \cdot)) = 0$: the Gâteaux differential of the length of a curve c is

$$D \text{len}(c; h) = \int_c (D_s c \cdot D_s h) \, ds \ ;$$

substituting the value of $D_s C(t, \cdot))$ from (63),

$$\partial_t \text{len}(C(t, \cdot)) = D(\text{len}(C))(\partial_t C)$$
$$= \frac{1}{\lambda L^2} \int \left\langle D_s C, \left(D_s C \left\langle (\text{avg}_c(C) - v), (C - \text{avg}_c(C)) \right\rangle + \alpha \right) \right\rangle ds =$$
$$= \frac{1}{\lambda L^2} \left\langle (\text{avg}_c(C) - v), \int (C - \text{avg}_c(C)) \, ds \right\rangle + \int (D_s C \cdot \alpha) \, ds = 0 \ .$$

We now present the proof of Theorem 10.29.

Proof. We will first of all prove existence and uniqueness of the gradient flow for small time, using the *Cauchy–Lipschitz theorem*[24] in the space M of immersed curves, seen as an open subset of the C^1 *Banach space* (see Definition 10.24). To this end we will prove that $c \mapsto \nabla_{\tilde{H}^1} E(c)$ is a locally Lipschitz functional from C^1 into itself. Afterward, we will directly prove that the solution exists for all times.

So, let us fix $c_1, c_2 \in C^1 \cap M$ that are two curves near c_0. Let us fix $\varepsilon > 0$ by

$$\varepsilon := \inf_\theta |c_0'(\theta)|/2 \ .$$

By "near" we mean that, in all of the following, we will require that $\|c_i - c_0\|_1 < \varepsilon$, for $i = 1, 2$. We will need the following (easy to prove) facts. (In the following, the index i will represent both 1, 2).

---

[24] A.k.a. as the *Picard–Lindelöf theorem*.

Metrics of Curves in Shape Optimization and Analysis

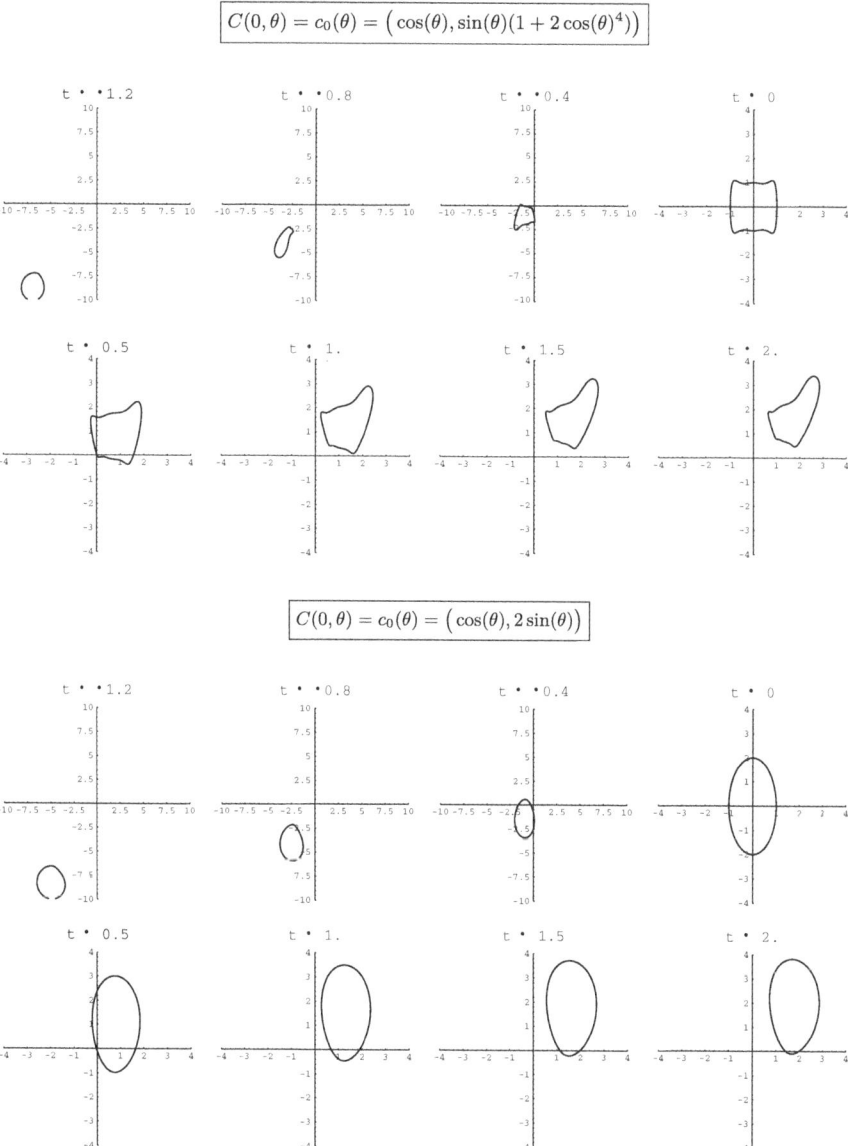

Fig. 12 $\tilde{H}^1$ gradient descent flow of the centroid energy. See 10.20 and 10.29. Two numerical simulation of the $\tilde{H}^1$ gradient descent flow of the *centroid energy* $E(c) = \frac{1}{2}|\text{avg}_c(c) - v|^2$, for two different initial curves c_0. The constants were set to $\lambda = 1/(4\pi^2)$, $v = (2, 2)$. Steps for numerical discretization were set to $\Delta t = 1/30$, $\Delta\theta = \pi/64$ (on $[0, 2\pi]$). Note that plots for $t < 0$ are shown in a smaller scale

- $\inf_\theta |c_i'(\theta)| \geq \varepsilon$, so all curves in the neighborhood are immersed.
- Setting $a_0 := \|c_0\|_1 + \varepsilon$, we have $\|c_i\|_1 \leq a_0$.
- By (ii) in Lemma 10.25,

$$\operatorname{len}(c_0) - 2\pi\|c_0 - c_i\|_1 \leq \operatorname{len}(c_i) \leq \operatorname{len}(c_0) + 2\pi\|c_0 - c_i\|_1$$

and in particular

$$2\pi\varepsilon \leq \operatorname{len}(c_i) \leq \operatorname{len}(c_0) + 2\pi\varepsilon .$$

Let $f_i = \nabla_{\tilde{H}^1} E(c_i)$ be the gradient, whose formula was expressed in (62). We decompose f_i using the relation $T_c M = \mathbb{R}^n \oplus D_c M$, as done previously:

$$\overline{f}_i = \operatorname{avg}_{c_i}(f_i) \in \mathbb{R}^n \quad , \quad \tilde{f}_i = f - \operatorname{avg}_{c_i}(f_i) \in D_{c_i} M$$

and obtain

$$\overline{f}_i = \operatorname{avg}_{c_i}(c_i) - v$$
$$\tilde{f}_i = \frac{1}{\lambda L_i^2} \mathcal{P}_{c_i} \Pi_{c_i} h_i, \text{ where}$$
$$h_i := D_s c_i \langle (\operatorname{avg}_{c_i}(c_i) - v), (c_i - \operatorname{avg}_{c_i}(c_i)) \rangle , \tag{74}$$

by rewriting (63) in the notation of this proof.

We will then exploit all the inequalities in the lemmas to prove that

$$|\overline{f}_1 - \overline{f}_2| \leq a_1 \|c_1 - c_2\|_1 \quad , \quad \|\tilde{f}_1 - \tilde{f}_2\|_1 \leq a_2 \|c_1 - c_2\|_1$$

for two constants $a_1, a_2 > 0$ and all c_1, c_2 near c_0; this effectively proves that

$$|f_1 - f_2| \leq (a_1 + a_2)\|c_1 - c_2\|_1$$

that is, $\nabla_{\tilde{H}^1} E(c)$ is a locally Lipschitz functional.

The first term is dealt with (iv) in Lemma 10.25, whence we obtain

$$|\overline{f}_1 - \overline{f}_2| = \left| \fint_{c_1} c_1(s)\,ds - \fint_{c_2} c_2(s)\,ds \right| \leq a_1 \|c_1 - c_2\|_1 \text{ with } a_1 := \frac{2a_0^2}{\varepsilon^2} .$$

By repeated applications of the inequalities listed in Lemma 10.25, we prove that, in the designed neighborhood, the following inequalities hold

$$\|h_2 - h_1\|_0 \leq a_3 \|c_2 - c_1\|_1$$
$$\|h_i\| \leq a_4$$

for two constants $a_3, a_4 > 0$. So we can choose constants $P, Q > 0$ such that the inequality (68) holds uniformly in the neighborhood, and then

$$\|\mathcal{P}_{c_1} \Pi_{c_1} h_1 - \mathcal{P}_{c_2} \Pi_{c_2} h_2\|_1 \le P \|c'_1 - c'_2\|_0 + Q \|h_2 - h_1\|_0 \le$$
$$\le (P + Qa_3)\|c_1 - c_2\|_1$$

By using (i) and (ii) from Lemma 10.25 once again, we can conclude that

$$\|\tilde{f}_1 - \tilde{f}_2\|_1 \le a_2 \|c_1 - c_2\|_1$$

for a constant $a_2 > 0$. The Cauchy–Lipschitz theorem is now invoked to guarantee that the gradient descent flow does exist and is unique for small times. Let then $C(t, \theta)$ be the solution, which will exist for $t \in (t^-, t^+)$, the maximal interval.

In the following, given any $g = g(t, \theta)$ we will simply write $\|g\|_0$ instead of

$$\|g(t, \cdot)\|_0 = \sup_\theta |g(t, \theta)|$$

and similarly

$$\|g\|_1 = \|g\|_0 + \|\partial_\theta g\|_0 = \sup_\theta (|g(t, \theta)| + |\partial_\theta g(t, \theta)|) ;$$

we will also write $\text{len}(C) = \text{len}(C(t, \cdot))$ for the length of the curve at time t.

We want to prove that the maximal interval is actually $\mathbb{R}$, that is, $t^+ = -t^- = \infty$. The base and rough idea of the proof is assuming that t^+ or t^- are finite, and derive a contradiction by showing that $\nabla E(C)$ does not *blow up* when $t \searrow t^-$ or $t \nearrow t^+$.

More precisely, we will show that, if $t^- > -\infty$ then

$$\limsup_{t \searrow t^-} \|\nabla E(C)\|_1 < \infty \qquad (**)$$

this implies that the flow $C(t, \cdot)$ admits a limit (inside the Banach space C^1) as $t \to t^-$, and then it may continued (contradicting the fact that t^- is the lowest time limit of existence of the flow). A similar result may be derived when $t \nearrow t^+$ (but we will omit the proof, that is actually simpler).

One key step in showing that (**) holds is to consider the two fundamental quantities

$$I(t) := \inf_\theta |\partial_\theta C(t, \theta)| , \ N(t) := \|C\|_1$$

and prove that

$$\liminf I(t) > 0 , \ \limsup N(t) < \infty$$

when $t^- > -\infty$ and $t \searrow t^-$.[25] We proceed in steps.

- By Remark 10.30, we know that $\text{len}(C)$ is constant (for as long as the flow is defined), and then equal to $\text{len}(c_0)$.
- At any fixed time, by (43),

$$|C - \text{avg}_c(C)| \leq \text{len}(C)/2 = \text{len}(c_0)/2 \tag{75}$$

where $\text{avg}_c(C)$ is the center of mass of $C(t, \cdot)$; and then

$$|C| \leq |C - \text{avg}_c(C)| + |v - \text{avg}_c(C)| + |v| \leq \text{len}(c_0)/2 + \sqrt{2E(C)} + |v| \tag{76}$$

- During the flow, the value of the energy changes with rate

$$\partial_t E(C) = -\|\nabla_{\tilde{H}^1} E(C)\|_{\tilde{H}^1}^2 =$$

$$= -|\text{avg}_c(C) - v|^2 - \frac{1}{\lambda \, \text{len}(C)^2} \oint_C \langle \text{avg}_c(C) - v, C - \text{avg}_c(C) \rangle^2 \, ds$$

using (75), we can bound

$$|\partial_t E(C)| \leq E(C)(2 + 1/\lambda)$$

so from Gronwall inequality we obtain that $E(C)$ does not blow up; consequently by (76), $\|C\|_0$ does not blow up as well.

- Let

$$H := D_s C \langle (\text{avg}_c(C) - v), (C - \text{avg}_c(C)) \rangle ,$$

from the above we obtain that H does not blow up as well, since

$$\|H(t, \cdot)\|_0 \leq \text{len}(c_0) \sqrt{2E(C)} \tag{77}$$

- The parameterization of the curves changes according to the law

$$\partial_t \log(|\partial_\theta C|^2) = 2 \langle D_s \partial_t C, D_s C \rangle = \frac{2H \cdot D_s C}{\lambda \, \text{len}(c_0)^2}$$

so

$$|\partial_t \log(|\partial_\theta C|^2)| \leq \frac{2\sqrt{2E(C)}}{\lambda \, \text{len}(c_0)}$$

---

[25] Actually, by tracking the first part of the proof in detail, it possible to prove that all other constants a_1, a_2, a_3, a_4, P, Q may be bounded in terms of these two quantities $I(t), N(t)$.

this proves that
$$\liminf_{t \searrow t^-} I(t) > 0$$

and
$$\limsup_{t \searrow t^-} \sup_\theta |\partial_\theta C(t,\theta)| < \infty.$$

(As a consequence, $\partial_\theta C(t,\theta) \neq 0$ at all times: curves will always be immersed)
- So by (67)
$$\|\mathcal{P}_C \Pi_C H\|_1 \leq (2\pi + 2)\|C'\|_0 \|H\|_0$$
does not blow up as well.
- Since both $\|C\|_0$ and $\|\partial_\theta C\|_0$ do not blow up, then $\|C\|_1$ does not blow up.
- Decomposing $F = -\partial_t C = \nabla_{\tilde{H}^1} E(C)$ (as was done in (74)) we write
$$\|\tilde{F}\|_1 = \frac{1}{\lambda \operatorname{len}(c_0)^2} \|\mathcal{P}_C \Pi_C H\|_1$$

and from all above $\|\tilde{F}\|_1$ does not blow up; but also
$$|\overline{F}| = |\operatorname{avg}_C(C) - v| \leq \sqrt{2E(C)}$$
as well; but then $\|\nabla_{\tilde{H}^1} E(C)\|_1$ itself does not blow up, as we wanted to prove.

Existence of Flow for Geodesic Active Contour

Theorem 10.31. *Let once again*
$$E(c) = \int_c \phi(c(s))\, ds$$
be the geodesic active contour *energy. Suppose that $\phi \in C^{1,1}_{loc}$ (that is, $\phi \in C^1$ and its derivative is Lipschitz on compact subsets of $\mathbb{R}^n$), let the initial curve be $c_0 \in C^1$; then the gradient flow*
$$\frac{dc}{dt} = -\nabla_{\tilde{H}^1} E(c)$$
exists and is unique in C^1 for small times.

Proof. We rapidly sketch the proof, that is quite similar to the proof of the previous theorem. We show that $\nabla_{\tilde{H}^1} E(c)$ is locally Lipschitz in the C^1 Banach space (see Definition 10.24). Indeed, since $\phi \in C^{1,1}_{\text{loc}}$, the maps

$$c \mapsto \nabla \phi(c)$$
$$c \mapsto \phi(c) D_s c$$

are locally Lipschitz as maps from C^1 to C^0, so (using Lemmas 10.25 and 10.26) we obtain that

$$c \mapsto \text{avg}_c(\nabla \phi(c))$$

is locally Lipschitz from C^1 to $\mathbb{R}^n$, and

$$c \mapsto \mathcal{P}_c \mathcal{P}_c \Pi_c (\nabla \phi(c))$$
$$c \mapsto \mathcal{P}_c \Pi_c (\phi(c) D_s c)$$

are locally Lipschitz as maps from C^1 to C^1; combining all together (and using the Lemma 10.25 again)

$$\nabla_{\tilde{H}^1} E(c) = L \text{avg}_c(\nabla \phi(c)) - \frac{1}{\lambda L} \mathcal{P}_c \mathcal{P}_c \Pi_c (\nabla \phi(c)) + \frac{1}{\lambda L} \mathcal{P}_c \Pi_c (\phi(c) D_s c)$$

is locally Lipschitz as maps from C^1 to C^1; so by the *Cauchy–Lipschitz theorem* we know that the gradient flow exists for small times.

Remark 10.32 (Gradient descent of curve length). Of particular interest is the case when $\phi = 1$, that is $E = L$, the length of the curve. In this case, we already know that the H^0 flow is the *geometric heat flow*. By 10.15, the $\tilde{H}^1$ gradient instead reduces to

$$\nabla_{\tilde{H}^1} L = \frac{c - \text{avg}_c(c)}{\lambda L} . \tag{78}$$

So the $\tilde{H}^1$ gradient flow constitutes a simple rescaling of the curve about its centroid (!).

It is interesting to notice that the H^1 and $\tilde{H}^1$ gradient flows are well-posed for both ascent and descent while the H^0 gradient flow is only well-posed for descent. This is related to the fact that the H^0 gradient descent flow smooths the curve, whereas the $\tilde{H}^1$ gradient descent (or ascent) has neither a beneficial nor a detrimental effect on the regularity of the curve.

10.7 Regularization of Energy vs Regularization of Flow/Metric

Imagine an energy E minimized on curves (numerically sampled to p points). Suppose it is not satisfactory in applications (it may be ill posed, or not robust to noise). We have two solutions available.

- Add a regularization term $R(c)$ and minimize $E(c) + \varepsilon R(c)$ by H^0 gradient descent. The numerical complexity of computing $\nabla_{H^0} R$ is of order $O(p)$.
- Minimize $E(c)$ by $\tilde{H}^1$ gradient descent. The numerical complexity of computing $\nabla_{\tilde{H}^1} E$ using the equations in Proposition 10.10 is of order $O(p)$.

The numerical complexity of the two approaches is similar; but in the second case we are minimizing the original energy. This can bring evident benefits, as shown in the following example (originally presented in the *2006 IMA conference on shape spaces*).

Example 10.33. We consider a *region-based segmentation* of a synthetic noisy image, where the energy is the *Chan-Vese* energy

$$E(c) = \int_{c_{in}} (I - u)^2 \, dA + \int_{c_{out}} (I - v)^2 \, dA$$

plus the regularization terms $\alpha \operatorname{len}(c)$ for H^0 flows.

When flowing according to $-\nabla_{H^0} E + \alpha \kappa N$ and a small value of $\alpha > 0$, we obtain the following evolution of the curve, where the curve gets trapped in a local minima due to noise.

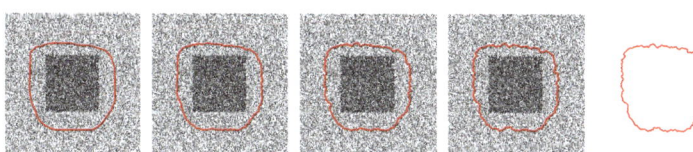

For a larger value of α, the regularization term forces the curve away from the square shape.

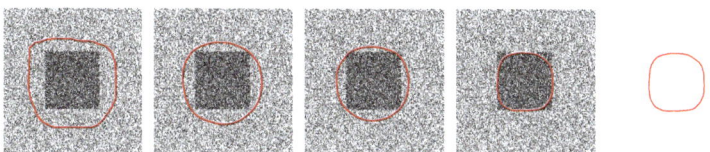

When evolving using $-\nabla_{\tilde{H}^1} E$, the curve captures the correct shape on coarse scale, and then segments the noisy boundary of the square further.

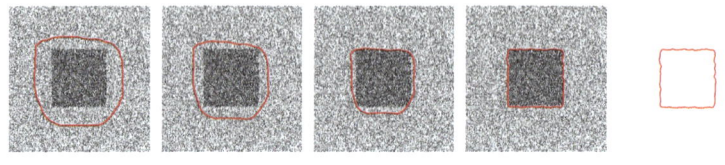

Robustness with respect to Local Minima Due to Noise

The concept of *local* depends on the norm.

Definition 10.34. A curve c_0 is a *local minimum* of an energy E iff $\exists \epsilon > 0$ such that if $\|h\|_{c_0} < \epsilon$ then $E(c_0) \le E(c_0 + h)$.

Note that Sobolev-type norms dominate the H^0-type norm:

$$\|h\|_{H^0} \le \|h\|_{\text{Sobolev}},$$

and the norms are *not* equivalent. As a result the *neighborhood* of critical points in Sobolev-type space is different than in H^0 space.

We present an experimental demonstration of what "local" means (originally presented in [57]).

Example 10.35. 1. We initialize a contour in a noisy image.
2. We run the H^0 gradient flow on the energy

$$E(c) = E_{cv}(c) + \alpha \operatorname{len}(c),$$

where E_{cv} is the *Chan-Vese* energy. Let us call the converged contour c_0; it is shown in the picture on the right.

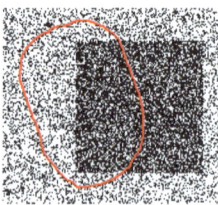

3. We adjust c_0 at **one sample point** by **one pixel**, and call the modification $\hat{c}_0$. Note: $\hat{c}_0$ is an H^0 (but also "rigidified") local perturbation of c_0.
4. We run and compare H^0, "rigidified," and Sobolev gradient flows initialized with $\hat{c}_0$.

The results are in Fig. 13. As we can see, the SAC evolution can escape from the local minimum that is induced by noise. Surprisingly, in the numerical experiments, the **same** result holds for SAC even without perturbing the local minimum: this is due to the effect of numerical noise.

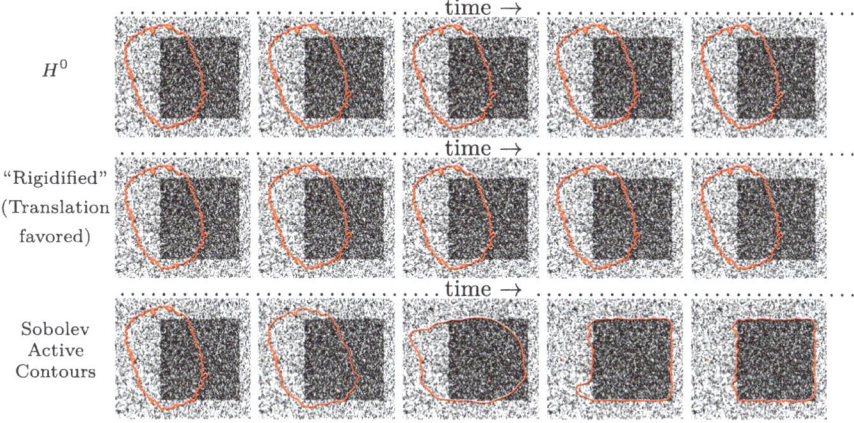

Fig. 13 Comparison of minimization flows when initialized near a local minimum; see Example 10.35 (From [57] ©2008 IEEE. Reproduced with permission)

Many other examples, on synthetic and on real images, are present in [56–58].

10.8 New Shape Optimization Energies

Most of what follows was originally presented in [55, 58].

There are many useful active contour energies that cannot be optimized using H^0 gradient flow. There are mainly two reasons why this happens.

- Some energies result in *ill-posed H^0* flows.
- Some energies result in high-order PDE and are difficult to implement numerically.

In many cases, we can optimize these energies using Sobolev active contours and avoid ill-posed problems and reduce the order of PDEs.

We remark that other gradients (e.g. the "rigidified" active contours of Charpiat et al. [10]) cannot be used in the examples we will present: the minimization flows still are ill-posed or high order. Similarly global methods (e.g. graph cuts) cannot deal with these types of energies.

Average Weighted Length

A simple example that falls in the first category is the **average weighted length** energy that we presented in (7) in the introduction:

$$E_1(c) = \oint_c \phi(c(s))\,\mathrm{d}s = \mathrm{avg}_c(\phi) \, ;$$

this energy was introduced since it does not present the *short length bias* (that was discussed in Sect. 2.4).

Let $L_\phi = \int_c \phi \, ds$ and $L = \text{len}(c)$ so that $E_1(c) = L_\phi/L$, then the gradient is

$$\nabla E_1 = \frac{1}{L} \nabla L_\phi - \frac{L_\phi}{L^2} \nabla L.$$

We already presented all the calculus needed to compute the H^0 and $\tilde{H}^1$ gradients of E_1 (see Sect. 2.4 and Example 10.21); let us summarize all the results.

- In the case of H^0 gradient descent flow, the second term is ill-posed, since $L_\phi/L^2 > 0$ and $\nabla_{H^0} L$ is the driving term in the *geometric heat flow*.

  This is also clear from the explicit formula

  $$\nabla_{H^0} E_1(c) = [\nabla \phi \cdot N - \kappa(\phi - \text{avg}_c(\phi))] N \qquad (79)$$

  since $\text{avg}_c(\phi)$ is the *average* value of ϕ, therefore $\phi - \text{avg}_c(\phi)$ is negative on roughly half of the curve; so the second term tries to increase length on half of curve using a *geometric heat flow* (and this is ill posed); whereas the first term $\nabla \phi$ does not stabilize the flow.

- In the case of the Sobolev gradient, we saw in the proof of Theorem 10.31 that the gradients $\nabla_{\tilde{H}^1} L_\phi$ and $\nabla_{\tilde{H}^1} L$ are locally Lipschitz in C^1, so, by using Lemma 10.25 we conclude that the gradient $\nabla_{\tilde{H}^1} E_1$ is locally Lipschitz as well, hence the gradient flow is well defined.

New Edge-Based Active Contour Models

The *average weighted length* idea can be used to build new energies with interesting applications in tracking.

Let us call

$$E_{\text{old}}(c) = \int_c \phi(c(s)) \, ds$$

the traditional edge-based active contour energy.

When using a Sobolev-type norm, we can consider *non-shrinking* edge-based models, as in the following examples.

Example 10.36. Let us consider the energy

$$E_{\text{new}}(c) = \int_c \phi(c(s)) \left(L^{-1} + \alpha L \kappa^2(s) \right) ds$$

where

- the first term is edge-detection without shrinking bias (nor regularity),

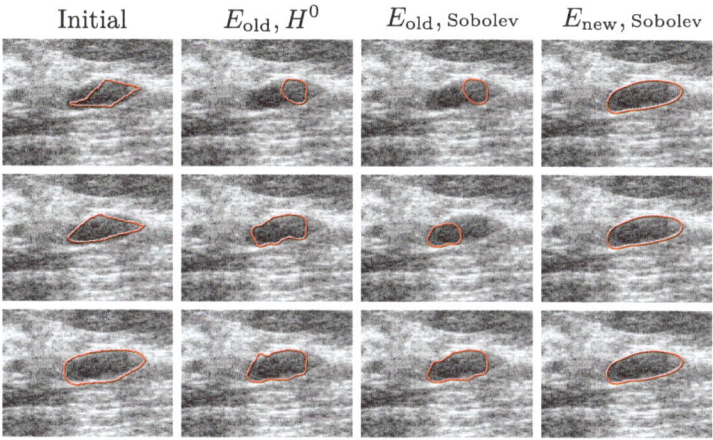

Fig. 14 Comparison of segmentations obtained with different energies, as explained in Example 10.36 (From [58] ©2008 IEEE. Reproduced with permission)

- and the second term is a kind of **elastic regularization** (that will be discussed also in the next section) that is moreover relaxed near edges;

the length terms render the whole energy *scale invariant*.

The frames in Fig. 14 show initialization and final results obtained with different norms and energies.

Example 10.37. Let us consider a **length increasing** edge-based model. In this case, we **maximize** the energy

$$E_{\text{inc}}(c) = \int_c \phi(c(s))\,\mathrm{d}s - \alpha \int_c \kappa^2(s)\,\mathrm{d}s$$

where $\phi > 0$ is high near edges. We compare numerical results with a typical **edge-based balloon model**

$$E_{bal}(c) = \int_c \phi(c(s))\,\mathrm{d}s - \alpha \int_R \phi(x)\,\mathrm{d}x$$

where R is the area enclosed by c. See Fig. 15.

10.9 New Regularization Methods

Typically, in active contour energies a *length penalty* is added to obtain regularity of the evolving contour:

$$E(c) = E_{\text{data}}(c) + \alpha \operatorname{len}(c).$$

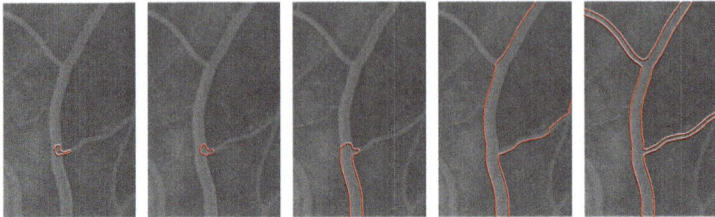

Fig. 15 *Left* to *right* we see the initial contour, the minima of E_{bal} for $\alpha = 0.2, 0.25, 0.4$ using H^0; the maximum for E_{inc} with $\alpha = 0.1$ using $\tilde{H}^1$ flowing. See Example 10.37 (From [58] ©2008 IEEE. Reproduced with permission)

It is important to note that this *length penalty* will regularize the curve only if the curve evolution is derived as a H^0 gradient descent flow. If another curve metric is used to derive the gradient descent flow, then *a priori* the *length penalty* may have no regularizing effect on the curve regularity (cf. 10.32).

Elastic Regularization

We can now consider an alternative approach for regularity of curve:

$$E(c) = E_{\text{data}}(c) + \alpha \operatorname{len}(c) \int_c \kappa^2(s)\, ds$$

where $\alpha > 0$ is a fixed constant. This energy favors regularity of curve, but this regularization does not rely on properties of the metric; and the regularization is *scale invariant*. Note that the H^0 gradient flow of this energy is ill-posed.

The numerical results (originally presented in [55]) are in Fig. 16. In all frames, the final limit of the gradient descent flow is shown; between different frames, the value of α is increased.

11 Mathematical Properties of the Riemannian Space of Curves

In this section we study some mathematical properties, and add some final remarks, regarding the Riemannian manifold of geometric curves, when this is endowed with the metrics that were presented previously.

11.1 Charts

Let again

$$M = M_{i,f} = \operatorname{Imm}_f(S^1, \mathbb{R}^n)$$

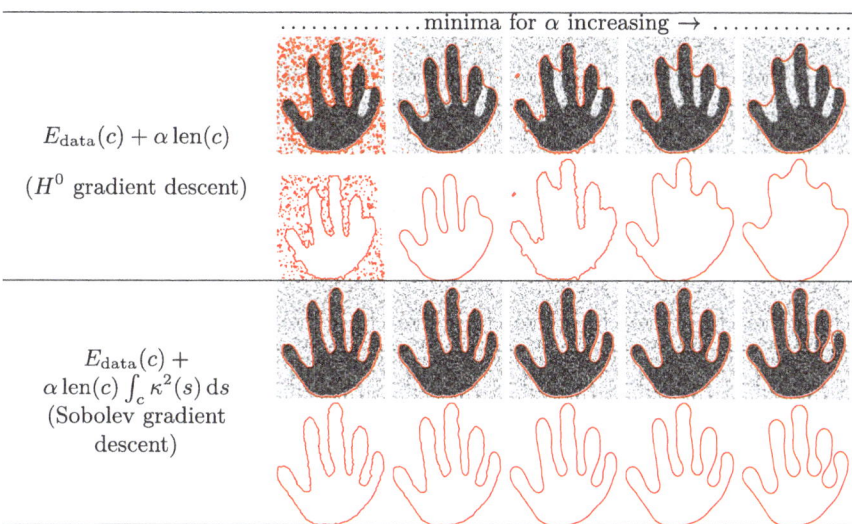

Fig. 16 Elastic regularization (From [58] ©2008 IEEE. Reproduced with permission)

be the space of all smooth freely-immersed curves.

We already remarked in Remark 4.4 that, if the topology on M is strong enough then the immersed curves are an open subset of all functions. We moreover represent both curves $c \in M$ and deformations $h \in T_c M$ as functions $S^1 \to \mathbb{R}^n$; this is a special structure that is not usually present in abstract manifolds: so we can easily define "charts" for M.

Proposition 11.1 (Charts in $M_{i,f}$). *Given a curve c, there is a neighborhood U_c of $0 \in T_c M$ such that for $h \in U_c$, the curve $c + h$ is still immersed; then this map $h \mapsto c + h$ is the simplest natural candidate to be a chart of $\Phi_c : U_c \to M$. Indeed, if we pick another curve $\tilde{c} \in M$ and the corresponding $U_{\tilde{c}}$ such that $U_{\tilde{c}} \cap U_c \neq \emptyset$, then the equality $\Phi_c(h) = c + h = \tilde{c} + \tilde{h} = \Phi_{\tilde{c}}(\tilde{h})$ can be solved for h to obtain $h = (\tilde{c} - c) + \tilde{h}$.*

The main goal of this section is to study the manifold

$$B = B_{i,f} := M/\mathrm{Diff}(S^1)$$

of geometric curves. Since B is an abstract object, we will actually work with M in everyday calculus. To recombine the two needs, we will identify an unique family of "small deformations" inside M, which has a specific meaning in B. A common choice is to restrict the family of infinitesimal motions h to those such that $h(\theta)$ is orthogonal to the curve, that is, to $c'(\theta)$.

Proposition 11.2 (Charts in $B_{i,f}$). *Let Π be the projection from M to the quotient B. Let $[c] \in B$: we pick a curve c such that $\Pi(c) = [c]$. We represent the tangent space $T_{[c]}B$ as the space of all $k : S^1 \to \mathbb{R}^n$ such that $k(s)$ is orthogonal to $c'(s)$. We choose $U_{[c]} \subset T_{[c]}B$ a neighborhood of 0; then the chart is defined by*

$$\Psi_{[c]} : U_{[c]} \to B \,, \ \Psi_{[c]}(k) := \Pi(c(\cdot) + k(\cdot))$$

that is, it moves $c(u)$ in direction $k(u)$.

If $U_{[c]}$ is small enough, then this chart is a smooth diffeomorphism; and, if we pick another curve $\tilde{c} \in M$ and the corresponding $U_{[\tilde{c}]}$ such that $U_{[\tilde{c}]} \cap U_{[c]} \neq \emptyset$, then the equality

$$\Psi_{[c]}(k) = \Psi_{[\tilde{c}]}(\tilde{k})$$

can be solved for k (this is not so easy to prove: see [38], or 4.4.7 and 4.6.6 in [24]).

11.2 Reparameterization to Normal Motion

In the preceding proposition we decided to use "orthogonal motion" as distinguished chart for the manifold B. This choice leads also to a "lifting".

Lemma 11.3. *Given any smooth homotopy C of immersed curves, there exists a reparameterization given by a parameterized family of diffeomorphisms $\Phi : [0, 1] \times S^1 \to S^1$, so that setting*

$$\tilde{C}(t, \theta) := C(t, \Phi(t, \theta)) \,;$$

we have that $\partial_t \tilde{C}$ is orthogonal to $\partial_\theta \tilde{C}$ at all points; more precisely,

$$\pi_{\tilde{T}} \partial_t \tilde{C} = 0 \quad , \quad \pi_{\tilde{N}} \partial_t \tilde{C} = \pi_N \partial_t C$$

(where the last equality is up to the reparameterization Φ).

For the proof, see Thm. 3.10 in [64] or Sect. 2.5 in [38].

This explains what is commonly done in the *level set method*, where the tangent part of the flow is discarded.

It is unclear that this choice is actually the best possible choice for *computer vision* application. Consider the following example.

Example 11.4. Suppose that $c(\theta) = (\cos(\theta), \sin(\theta))$ is a planar circle. Let $C(t, \theta) = c(\theta) + vt$ be an uniform translation of c. Let $\tilde{C}$ be as in the previous proposition; Fig. 17 shows the two motions. The two motions C and $\tilde{C}$ coincide in B but are represented differently in M. Which one is best? Both "translation" and "orthogonal motions" seem a natural idea at first glance.

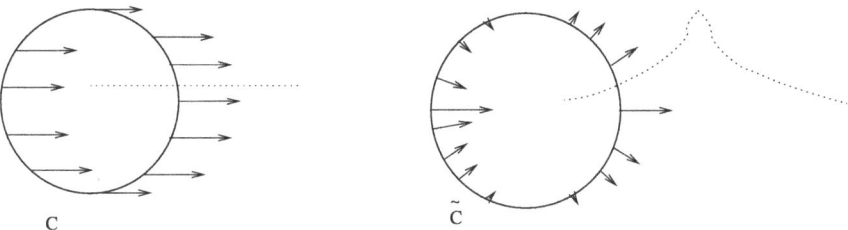

Fig. 17 Translation of a circle as a normal-only motion, by Proposition 11.3. The *dotted line* represents the trajectory of a point on the curve

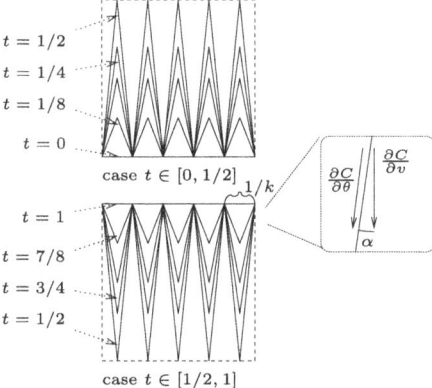

Fig. 18 Zig-zag homotopy

11.3 The H^0 Distance is Degenerate

In Sect. C in [64], or Sect. 3.10 in [38], the following theorem is proved.

Theorem 11.5. *The H^0-induced distance is degenerate: the distance between any two curves vanishes.*

This results is generalized in [37] to L^2-type metrics of submanifolds of any codimension.

Here we will sketch the main idea of the proof for a very simple case: it is possible to connect two segments

$$c_0(u) = (u, 0), \quad c_1(u) = (u, 1)$$

with a family of "zigzag" homotopies $C^k(t, \theta)$ so that the H^0 action is infinitesimal when $k \to \infty$. A snapshot of the curves along the homotopy, for $t = 0, 1/8 \ldots 1$ and $k = 5$, is in Fig. 18.

In the following, let $C = C^k$ for k large. We recall that the H^0 action is

$$\mathbb{E}(C) = \int_0^1 \int_C |\partial_t C|^2 \, ds \, dt = \int_0^1 \int_0^1 |\partial_t C|^2 |\partial_\theta C| \, d\theta \, dt \, ;$$

and we also define

$$\mathbb{E}_\perp(C) = \int_0^1 \int_C |\pi_N \partial_t C|^2 \, ds \, dt = \int_0^1 \int_0^1 |\pi_N \partial_t C|^2 |\partial_\theta C| \, d\theta \, dt \quad (80)$$

(this "action" will be explained in Sect. 11.10). The argument goes as follows.

1. The angle α is (the absolute value of) the angle between $\partial_\theta C$ and the vertical direction (that is also the direction of $\partial_t C$). Note that
   - $\alpha = \alpha(k, t)$, and
   - at $t = 1/2$ it achieves the maximum $\alpha(k, 1/2) = \arctan(1/(2k))$, and also
   - $|\partial_\theta C| = 1/\sin(\alpha)$.
2. Now

$$|\pi_N \partial_t C|^2 |\partial_\theta C| = |\partial_t C|^2 |\partial_\theta C| \sin^2(\alpha) \sim \sin(\alpha)$$

so if k is large, then the angle α is small and then $\mathbb{E}_\perp(C) < \varepsilon \sim 1/k$.
3. We now smooth C just a bit to obtain $\hat{C}$, so that $\mathbb{E}_\perp(\hat{C}) < 2\varepsilon$.
4. We then apply Proposition 11.3 to $\hat{C}$ to finally obtain $\tilde{C}$, that moves only in the normal direction, so

$$\mathbb{E}_\perp(\hat{C}) = \mathbb{E}_\perp(\tilde{C}) = \mathbb{E}(\tilde{C}) < 2\varepsilon$$

as we wanted to show.

11.4 Existence of Critical Geodesics for H^j

In contrast, it is possible to show that a Sobolev-type metric does admit critical geodesics.

Theorem 11.6 (4.3 in Michor and Mumford [39]). *Consider the Sobolev type metric*

$$\langle h, k \rangle_{G_c^n} = \int_c \langle h, k \rangle + \langle D_s^n h, D_s^n k \rangle \, ds$$

with $n \geq 1$. Let $k \geq 2n + 1$; suppose c, h are in the (usual and parametric) H^k Sobolev space (that is, c, h admit a k-th (weak)derivative, which is square integrable). Then it is possible to shoot the geodesic, starting from c and in direction h, for short time.

In Sect. 4.8 in [39] it is suggested that the theorem may similarly hold for Sobolev metrics with length-dependent scale factors.

11.5 Parameterization Invariance

Let M be the manifold of (freely)immersed curves. Let $\langle h_1, h_2 \rangle_c$ be a Riemannian metric, for $h_1, h_2 \in T_c M$; let $\|h\|_c = \sqrt{\langle h, h \rangle_c}$ be its associated norm. Let $\mathbb{E}(\gamma) := \int_0^1 \|\dot\gamma(v)\|^2_{\gamma(v)} \, dv$ be the *action*.

Let $C : [0, 1] \times S^1 \to S^1$ be a smooth homotopy. We recall the definition that we saw in Sect. 4.6, and add a second stronger version.

Definition 11.7. A Riemannian metric is

- **curve-wise parameterization invariant** when the metric does not depend on the parameterization of the curve, that is $\|\tilde h\|_{\tilde c} = \|h\|_c$ when $\tilde c(t) = c(\varphi(t))$ and $\tilde h(t) = h(\varphi(t))$;
- **homotopy-wise parameterization invariant** if, for any $\varphi : [0, 1] \times S^1 \to S^1$ smooth, $\varphi(t, \cdot)$ a diffeomorphism of S^1 for all fixed t, let $\tilde C(t, \theta) = C(t, \varphi(t, \theta))$, then $\mathbb{E}(\tilde C) = \mathbb{E}(C)$.

It is not difficult to prove that the second condition implies the first. There is an important remark to note: a homotopy-wise-parameterization-invariant Riemannian metric cannot be a proper metric.

Proposition 11.8. *Suppose that a Riemannian metric is* homotopy-wise parameterization invariant; *if $h_1, h_2 \in T_c M$ and h_1 is tangent to c at all points, then $\langle h_1, h_2 \rangle_c = 0$. So the Riemannian metric in this case is actually a* **semimetric** *(and $\|\cdot\|_c$ is not a norm, but rather a* seminorm *in $T_c M$).*

Proof. Let $C(t, u) = c(\varphi(t, u))$ with $\varphi(t, \cdot)$ be a time-varying family of diffeomorphisms of S^1. So $\mathbb{E}(C) = \mathbb{E}(c) = 0$, that is

$$\int_0^1 \|c' \partial_t \phi\|^2_c \, dt = 0$$

so $\|c' \partial_t \phi\|_c = 0$; but note that, by choosing ϕ appropriately, we may represent any $h \in T_c M$ that is tangent to c as $h = c' \partial_t \phi$. By polarization, we obtain the assertion.

11.6 Standard and Geometric Distance

The **standard distance** $d(c_0, c_1)$ in M is the infimum of $\text{Len}(\gamma)$ where γ is any homotopy that connects $c_0, c_1 \in M$ (cf. Definition 3.25).

We are though interested in studying metrics and distances in the quotient space $B := M/\text{Diff}(S^1)$.

We suppose that the metric G is *curve-wise parameterization invariant*; then G may be projected to $B := M/\text{Diff}(S^1)$. So in the following we will use a different definition of "geometric distance".

Definition 11.9 (geometric distance). $d_G(c_0, c_1)$ is the infimum of the length $\text{Len}(C)$ in the class of all homotopies C connecting the curve c_0 to any reparameterization $c_1 \circ \phi$ of c_1.

This implements the quotienting formula that we saw in Definition 17 (in this case, the group is $\mathcal{G} = \text{Diff}(S^1)$): so we will use d_G as a distance on B.

At the same time, note that in writing $d_G(c_0, c_1)$ we are abusing notation: d_G is not a distance in the space M, but rather it is a *semidistance*, since the distance between c and a reparameterization $c \circ \phi$ is zero.

11.7 Horizontal and Vertical Space

Consider a metric G (curve-wise parameterization invariant) on M. Let Π once again be the projection from $M := \text{Imm}_f(S^1)$ to the quotient $B = B_{i,f} = M/\text{Diff}(S^1)$.

We present a list of definitions (see also the Fig. 19).

Definition 11.10.
- The **orbit** is $O_c = [c] = \{c \circ \phi \mid \phi\} = \Pi^{-1}(\{c\})$.
- The **vertical space** V_c is the tangent to O_c:

$$V_c := T_c O_c \subset T_c M$$

that can be explicitly written as

$$V_c := \{h = b(s)c'(s) \mid b : S^1 \to \mathbb{R}\}$$

i.e. all the vector fields h where $h(s)$ is tangent to c.
- The **horizontal space** W_c is the orthogonal complement inside $T_c M$

$$W_c := V_c^\perp .$$

Note that W_c depends on G, but V_c does not.

Figure 20 may also help in understanding. The whole orbit O_c is projected to $[c]$. The vertical space V_c is the kernel of $D\Pi_c$, so it is projected to 0.

Fig. 19 The action of Diff(S^1) on M. The orbits O_c are *dotted*, the spaces W_c and V_c are *dashed*

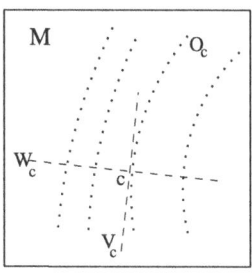

Fig. 20 The vertical and horizontal spaces, and the projection Π from M to B

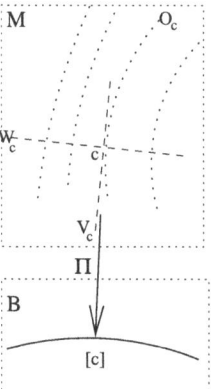

11.8 From Curve-Wise Parameterization Invariant to Homotopy-Wise Parameterization Invariant

In the following sections, we just aim to give an overview of the theory; we will not dwell in depicting the most general hypotheses such that all presented results do hold for a generic metric G; we will though, case by case, present references to how the results can be proven for the metrics discussed in this paper, and in particular the Sobolev-type metrics.

We suppose that this theorem holds.

Theorem 11.11. *Given any $h \in T_c M$, there is an unique minimum point for*

$$\min_{x \in W_c} \|x - h\|_G$$

*and the minimum is called the **horizontal projection** of h.*

This theorem is verified when G is a Sobolev-type metrics, see Sect. 4.5 in [39]; and it is trivially verified by H^0.

Definition 11.12. The **horizontally projected metric** $G^\perp$ is defined by

$$\langle h, k \rangle_{G^\perp, c} := \langle \tilde{h}, \tilde{k} \rangle_{G, c} \tag{81}$$

where $\tilde h, \tilde k$ are the projections of h, k to W_c.

Proposition 11.13. *Equivalently the norm $\|h\|_{G^\perp}$ can be defined by*

$$\|h\|_{G^\perp} = \inf_b \|h + bc'\|_G \tag{82}$$

where the infimum is in the class of smooth $b : S^1 \to \mathbb{R}$.

Proof. Indeed by the projection theorem,

$$\inf_b \|h + bc'\|_G = \|\tilde h\|_G$$

where $\tilde h$ is the projections of h to W_c. By polarization we obtain (81).

It is easy to prove that $G^\perp$ is *curve-wise parameterization invariant*, but moreover

Proposition 11.14. $G^\perp$ *is* homotopy-wise parameterization invariant.

Proof. Let $\tilde C(t, \theta) = C(t, \varphi(t, \theta))$, then

$$\partial_t \tilde C = \partial_t C + C'\varphi'$$

so at any given time t

$$\|\partial_t \tilde C(t, \cdot)\|_{G^\perp} = \|\partial_t C + C'\varphi'\|_{G^\perp} = \|\partial_t C\|_{G^\perp}$$

by (82), *where the terms RHS are evaluated at* $(t, \varphi(t, \cdot))$; *since $G^\perp$ is* curve-wise parameterization invariant, *then*

$$\|\partial_t \tilde C(t, \cdot)\|_{G^\perp} = \|\partial_t C(t, \cdot)\|_{G^\perp}.$$

Consequently,

Corollary 11.15. *G is homotopy-wise parameterization invariant if and only if*

$$\|h\|_G = \|h + bc'\|_G$$

for all b.

Summarizing all above theory, we obtain a method to understand/study/design the metric G on B, following these steps:

1. choose a metric G on M that is curve-wise parameterization invariant
2. G generates the horizontal space $W = V^\perp$
3. project G on W to define $G^\perp$ that is homotopy-wise parameterization invariant.

Horizontal $G^\perp$ as Length Minimizer

As we defined in Definition 11.9, to define the distance we minimize the length of the paths $\gamma : [0, 1] \to M$ that connect a curve c_0 to a reparameterization $c_1 \circ \phi$ of the curve c_1. Consider the following sketchy example.

Example 11.16. Consider two paths γ_1 and γ_2 connecting c_0 to reparameterizations of c_1. Recall that the space W_c is orthogonal to the orbits, the space V_c is tangent to the orbits. The path γ_1 (that moves in some tracts along the orbits, with $\dot\gamma_1 \in V_\gamma$) is longer than the path γ_2.

The above example explains the following lemma.

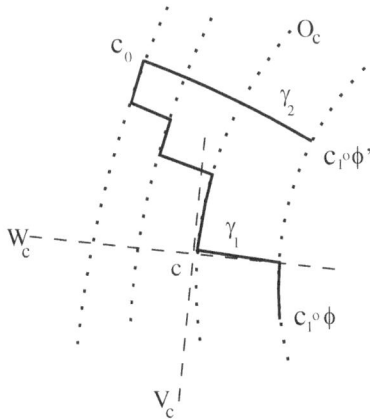

Lemma 11.17. *Let us choose G to be a curve-wise parameterization invariant metric (so, by Proposition 11.8, it cannot be homotopy-wise parameterization invariant).*

1. *Let $\tilde{C}(t, \theta) = C(t, \varphi(t, \theta))$ where $\varphi(t, \cdot)$ is a diffeomorphism for all fixed t. Minimize*

$$\min_\varphi \mathbb{E}_G(\tilde{C})$$

Then at the minimum $\tilde{C}^*$, $\partial_t \tilde{C}^*$ is horizontal at every point, and

$$\mathbb{E}_G(\tilde{C}^*) = \mathbb{E}_{G^\perp}(\tilde{C}^*).$$

2. *So the distance can be equivalently computed as the infimum of $\mathrm{Len}_{G^\perp}$ or of Len_G; and distances are equal, $d_{G^\perp} = d_G$.*

In practice, when computing minimal geodesic (possibly by numerical methods), it is usually best to minimize $\mathbb{E}_G$, since it also penalizes the vertical part of the motion, and hence the minimization will produce a "smoother" geodesic.

The proof of the above lemma is based on the validity of *the lifting Lemma*:

Lemma 11.18 (lifting Lemma). *Given any smooth homotopy C of immersed curves, there exists a reparameterization given by a parameterized family of diffeomorphisms $\Phi : [0, 1] \times S^1 \to S^1$, so that setting*

$$\tilde{C}(t, \theta) := C(t, \Phi(t, \theta))$$

we have that $\partial_t \tilde{C}$ is in the horizontal space W_C at all times t.

For the metric H^0, this reduces to Lemma 11.3 in this paper; for Sobolev-type metrics, the proof is in Sect. 4.6 in [39].

11.9 A Geometric Gradient Flow is Horizontal

Proposition 11.19. *If E is a geometric energy of curves, in the sense that E is invariant with respect to reparameterizations $\phi \in \mathrm{Diff}^+(S^1)$; then ∇E is horizontal.*

Proof. Let c be a smooth curve; suppose for simplicity (and with no loss of generality) that it has $\mathrm{len}(c) = 2\pi$ and that it is parameterized by arc parameter, so that $|c'| \equiv 1$ and $c' = D_s c$. Let $b : S^1 \to \mathbb{R}$ be smooth, and define $\phi : \mathbb{R} \times S^1 \to S^1$ by solving the b flow, that is the ODE

$$\partial_t \phi(t, \theta) = b(\phi(t, \theta))$$

with initial data $\phi(0, \theta) = \theta$. Note that, for all t, $\phi(t, \cdot) \in \mathrm{Diff}^+(S^1)$. Let $C(t, \theta) = c(\phi(t, \theta))$, note that

$$\partial_t C(t, \theta) = b(\phi(t, \theta)) c'(\phi(t, \theta))$$

then $E(C) = E(c)$, so that (deriving and setting $t = 0$)

$$\partial_t E(C) = 0 = \langle \nabla E(c), bc' \rangle$$

and we conclude by arbitrariness of b.

11.10 Horizontality According to H^0

Proposition 11.20. *The horizontal space W_c w.r.to H^0 is*

$$W_c := \{h : h(s) \perp c'(s) \forall s\}$$

that is the space of vector fields orthogonal to the curve.

Metrics of Curves in Shape Optimization and Analysis

So Proposition 11.19 guarantees that the H^0 gradients of geometric energies have only a normal component with respect to the curve—and this couples well with *level set methods*.

The horizontal version of H^0 is

$$\langle h_1, h_2 \rangle_{H^{0,\perp}} := \int_c \pi_N h_1 \cdot \pi_N h_2 \, ds \tag{83}$$

where $\pi_N h_1(s)$ is the component of $h_1(s)$ that is orthogonal to c' (as was defined in Sect. 7.1); note that the action of this (semi)metric was already presented in (80).

So a good paradigm is to think at the metric H^0 as

$$\int_c h_1 \cdot h_2 \, ds \quad \text{on } M,$$
$$\int_c \pi_N h_1 \cdot \pi_N h_2 \, ds \quad \text{on } B.$$

Remark 11.21. The horizontal space W_c is isometric (thru $D\Pi_c$) with the tangent space $T_{[c]}B$: this implies that we may decide to use W_c and Π_c as a "local chart" for B. If the metric is $G = H^0$, this brings us back the chart 11.2; with other metrics, this process instead defines a novel choice of chart.

The same reasoning above holds for the H^A metric (from Sect. 9.2), and for the Finsler metrics F^1 and F^∞ (from Sect. 7); in this latter case, we already presented the horizontally projected version of the metrics.

For all above reasons, it is common thinking that *a good movement of a curve is a movement that is orthogonal to the curve*. But, when using a different metric (as for example, the Sobolev-type metrics) the above is not true anymore.

11.11 Horizontality According to H^j

Horizontality according to H^j is a complex matter. To study the horizontally projected metric, we need to express the solution of

$$\inf_b \|h + bc'\|_{H^j}$$

in closed form, and this needs (pseudo)differential operators: see [39] for details. We just remark that in this case, *a good movement of a curve is **not necessarily** a movement that is orthogonal to the curve.*

11.12 Horizontality is for Any Group Action

In the above we studied the horizontality properties of the quotient $B = M/\text{Diff}(S^1)$. But the theory works in general for any other quotient. So we may apply the same study and ideas to further quotients (such as $B/E(n)$).

11.13 Momenta

What follows comes from Sect. 2.5 in [39].

Consider again the action of a group $\mathcal{G}$ on a Riemannian manifold M (as defined in Sect. 3.8). Most groups that act on curves are smooth differentiable manifolds themselves, and the group operations are smooth: so they are **Lie groups**. In our cases of interest, the actions are smooth as well.

We want to study some differentiable properties of the action $g \cdot m$; using the rule (16), we see that we can restrict to studying the case $e \cdot m$, where $e \in \mathcal{G}$ is the identity element.

Conservation of Momenta

Given a curve $c \in M$, and a tangent vector $\xi \in T_e\mathcal{G}$ (where $T_e\mathcal{G}$ is the **Lie algebra** of $\mathcal{G}$), we derive the action, for fixed c and e "moving" in direction ξ; the result of this derivative is a $\zeta = \zeta_{\xi,c} \in T_c M$ (depending linearly on ξ). This direction ζ is intuitively *the infinitesimal motion of c, when we infinitesimally act in direction ξ on c*.

To exemplify the above process, we provide a simple toy example.

Example 11.22 (Rotation action on the plane). The group of 2D rotations $\mathcal{G} = \mathbb{R}/(2\pi)$ (the real line modulus 2π, with the $+$ operation) acts on the plane, as follows

$$\mathcal{G} \times \mathbb{R}^2 \to \mathbb{R}^2$$
$$g, m \mapsto g \cdot m = R_g m$$

with

$$R_g := \begin{pmatrix} \cos g & -\sin g \\ \sin g & \cos g \end{pmatrix}.$$

Since $T_e\mathcal{G} = \mathbb{R}$, there is only one direction in the tangent to $\mathcal{G}$, hence the derivative in direction $\xi = 1$ is just the standard derivative d/dg; similarly $T\mathbb{R}^2 = \mathbb{R}^2$; by deriving the action in point $g = e = 0$ (the identity element), we obtain that

$$\zeta_{1,m} = Jm$$

with

$$J := \begin{pmatrix} 0 & -1 \\ 1 & 0 \end{pmatrix}.$$

Note that Jm is the vector (normal to m) obtained by rotating m of an angle $\pi/2$ counterclockwise. The physical interpretation is that, when rotating the plane in counterclockwise direction with speed $\xi = 1$, each point m will move with velocity $\zeta = Jm$.

In the above toy example, the result ζ is vector; but in the case of immersed curves M, since $\zeta \in T_c M$, then $\zeta = \zeta(\theta) : S^1 \to \mathbb{R}^n$ is a vector field.

We then recall a very interesting condition by Emmy Noether

Theorem 11.23. *Suppose that the Riemannian metric $\langle \cdot, \cdot \rangle_c$ on M is invariant w.r.to the action of $\mathcal{G}$: this is equivalent to saying that the action is an isometry. Let $\gamma(t)$ be a critical geodesic: then*

$$\left\langle \zeta_{\xi, \gamma(t)}, \dot{\gamma}(t) \right\rangle_{\gamma(t)} \tag{84}$$

is constant in t (for any choice of $\xi \in T_e \mathcal{G}$).

The quantity (84) is called "the **momentum** of the action".

As an alternative interpretation, note that the vectors ζ that are obtained by deriving the action are exactly all the vectors in the vertical spaces V_c of the corresponding action $\mathcal{G}$ (since ζ are infinitesimal motions inside the orbit). Recall that $h \in T_c M$ is horizontal (that is, $h \in W_c$) iff it is orthogonal to V_c $\langle \zeta, h' \rangle = 0$ for all $\zeta \in V_c$. So, as corollary of Emmy Noether's theorem, we obtain that

Corollary 11.24. *If a geodesic is shot in a horizontal direction $\dot{\gamma}(0)$, then the geodesic will be horizontal for all subsequent times.*

The following are examples of momenta that are related to the actions on curves (that we saw in Example 4.8).

Example 11.25. • The *rescaling* group is represented by $\mathbb{R}^+$, that is one dimensional, so there is only one tangent direction $\xi = 1$ in $\mathbb{R}^+$. The action is $l, c \mapsto lc$, so there is only one direction, that is $\zeta = c$.
- The *translation* group is represented by $\mathbb{R}^n$, that is a vector space, so the tangent directions are $\xi \in \mathbb{R}^n$; the action is $\xi, c \mapsto \xi + c$; then $\zeta = \xi$ (and note that this is constant in θ).
- The *rotation* group is represented by orthogonal matrixes, so we set $\mathcal{G} = O(n)$; the group identity e is the identity matrix; the action is matrix-vector multiplication $A, c(\theta) \mapsto Ac(\theta)$; the tangent $T_e \mathcal{G}$ is the set of the antisymmetric matrixes $B \in \mathbb{R}^{n \times n}$; then $\zeta = Bc$.
- The *reparameterization* group is $\mathcal{G} = \text{Diff}(S^1)$; a tangent vector is a scalar field $\xi : S^1 \to \mathbb{R}$; the action is the composition $\phi, c \mapsto c \circ \phi$; we in the end have that

$$\zeta(\theta) = \xi(\theta) c'(\theta)$$

(where $c' = D_\theta c$) that is, ζ is a generic vector field parallel to the curve.

For the $\tilde{H}^1$ metric, this implies the following properties of a geodesic path $\gamma(t)$.

Proposition 11.26. • *Scaling momentum:*

$$\langle \gamma, \dot\gamma \rangle_{\tilde H^1} = \mathrm{avg}_c(\gamma) \cdot \mathrm{avg}_c(\dot\gamma) + \lambda L^2 \oint D_s\gamma \cdot D_s\dot\gamma \, \mathrm{d}s = \text{constant.}$$

- *Linear momentum:* $\mathrm{avg}_c(\dot\gamma)$ *is constant, since, for all* $\xi \in \mathbb{R}^n$,

$$\langle \xi, \dot\gamma \rangle_{\tilde H^1} = \xi \cdot \mathrm{avg}_c(\dot\gamma) = \text{constant;}$$

since ξ is arbitrary, this means that $\mathrm{avg}_c(\dot\gamma)$ is constant in t.
- *Angular momentum: for any antisymmetric matrix $B \in \mathbb{R}^{n\times n}$,*

$$\langle B\gamma, \dot\gamma \rangle_{\tilde H^1} = (B\mathrm{avg}_c(\gamma)) \cdot \mathrm{avg}_c(\dot\gamma) + \lambda L^2 \oint (BD_s\gamma) \cdot (D_s\dot\gamma) \, \mathrm{d}s = \text{constant.}$$

- *Reparameterization momentum: for any scalar field $\xi : S^1 \to \mathbb{R}$, setting*

$$\zeta(\theta, t) = \xi(\theta)\gamma'(\theta, t)$$

we get

$$\langle \zeta, \dot\gamma \rangle_{\tilde H^1} = \mathrm{avg}_c(\xi\gamma') \cdot \mathrm{avg}_c(\dot\gamma) + \lambda L^2 \oint D_s(\xi\gamma') \cdot D_s\dot\gamma = \text{constant;}$$

integrating by parts,

$$\mathrm{avg}_c(\xi\gamma') \cdot \mathrm{avg}_c(\dot\gamma) - \lambda L^2 \oint (\xi\gamma') \cdot D_{ss}\dot\gamma = \text{constant;}$$

since ξ is arbitrary, this means that

$$\gamma' \cdot \mathrm{avg}_c(\dot\gamma) - \lambda L^2 \gamma' \cdot D_{ss}\dot\gamma$$

is constant in t, for any $\theta \in S^1$.

References

1. L. Ambrosio, G. Da Prato, A.C.G. Mennucci, An introduction to measure theory and probability, 2007. http//dida.sns.it/dida2/cl/07-08/folde2/pdf0
2. T.M. Apostol, *Mathematical Analysis* (Addison Wesley, Reading, 1974)
3. C.J. Atkin, The Hopf-Rinow theorem is false in infinite dimensions. Bull. Lond. Math. Soc. 7(3), 261–266 (1975) doi: 10.1112/blms/7.3.261
4. H. Brezis, *Analisi Funzionale* (Liguori Editore, Napoli, 1986). Italian translation of *Analyse fonctionelle* (Masson, Paris, 1983)

5. J. Canny, A computational approach to edge detection. IEEE Trans. Pattern Anal. Mach. Intell. **8**(6), 679–698 (1986) ISSN 0162-8828.
6. V. Caselles, F. Catte, T. Coll, F. Dibos, A geometric model for edge detection. Num. Mathematik **66**, 1–31 (1993)
7. V. Caselles, R. Kimmel, G. Sapiro, Geodesic active contours, in *Proceedings of the IEEE International Conference on Computer Vision*, Cambridge, MA, June 1995, pp. 694–699
8. T. Chan, L. Vese, Active contours without edges. IEEE Trans. Image Process. **10**(2), 266–277 (2001)
9. G. Charpiat, O. Faugeras, R. Keriven, Approximations of shape metrics and application to shape warping and empirical shape statistics. Found. Comput. Math. (2004) doi: 10.1007/s10208-003-0094-xgg819. INRIA report 4820 (2003)
10. G. Charpiat, R. Keriven, J.P. Pons, O. Faugeras, Designing spatially coherent minimizing flows for variational problems based on active contours, in *ICCV* (2005). doi: 10.1109/ICCV.2005.69
11. G. Charpiat, P. Maurel, J.-P. Pons, R. Keriven, O. Faugeras, Generalized gradients: Priors on minimization flows. Int. J. Comp. Vis. (2007). doi: 10.1007/s11263-006-9966-2
12. Y. Chen, H. Tagare, S. Thiruvenkadam, F. Huang, D. Wilson, K. Gopinath, R. Briggs, E. Geiser, Using prior shapes in geometric active contours in a variational framework. Int. J. Comp. Vis. **50**(3), 315–328 (2002)
13. D. Cremers, S. Soatto, A pseuso distance for shape priors in level set segmentation, in *2nd IEEE Workshop on Variational, Geometric and Level Set Methods in Computer Vision*, Nice, ed. by N. Paragios, Oct 2003, pp. 169–176
14. M.P. do Carmo, *Riemannian Geometry* (Birkhäuser, Boston, 1992)
15. A. Duci, A.C.G. Mennucci, Banach-like metrics and metrics of compact sets . arXiv preprint arXiv:0707.1174 (2007)
16. A. Duci, A.J. Yezzi, S.K. Mitter, S.Soatto, Shape representation via harmonic embedding, in *International Conference on Computer Vision (ICCV03)*, vol. 1, Washington, DC, pp. 656–662 (IEEE Computer Society, Silver Spring, 2003). ISBN 0-7695-1950-4. doi: 10.1109/ICCV.2003.1238410
17. A. Duci, A.J. Yezzi, S. Soatto, K. Rocha, Harmonic embeddings for linear shape. J. Math Imag. Vis **25**, 341–352 (2006). doi: 10.1007/s10851-006-7249-8
18. J. Eells, K.D. Elworthy, Open embeddings of certain Banach manifolds. Ann. Math. (2) **91**, 465–485 (1970)
19. I. Ekeland, The Hopf-Rinow theorem in infinite dimension. J. Differ. Geom. **13**(2), 287–301 (1978)
20. A.T. Fomenko, *The Plateau Problem*. Studies in the Development of Modern Mathematics (Gordon and Breach, New York, 1990)
21. M. Gage, R.S. Hamilton, The heat equation shrinking convex plane curves. J. Differ. Geom. **23**, 69–96 (1986)
22. J. Glaunès, A. Trouvé, L. Younes, Modeling planar shape variation via Hamiltonian flows of curves, in *Analysis and Statistics of Shapes*, ed. by A. Yezzi, H. Krim. Modeling and Simulation in Science, Engineering and Technology, chapter 14 (Birkhäuser, Basel, 2005)
23. M. Grayson, The heat equation shrinks embedded planes curves to round points. J. Differ. Geom. **26**, 285–314 (1987)
24. R.S. Hamilton, The inverse function theorem of Nash and Moser. Bull. Am. Math. Soc. (N.S.) **7**(1), 65–222 (1982) ISSN 0273-0979.
25. J. Itoh, M. Tanaka, The Lipschitz continuity of the distance function to the cut locus. Trans. A.M.S. **353**(1), 21–40 (2000)
26. H. Karcher, Riemannian center of mass and mollifier smoothing. Comm. Pure Appl. Math. **30**, 509–541 (1977)
27. M. Kass, A. Witkin, D. Terzopoulos, Snakes: active contour models. Int. J. Comp. Vis. **1**, 321–331 (1987)
28. S. Kichenassamy, A. Kumar, P. Olver, A. Tannenbaum, A. Yezzi, Gradient flows and geometric active contour models, in *Proceedings of the IEEE International Conference on Computer Vision*, Cambridge, MA (1995), pp. 810–815. doi:10.1109/ICCV.1995.466855

29. E. Klassen, A. Srivastava, W. Mio, S.H. Joshi, Analysis of planar shapes using geodesic paths on shape spaces. IEEE Trans. Pattern Anal. Mach. Intell. **26**, 372–383 (2004). ISSN 0162-8828. doi: 10.1109/TPAMI.2004.1262333
30. W. Klingenberg, *Riemannian Geometry* (W. de Gruyter, Berlin, 1982)
31. S. Lang, *Fundamentals of Differential Geometry* (Springer, Berlin, 1999)
32. M. Leventon, E. Grimson, O. Faugeras, Statistical shape influence in geodesic active contours, *IEEE Conference on Computer Vision and Pattern Recognition*, Hilton Head Island, SC, vol. 1 (2000), pp. 316–323. doi:10.1109/CVPR.2000.855835
33. Y. Li, L. Nirenberg, The distance function to the boundary, Finsler geometry and the singular set of viscosity solutions of some Hamilton-Jacobi equations. Comm. Pure Appl. Math. **58**, 85–146 (2005)
34. R. Malladi, J. Sethian, B. Vemuri, Shape modeling with front propagation: a level set approach. IEEE Trans. Pattern Anal. Mach. Intell. **17**, 158–175 (1995)
35. A.C.G. Mennucci, A. Yezzi, G. Sundaramoorthi, Properties of Sobolev Active Contours. Interf. Free Bound. **10**, 423–445 (2008)
36. A.C.G. Mennucci, On asymmetric distances, 2nd version, preprint, 2004. http://cvgmt.sns.it/papers/and04/
37. P.W. Michor, D. Mumford, Vanishing geodesic distance on spaces of submanifolds and diffeomorphisms. Documenta Math. **10**, 217–245 (2005). http://www.univie.ac.at/EMIS/journals/DMJDMV/vol-10/05.pdf
38. P.W. Michor, D. Mumford, Riemannian geometris of space of plane curves. J. Eur. Math. Soc. (JEMS) **8**, 1–48 (2006)
39. P.W. Michor, D. Mumford, An overview of the Riemannian metrics on spaces of curves using the Hamiltonian approach. Appl. Comput. Harmonic Anal. **23**, 76–113 (2007). doi: 10.1016/j.acha.2006.07.004. http://www.mat.univie.ac.at/~michor/curves-hamiltonian.pdf
40. W. Mio, A. Srivastava, Elastic-string models for representation and analysis of planar shapes, in *Proceedings of the 2004 IEEE Computer Society Conference on Computer Vision and Pattern Recognition*, 2004 (CVPR 2004), vol. 2 (2004), pp. 10–15. doi:10.1109/CVPR.2004.1315138
41. W. Mio, A. Srivastava, Elastic-string models for representation and analysis of planar shapes, in *Conference on Computer Vision and Pattern Recognition (CVPR)*, June 2004. http://stat.fsu.edu/~anuj/pdf/papers/CVPR_Paper_04.pdf
42. D. Mumford, J. Shah, Boundary detection by minimizing functionals, in *Proceedings CVPR 85: IEEE Computer Society Conference on Computer Vision and Pattern Recognition*, June 19–23, 1985, San Francisco, CA, 1985
43. D. Mumford, J. Shah, Optimal approximations by piecewise smooth functions and associated variational problems. Comm. Pure Appl. Math. **42**, 577–685 (1989)
44. S. Osher, J. Sethian, Fronts propagating with curvature-dependent speed: algorithms based on the Hamilton-Jacobi equations. J. Comp. Phys. **79**, 12–49 (1988)
45. T.R. Raviv, N. Kiryati, N. Sochen, Unlevel-set: geometry and prior-based segmentation, in *Proceedings of European Conference on Computer Vision*, 2004 *(Computer Vision-ECCV 2004)*, ed. by T. Pajdla (Springer, Berlin, 2004). http://dx.doi.org/10.1007/978-3-540-24673-2_5
46. R. Ronfard, Region based strategies for active contour models. Int. J. Comp. Vis. **13**(2), 229–251 (1994). http://perception.inrialpes.fr/Publications/1994/Ron94
47. M. Rousson, N. Paragios, Shape priors for level set representations, in *Proceedings of the European Conference on Computer Vision*, vol. 2 (2002), pp. 78–93
48. W. Rudin, *Functional Analysis* (McGraw-Hill, New York, 1973)
49. W. Rudin, *Real and Complex Analysis* (McGraw-Hill, New York, 1987)
50. J. Shah, H^0 type Riemannian metrics on the space of planar curves. Q. Appl. Math. **66**, 123–137 (2008)
51. S. Soatto, A.J. Yezzi, DEFORMOTION: deforming motion, shape average and the joint registration and segmentation of images. ECCV (3), 32–57 (2002)
52. A. Srivastava, S.H. Joshi, W. Mio, X. Liu, Statistical shape analysis: clustering, learning, and testing. IEEE Trans. Pattern Anal. Mach. Intell. **27**, 590–602 (2005). ISSN 0162-8828. doi: 10.1109/TPAMI.2005.86

53. G. Sundaramoorthi, A. Yezzi, A.C.G. Mennucci, Sobolev active contours, in *VLSM*, ed. by N. Paragios, O.D. Faugeras, T. Chan, C. Schnörr. Lecture Notes in Computer Science, vol. 3752 (Springer, Berlin, 2005), pp. 109–120. ISBN 3-540-29348-5. doi: 10.1007/11567646_10
54. G. Sundaramoorthi, J.D. Jackson, A. Yezzi, A.C.G. Mennucci, Tracking with Sobolev active contours, in *Conference on Computer Vision and Pattern Recognition (CVPR06)* (IEEE Computer Society, Silver Spring, 2006). ISBN 0-7695-2372-2. doi: 10.1109/CVPR.2006.314
55. G. Sundaramoorthi, A. Yezzi, A.C.G. Mennucci, G. Sapiro, New possibilities with Sobolev active contours, in *Scale Space Variational Methods 07* (2007). http://ssvm07.ciram.unibo.it/ssvm07_public/index.html. "Best Numerical Paper-Project Award"; also [58]
56. G. Sundaramoorthi, A. Yezzi, A.C.G. Mennucci, Sobolev active contours. Int. J. Comp. Vis. (2007). doi: 10.1007/s11263-006-0635-2
57. G. Sundaramoorthi, A. Yezzi, A.C.G. Mennucci, Coarse-to-fine segmentation and tracking using Sobolev Active Contours. IEEE Trans. Pattern Anal. Mach. Intell. (TPAMI) (2008). doi: 10.1109/TPAMI.2007.70751
58. G. Sundaramoorthi, A. Yezzi, A.C.G. Mennucci, G. Sapiro, New possibilities with Sobolev active contours. Int. J. Comp. Vis. (2008). doi: 10.1007/s11263-008-0133-9
59. A. Trouvé, L. Younes, Local geometry of deformable templates. SIAM J. Math. Anal. **37**(1), 17–59 (electronic) (2005). ISSN 0036-1410
60. A. Tsai, A. Yezzi, W. Wells, C. Tempany, D. Tucker, A. Fan, E. Grimson, A. Willsky, Model-based curve evolution technique for image segmentation, in *Proceedings of the 2001 IEEE Computer Society Conference on Computer Vision and Pattern Recognition, 2001* (CVPR 2001), vol. 1, Dec 2001, pp. I-463, I-468 . doi: 10.1109/CVPR.2001.990511
61. A. Tsai, A. Yezzi, A.S. Willsky, Curve evolution implementation of the mumford-shah functional for image segmentation, denoising, interpolation, and magnification. IEEE Trans. Image Process. **10**(8), 1169–1186 (2001)
62. L.A. Vese, T.F. Chan, A multiphase level set framework for image segmentation using the mumford and shah model. Int. J. Comp. Vis. **50**(3), 271–293 (2002)
63. A. Yezzi, A.C.G. Mennucci, Geodesic homotopies, in *EUSIPCO04* (2004). http://www.eurasip.org/content/Eusipco/2004/defevent/papers/cr1925.pdf
64. A. Yezzi, A.C.G. Mennucci, Metrics in the space of curves. arXiv (2004)
65. A. Yezzi, A.C.G. Mennucci, Conformal metrics and true "gradient flows" for curves, in *International Conference on Computer Vision (ICCV05)* (2005), pp. 913–919. doi: 10.1109/ICCV.2005.60. URL http://research.microsoft.com/iccv2005/
66. A. Yezzi, A. Tsai, A. Willsky, A statistical approach to snakes for bimodal and trimodal imagery, in *The Proceedings of the Seventh IEEE International Conference on Computer Vision, 1999*, vol. 2, October 1999, pp. 898, 903. doi:10.1109/ICCV.1999.790317
67. L. Younes, Computable elastic distances between shapes. SIAM J. Appl. Math. **58**(2), 565–586 (1998). doi: 10.1137/S0036139995287685
68. L. Younes, P.W. Michor, J. Shah, D. Mumford, A metric on shape space with explicit geodesics. Atti Accad. Naz. Lincei Cl. Sci. Fis. Mat. Natur. Rend. Lincei (9) Mat. Appl. **19**(1), 25–57 (2008). ISSN 1120-6330. doi: 10.4171/RLM/506
69. C.T. Zahn, R.Z. Roskies, Fourier descriptors for plane closed curves. IEEE Trans. Comput. **21**(3), 269–281 (1972). ISSN 0018-9340. doi: 10.1109/TC.1972.5008949
70. S.C. Zhu, T.S. Lee, A.L. Yuille, Region competition: Unifying snakes, region growing, energy/bayes/MDL for multi-band image segmentation, in *ICCV* (1995), p. 416. citeseer.ist.psu.edu/zhu95region.html

LECTURE NOTES IN MATHEMATICS

Edited by J.-M. Morel, B. Teissier; P.K. Maini

Editorial Policy (for Multi-Author Publications: Summer Schools / Intensive Courses)

1. Lecture Notes aim to report new developments in all areas of mathematics and their applications - quickly, informally and at a high level. Mathematical texts analysing new developments in modelling and numerical simulation are welcome. Manuscripts should be reasonably selfcontained and rounded off. Thus they may, and often will, present not only results of the author but also related work by other people. They should provide sufficient motivation, examples and applications. There should also be an introduction making the text comprehensible to a wider audience. This clearly distinguishes Lecture Notes from journal articles or technical reports which normally are very concise. Articles intended for a journal but too long to be accepted by most journals, usually do not have this "lecture notes" character.

2. In general SUMMER SCHOOLS and other similar INTENSIVE COURSES are held to present mathematical topics that are close to the frontiers of recent research to an audience at the beginning or intermediate graduate level, who may want to continue with this area of work, for a thesis or later. This makes demands on the didactic aspects of the presentation. Because the subjects of such schools are advanced, there often exists no textbook, and so ideally, the publication resulting from such a school could be a first approximation to such a textbook. Usually several authors are involved in the writing, so it is not always simple to obtain a unified approach to the presentation.

   For prospective publication in LNM, the resulting manuscript should not be just a collection of course notes, each of which has been developed by an individual author with little or no coordination with the others, and with little or no common concept. The subject matter should dictate the structure of the book, and the authorship of each part or chapter should take secondary importance. Of course the choice of authors is crucial to the quality of the material at the school and in the book, and the intention here is not to belittle their impact, but simply to say that the book should be planned to be written by these authors jointly, and not just assembled as a result of what these authors happen to submit.

   This represents considerable preparatory work (as it is imperative to ensure that the authors know these criteria before they invest work on a manuscript), and also considerable editing work afterwards, to get the book into final shape. Still it is the form that holds the most promise of a successful book that will be used by its intended audience, rather than yet another volume of proceedings for the library shelf.

3. Manuscripts should be submitted either online at www.editorialmanager.com/lnm/ to Springer's mathematics editorial, or to one of the series editors. Volume editors are expected to arrange for the refereeing, to the usual scientific standards, of the individual contributions. If the resulting reports can be forwarded to us (series editors or Springer) this is very helpful. If no reports are forwarded or if other questions remain unclear in respect of homogeneity etc, the series editors may wish to consult external referees for an overall evaluation of the volume. A final decision to publish can be made only on the basis of the complete manuscript; however a preliminary decision can be based on a pre-final or incomplete manuscript. The strict minimum amount of material that will be considered should include a detailed outline describing the planned contents of each chapter.

   Volume editors and authors should be aware that incomplete or insufficiently close to final manuscripts almost always result in longer evaluation times. They should also be aware that parallel submission of their manuscript to another publisher while under consideration for LNM will in general lead to immediate rejection.

4. Manuscripts should in general be submitted in English. Final manuscripts should contain at least 100 pages of mathematical text and should always include
   - a general table of contents;
   - an informative introduction, with adequate motivation and perhaps some historical remarks: it should be accessible to a reader not intimately familiar with the topic treated;
   - a global subject index: as a rule this is genuinely helpful for the reader.

   Lecture Notes volumes are, as a rule, printed digitally from the authors' files. We strongly recommend that all contributions in a volume be written in the same LaTeX version, preferably LaTeX2e. To ensure best results, authors are asked to use the LaTeX2e style files available from Springer's web-server at
   ftp://ftp.springer.de/pub/tex/latex/svmonot1/ (for monographs) and
   ftp://ftp.springer.de/pub/tex/latex/svmultt1/ (for summer schools/tutorials).
   Additional technical instructions, if necessary, are available on request from:
   lnm@springer.com.

5. Careful preparation of the manuscripts will help keep production time short besides ensuring satisfactory appearance of the finished book in print and online. After acceptance of the manuscript authors will be asked to prepare the final LaTeX source files and also the corresponding dvi-, pdf- or zipped ps-file. The LaTeX source files are essential for producing the full-text online version of the book. For the existing online volumes of LNM see:
   http://www.springerlink.com/openurl.asp?genre=journal&issn=0075-8434.
   The actual production of a Lecture Notes volume takes approximately 12 weeks.

6. Volume editors receive a total of 50 free copies of their volume to be shared with the authors, but no royalties. They and the authors are entitled to a discount of 33.3 % on the price of Springer books purchased for their personal use, if ordering directly from Springer.

7. Commitment to publish is made by letter of intent rather than by signing a formal contract. Springer-Verlag secures the copyright for each volume. Authors are free to reuse material contained in their LNM volumes in later publications: a brief written (or e-mail) request for formal permission is sufficient.

Addresses:
Professor J.-M. Morel, CMLA,
École Normale Supérieure de Cachan,
61 Avenue du Président Wilson, 94235 Cachan Cedex, France
E-mail: morel@cmla.ens-cachan.fr

Professor B. Teissier, Institut Mathématique de Jussieu,
UMR 7586 du CNRS, Équipe "Géométrie et Dynamique",
175 rue du Chevaleret, 75013 Paris, France
E-mail: teissier@math.jussieu.fr

For the "Mathematical Biosciences Subseries" of LNM:

Professor P. K. Maini, Center for Mathematical Biology,
Mathematical Institute, 24-29 St Giles,
Oxford OX1 3LP, UK
E-mail: maini@maths.ox.ac.uk

Springer, Mathematics Editorial I,
Tiergartenstr. 17,
69121 Heidelberg, Germany,
Tel.: +49 (6221) 4876-8259
Fax: +49 (6221) 4876-8259
E-mail: lnm@springer.com

The manufacturer's authorised representative in the EU is Springer Nature Customer Service Centre GmbH, Europaplatz 3, 69115 Heidelberg, Germany. If you have any concerns regarding our products, please contact ProductSafety@springernature.com

Printed and bound by CPI Group (UK) Ltd, Croydon, CR0 4YY

23/03/2026

02076676-0005